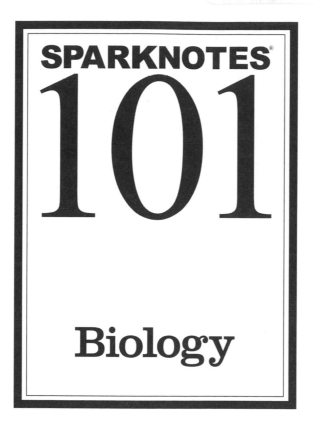

SPARKNOTES®
101

Biology

SPARK PUBLISHING

SPARKNOTES is a registered trademark of SparkNotes LLC.

Spark Publishing
A Division of Barnes & Noble
120 Fifth Avenue
New York, NY 10011
www.sparknotes.com

Please submit all comments and questions or report errors to www.sparknotes.com/errors.

Library of Congress Cataloging-in-Publication Data
Sparknotes 101. Biology.
     p. cm.
 ISBN-13: 978-1-4114-0337-6
 ISBN-10: 1-4114-0337-1
 1. Biology—Textbooks. I. Title: Biology.
QH308.2.S68 2008

570—dc22

2007029603

Printed and bound in the United States

10   9   8   7

Contents

Chapter 19

Chapter 20

Chapter 21

Chapter 22

Acknowledgments

SparkNotes would like to thank the following writers and contributors:

Shelley Wake

Daniella Swenton-Olson
University of New Mexico

Cynthia L. Tech
University of New Mexico

Franklin Bell
Mercersburg Academy Science Instructor,
AP Biology & PreAP Science Consultant

Rachel Warren

A Note from SparkNotes

Welcome to the *SparkNotes 101* series! This book will help you succeed in your introductory college course for Biology.

Every component of this study guide has been designed to help you process the material more quickly and score higher on your exams. You'll see lots of headings, lists, and charts, and, most important, you won't see any long blocks of text. This format will allow you to quickly learn what you need to know.

We've included these features to help you get the most out of this book:

Introduction: Before diving into the major chapters, you may want to get a broader sense of Biology as an academic discipline. The Introduction discusses the unifying themes of Biology; lists some careers related to the field; explains the relationship between the study of biology and conservation; and summarizes the scientific method by which biologists, and all other scientists, conduct their research.

Chapters 1–25: Each chapter provides clarification of material included in your textbook, as well as the following:

- **Examples:** These clarify main points and illustrate biological concepts from the real world.

- **Key Terms:** Important terms are bolded throughout each chapter for quick review. Definitions for these terms are compiled in the glossary at the back of the book.

- **Sample Test Questions:** The Sample Test Questions, including true/false, multiple choice, and short answer, show you the kinds of questions you're most likely to encounter on a test and give you the chance to practice what you've learned. Answers are provided.

Glossary: Review new terms and refresh your memory at exam time.

Index: Turn to the index at the back of the book to look up specific concepts and terms.

Your textbook might be longer or look different from our study guide. Not to worry—we've got you covered. Everything you need is here. We've gone for concision to make your studying easier. The material is organized in a clear, logical way that won't overwhelm you and will give you everything you need to know to keep up in class.

We hope *SparkNotes 101: Biology* helps you, gives you confidence, and occasionally saves your butt! Your input makes us better. Let us know what you think or how we can improve this book at **www.sparknotes.com/comments.**

Introduction

Biology is the scientific study of life. Living things, called **organisms,** include everything from unicellular bacteria to highly complex multicellular plants and animals. All organisms can be defined according to a set of common characteristics:

- They are made of one or more **cells,** which are compartments that carry out the functions of life and are separated from the outside environment by a membrane.

- They use energy obtained from the environment to perform their life functions.

- They maintain stable internal conditions through a process called **homeostasis.**

- They reproduce, passing on the hereditary information encoded in DNA to subsequent generations.

Unifying Themes of Biology

Advances in biology have had a profound impact on the quality of human life throughout history. Modern improvements in medicine, nutrition, agriculture, and the maintenance of our fragile biosphere have all come about through research conducted in this field. Biologists study a wide variety of topics ranging from the movement of molecules within cells to the flow of energy between all organisms living in a single ecosystem. Despite the diversity of topics, the study of biology remains unified by eight themes:

- **Cell theory**

- **Hierarchical organization of life**

- **Heredity**

- **Evolution**

- **Regulation**

- **Structure and function**

- **Environmental interactions**

- **Energy flow**

CELL THEORY

Although cells were first described in the seventeenth century, cell theory originated in the nineteenth century. **Cell theory,** refined through advances in research, remains the central idea in all studies of biology. The three main tenets of cell theory are as follows:

1. All organisms are made up of one or more cells.

2. Cells are the smallest structures that perform the processes essential to life, including food consumption, waste production, and the reproduction of new cells.

3. All new cells arise from the division of existing cells.

HIERARCHICAL ORGANIZATION OF LIFE

All life exhibits a distinct hierarchical organization, in which small, relatively simple structures combine to form larger, more complex structures. The structures that make up life, from the smallest to largest, are as follows:

- **Atoms:** The smallest units that possess the characteristics of an **element,** the foundational unit of all matter. An element, such as hydrogen, is any substance that cannot be broken down to any other substance by chemical reaction.

- **Molecules:** A group of two or more atoms joined together. DNA is an example of a molecule.

- **Organelles:** Groups of molecules within a cell organized to perform specific cellular functions. A chloroplast is an example of an organelle.

- **Cells:** The smallest units that can carry out all of the processes of life. A cardiac muscle cell is an example of a cell.

- **Tissues:** Groups of cells organized to carry out a particular function within an organism. Cardiac muscle is an example of tissue.

- **Organs:** Groups of tissues organized to carry out a particular function. The heart is an example of an organ.

- **Organ systems:** Groups of organs designed to carry out a particular task. The circulatory system is an example.

- **Organisms:** Individual living things. Organisms may be single-celled or they may contain multiple organ systems working together. A salmon is an example of an organism.

- **Populations:** Groups of individuals of the same species living in the same geographic area. All of the salmon within a single river form a population.

- **Species:** Groups of populations that are capable of interbreeding. The chinook (*Oncorhynchus tshawytscha*) species of Pacific salmon is an example of a species.

- **Communities:** All of the species that live in the same geographic area. All animals, bacteria, fungi, protists, and insects that live in a section of a river form a community.

- **Ecosystems:** All living and nonliving components of a particular area. A river ecosystem would include all the life within the river, as well as the water, soil, and sunlight that the living organisms draw resources from.

- **Biosphere:** All of the ecosystems on Earth together constitute the biosphere.

HEREDITY

All organisms contain DNA, the genetic code of life. Biological information, in the form of genes, is inherited from parents in one generation by the offspring in the next. Heredity works to gradually bring about change in a species through interactions between individual organisms and their environment—other organisms in their population, community, ecosystem—over the course of time. Heredity forms the basis for the process of evolution.

EVOLUTION

Evolution, defined as the modification of populations over time, is perhaps the greatest unifying theme of biology. Every aspect of the living world, from organelles to the biosphere, can be described through the mechanisms of evolution over the billions of years of Earth's history. Biologists use evolution to explain the great diversity of life on Earth.

REGULATION

To survive and reproduce, all forms of life must regulate their internal, and sometimes external, environment. Regulation often occurs through feedback mechanisms, in which stimuli increases or decreases a biological response.

STRUCTURE AND FUNCTION

Correlations between the structure of a biological object, such as a protein, cell, or organ, and its function are found on all levels of biological organization, from enzymes to appendages to ecosystems. The leaves of most plants, for example, are flat and broad with a large surface area. This structure is ideal for maximizing the capture of the sun's light for use in photosynthesis. Exceptions to the general shape are observed in desert plants, where water, not sunlight, is limited. Desert plants minimize water loss by having absent, small, or ephemeral (short-lasting) leaves.

ENVIRONMENTAL INTERACTIONS

Organisms do not exist as independent entities in a static (unchanging) environment. Individuals interact with other organisms of their own species and those of the greater community. Organisms are also highly influenced by the dynamic environment in which they live. These combined interactions determine an organism's behavior over time, and the organism's behavior, in turn, shapes its environment.

ENERGY FLOW

Since energy can neither be created nor destroyed, organisms must acquire the energy they need to survive and reproduce from the environment. Biologists study energy as it flows through a food chain, cycles through an ecosystem, or is converted to different forms within the cells of an organism.

Careers in Biology

The field of biology has become increasingly important to the health of individuals, societies, and the Earth itself. The quality of human life in many countries has improved beyond measure over the last two centuries, in part because of biological discoveries such as vaccines, antibiotics, medicines, pesticides, herbicides, water purification methods, and food preservation methods. Careers that require a biology background are numerous and diverse. A select list of broad fields of study within biology includes the following:

CLINICAL MEDICINE

Careers in clinical medicine involve direct work in healing patients, both human and animal. Medical doctors (MDs), veterinarians (DVMs), physician assistants (PAs), nurses (RNs), pharmacists, and medical technicians are all medical professionals who must have, at the very least, a basic understanding of biology.

MEDICINAL RESEARCH

Medicinal researchers work to understand and, hopefully, cure disease. Medicinal researchers may development new drugs or treatments, or they may work in fields such as genetic engineering and stem cell research. Although work in medical research often takes place in a laboratory setting, these biologists also find themselves working directly in the field, finding new cures or studying a disease in its native habitat.

BIOLOGICAL RESEARCH

Biological researchers study the functions and structures in organisms from bacteria to animals. As with medical researchers, their work can be done either in a laboratory or in the field. They may work with fossils or live organisms. Many of these researchers are academics associated with colleges or universities where they teach. Alternately, they may work for government agencies as wildlife biologists, agricultural researchers, and the like.

Biology and Conservation

Understanding the biological world is the first step in preserving it. The explosion of the human population in the twentieth century and through the early part of the twenty-first has impacted the globe in many ways, some devastating. In the last two centuries, scientists have documented unprecedented extinction rates. Global warming can also be linked to human actions. Humans are radically shaping the Earth into one that may be unrecognizable in the next century.

Understanding and preserving our living world has become more important than ever to prevent the harmful consequences of human impact. Biologists are on the forefront of the movements that seek to understand and protect the biosphere for the generations that follow.

The Scientific Method

Biologists, like all scientists, conduct their research using the scientific method. The scientific method is a standardized way of making observations, gathering data, forming theories, testing predictions, and interpreting results.

Researchers make observations to describe and measure biological states and behaviors. After observing certain events repeatedly, researchers come up with a theory that explains these observations. A theory is an explanation that organizes separate pieces of information in a coherent way. Researchers generally develop a theory only after they have collected a great deal of evidence and make sure that their research results can be reproduced by others.

Biological research, like research in other fields, must meet certain criteria to be considered scientific. Research must be:

- **Replicable**

- **Falsifiable**

- **Precise**

- **Parsimonious**

REPLICABLE

Research is replicable when others can repeat it and get the same results. When biologists report what they have found through their research, they also describe in detail how they made their discoveries. This way, other biologists can repeat the research to see if they can replicate the findings.

After biologists conduct their research and make sure it's replicable, they develop a theory and translate the theory into a precise hypothesis. A hypothesis is a testable or observable prediction of what will happen given a certain set of conditions. If further tests or observations do not confirm the hypothesis, the biologist revises or rejects the original theory.

FALSIFIABLE

A good theory or hypothesis must also be falsifiable, which means that it must be stated in a way that makes it possible to reject it. In other words, other researchers have to be able to prove a theory or hypothesis wrong. Theories and hypotheses need to be falsifiable because all researchers can succumb to the confirmation bias. Researchers who display confirmation bias look for and accept evidence that supports what they want to believe and ignore or reject evidence that refutes their beliefs.

PRECISE

By stating hypotheses precisely, researchers ensure that they can replicate their own and others' research. To make hypotheses more precise, researchers state exactly how their research has been conducted.

PARSIMONIOUS

The principle of parsimony, also called Occam's razor, maintains that researchers should apply the simplest explanation possible to any set of observations. For instance, biologists try to explain results by using well-accepted theories instead of elaborate new hypotheses. Parsimony prevents researchers from inventing and pursuing outlandish theories.

THE CHEMISTRY OF LIFE

Atoms

Chemical Bonds

Water

Biological Macromolecules

Atoms

All matter is made up of **elements,** which are substances that cannot be broken down into small substances in a chemical reaction. An **atom,** the fundamental unit of matter, is the smallest unit of an element that displays all of that element's characteristics. Atoms contain the following subatomic particles:

- **Protons:** Positively charged particles located within the **atomic nucleus,** or the core of the atom

- **Neutrons:** Electrically neutral particles located within the atomic nucleus

- **Electrons:** Negatively charged particles orbiting in a cloud around the atomic nucleus

Atoms are described by their atomic number and mass number. **Atomic number** indicates the number of protons in an atom. **Mass number** indicates the combined number of protons and neutrons in an atom. An atom's chemical symbol, the written representation of an atom, includes atomic number, mass number, and any charge on the atom, allowing chemists to derive the number of protons, neutrons, and electrons.

> *EXAMPLE:* The chemical symbol $^{12}_{6}C$ refers to an atom of carbon with a mass number of 12, an atomic number of 6, and no charge. Because the mass is 12 and there are 6 protons, the atom must contain 6 neutrons (12 – 6 = 6). Because the atom has no charge, the number of electrons must equal the number of protons.

ATOMIC BEHAVIOR

An atom's behavior is largely determined by the number and arrangement, or electron structure, of electrons orbiting its nucleus. Electrons move about in regions called orbitals located at varying distances from the nucleus. Because it is impossible to pinpoint the location of an electron at any given time, orbitals represent the region where a specific electron is likely to be found. A single orbital cannot contain more than two electrons; therefore, the number of electrons in an atom will determine the number of orbitals.

All of an atom's electrons are arranged in orbitals located at different energy levels, called electron shells, around the nucleus. Shells closest to the nucleus require less energy to maintain than shells lying farther out. Electrons tend to move toward lower energy positions and will fill orbitals in the lowest electron shells before filling orbitals in electron shells farther out. Each electron shell has a maximum number of orbitals, and therefore electrons, that it can hold:

- **First electron shell:** maximum one orbital; two electrons

- **Second electron shell:** maximum four orbitals; eight electrons

- **Third electron shell:** maximum four orbitals; eight electrons

- **Fourth electron shell:** maximum nine orbitals; eighteen electrons

Valence Electrons

Valence electrons occupy the energy shell farthest from the nucleus. Atoms with a full complement of electrons in their outermost shells are more stable, and therefore less reactive, than atoms with incomplete outer electron shells. Atoms react with other atoms and form chemical bonds by sharing valence electrons, where one atom gains an electron and the other atom loses one, to create a full outer shell.

EXAMPLE: Sodium, ^{11}Na, has 11 electrons. Two electrons occupy the first shell, eight electrons occupy the second shell, and one electron, the valence electron, occupies the outermost third shell. To create a full outer shell of electrons, a sodium atom loses the single valence electron from the third shell. In contrast, chlorine, 17Cl, has seventeen electrons. Two electrons occupy the first shell, eight electrons occupy the second shell, and seven valence electrons occupy the outermost third shell. To create a full outer shell of electrons, a chlorine atom gains one electron.

CHAPTER 1
The Chemistry of Life

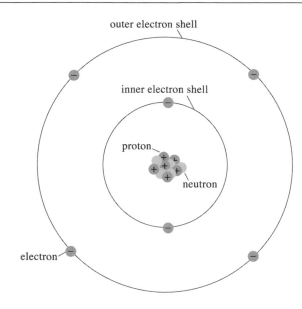

Carbon Atom

Ions and Isotopes

Although all atoms of the same element contain the same number of protons, they can have different numbers of electrons and neutrons. An **ion** is an atom that has acquired a positive or negative electric charge by gaining or losing electrons. There are two types of ions:

- A **cation** is an atom that has lost electrons and acquired a positive charge.

- An **anion** is an atom that has gained electrons and acquired a negative charge.

EXAMPLE: Potassium (K) can lose an electron to become a cation, K^+. Sulfur (S) can lose two electrons to become an anion, S^{-2}.

Isotopes are atoms of the same element that have different numbers of neutrons and therefore different atomic masses.

EXAMPLE: There are three naturally occurring carbon isotopes, each with a different atomic number: carbon-12 ($^{12}_6C$), carbon-13 ($^{13}_6C$), and carbon-14 ($^{14}_6C$). As represented in the atomic number, carbon-14 has one more neutron than carbon-13 and two more neutrons than carbon-12.

Radioactive isotopes are highly unstable and spontaneously decay by losing protons or neutrons along with energy. The rate of decay is constant for any given isotope. The **half-life** of a radioactive substance is a measure of the amount of time it takes for half of the atoms in that substance to decay. All living organisms are composed of carbon-14 and carbon-12 in a certain ratio, designated by the variable x. Once an organism dies, this ratio decreases as carbon-14, with a half-life of 5,730 years, and decays to form the nonradioactive isotope nitrogen-14 ($^{14}_7N$). Scientists measure the ratio of radioactive isotopes, such as $^{14}_6C$, against other elements present in a fossil to determine the approximate age of the fossil.

EXAMPLE: The age of a bone fossil that contains carbon-14 and carbon-12 in a ratio of $\frac{1}{8}x$ can be determined using the half-life of carbon-14. Because one-eighth is $\left(\frac{1}{2}\right)^3$, three half-lives have passed for carbon-14 present in this fossil. Three times 5,730 equals 17,190; therefore the fossil is approximately 17,190 years old.

Chemical Bonds

The interactions between the electrons of two or more atoms result in **chemical bonds. Electronegativity,** the strength of the attraction an atom has for its electrons, helps determine the nature of the chemical bonds that atom can form. The more electronegative an atom is, the greater the strength of its attraction for electrons. In general, an atom with a full or nearly full set of valence electrons has high electronegativity and will hold on to its electrons and may attract the electrons of other atoms. In contrast, atoms with few valence electrons have weak electronegativity and may lose their electrons to other atoms.

COVALENT BONDS

Two or more atoms share valence electrons to form **covalent bonds.** Covalent bonds result in the formation of **molecules,** which are strong, stable associations between two or more atoms. By sharing electrons, each atom in a covalent bond fills its outer electron shell. Atoms joined by covalent bonds may share one, two, or three electrons, resulting in single, double, or triple bonds.

The distribution of shared atoms in a covalent bond depends on the electronegativities of the atoms involved. Two terms describe the different ways atoms can share their electrons in a covalent bond:

*   **Nonpolar covalent bonds** form between atoms that have similar electronegativity. The electrons in a nonpolar covalent bond are shared equally between the two atoms.

*   **Polar covalent bonds** form between atoms that have different electronegativities. Valence electrons in a polar covalent bond will be more attracted to the atom of higher electronegativity, resulting in a partial negative charge on that atom and a partial positive charge on the other atom.

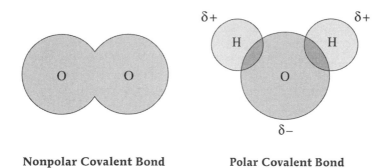

Nonpolar Covalent Bond   **Polar Covalent Bond**

IONIC BONDS

Ionic bonds form when electrons are transferred from an atom of low electronegativity to an atom of high electronegativity. The atom that has lost an electron becomes a positively charged anion, while the atom that has gained an electron becomes a negatively charged cation. Cations and anions are mutually attracted to one another by their charges. This mutual attraction results in the formation of a **crystal,** which is a highly regular and ordered solid whose atoms are arranged in repeating units.

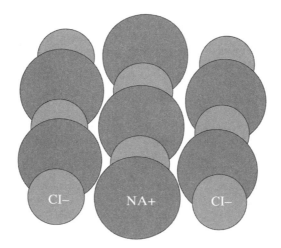

Ionic Bonds

HYDROGEN BONDS

Hydrogen bonds, the weakest of all chemical attractions, form when a hydrogen atom that is covalently bonded to an electronegative atom is attracted to another electronegative atom, generally either oxygen (O) or nitrogen (N).

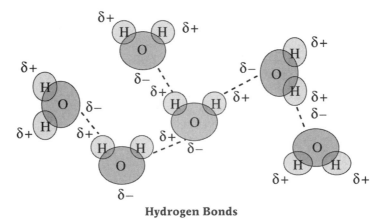

Hydrogen Bonds

Hydrogen bonds are only 5 to 10 percent as strong as covalent bonds. Nevertheless, hydrogen bonds play a very important role in biology. For example, hydrogen bonds are one of the main forces that give proteins their three-dimensional shapes.

Water

Water is the most abundant molecule present in all living organisms. All chemical reactions within an organism take place in the presence of water. Several characteristics unique to water contribute to its vital importance in the processes of life, such as its properties as a solvent and tendency to form ions.

WATER AS A SOLVENT

Solvents dissolve other molecules, called **solutes,** to form **solutions,** which are homogeneous mixtures of molecules.

The hydrogen bonds that hold water molecules make water a versatile solvent that can form solutions with polar **(hydrophilic)** molecules. Water cannot form solutions with nonpolar **(hydrophobic)** molecules.

EXAMPLE: In an aqueous solution of NaCl and water, the solvent is water and the solute is NaCl. NaCl is ionic, and therefore hydrophilic, and forms an aqueous solution in which water is the solvent and NaCl is the solute. Oil is nonpolar, and therefore hydrophobic. Oil will not dissolve in water.

WATER IONIZATION

Occasionally, water molecules spontaneously ionize, or break apart, into **hydroxide ion** (OH⁻) and **hydrogen ion** (H⁺), as illustrated in the following chemical equation.

$$H_2O \rightleftarrows OH^- + H^+$$

The **pH scale,** which ranges from 0 to 14, expresses the relative concentrations of OH⁻ and H⁺ in a solution. The pH value determines whether a solution is acidic, basic, or neutral:

- **Neutral solutions** have equal concentrations of OH⁻ and H⁺ and a pH of 7.

- **Acidic solutions** have a greater concentration of H⁺ and a pH of less than 7.

- **Basic solutions** have a greater concentration of OH⁻ and a pH of greater than 7.

Acids lower the pH of solutions, while **bases** raise the pH. **Buffers** are substances that reduce the effect of acids and bases on the pH of a solution.

Calculating pH

The pH of a given solution expresses the negative logarithm of the hydrogen ion concentration, as represented in the following equation.

$$pH = -\log[H^+]$$

For example, the $[H^+]$ of wine equals 10^{-3}. Because the negative logarithm (represented by the exponent in this equation) is three, wine pH = 3 and is therefore acidic.

Biological Macromolecules

All activities of an organism, including breathing, eating, and growing, are controlled by the interactions of the chemicals in its body. Although millions of different types of molecules make up an organism, most biologically important molecules fall into one of the following macromolecule groups: nucleic acids, proteins, carbohydrates, or lipids.

Biological macromolecules are large, carbon-based molecules formed within an organism to serve a variety of functions. Most biological macromolecules are **polymers,** which are long chains composed of many **monomers,** similarly structured subunits bonded together. A dehydration synthesis reaction, which requires the input of energy and results in the release of water, links together the monomers that form a macromolecule.

NUCLEIC ACIDS

Nucleic acids form from polymers of **nucleotides,** molecules composed of a phosphate group, a five-carbon sugar, and a nitrogenous base. Five different nitrogenous bases exist:

- **Adenine**

- **Guanine**

- **Cytosine**

- **Thymine**

- **Uracil**

DNA and RNA are nucleic acids that function in protein synthesis and the storage and transmission of genetic information.

PROTEINS

Proteins consist of one or more **polypeptides,** polymers of **amino acids** folded into complex three-dimensional shapes. An amino acid is a small molecule made up of a central carbon atom, an amino group, a carboxyl group, a hydrogen atom, and a functional group labeled "R." Twenty different amino acids exist, each formed with a different R group. Polypeptides form when amino acids bond together in long chains. The twenty different amino acids can produce a diverse range of proteins, including enzymes, hormones, cell receptors, antibodies, transport proteins, storage proteins, motor proteins, and structural proteins, which perform a wide range of vital tasks in organisms.

All proteins have either three or four structural levels:

- **Primary structure** refers to the sequence of amino acids that form the polypeptides.

- Hydrogen bonds in single groups in a polypeptide chain result in a folded region referred to as the **secondary structure.** Secondary structures include helices (coils) and sheets (pleated folds).

- The **tertiary structure** describes the folding of an entire polypeptide chain. Interactions between the R groups of the polypeptide chain determine the overall shape of the tertiary structure.

- Interaction between two or more polypeptides forms the **quaternary structure.** Since some proteins consist of a single polypeptide, not all proteins exhibit quaternary structure

Primary Structure   Secondary Structure   Tertiary Structure   Quaternary Structure

Protein Structure

CARBOHYDRATES

Carbohydrates include both **monosaccharides** (also called simple sugars or simple carbohydrates) and **polysaccharides** (also called complex carbohydrates). Monosaccharides and polysaccharides perform different functions in organisms:

- Monosaccharides, which include glucose, fructose, and lactose, provide cells with energy.

- Polysaccharides, which include glycogen, starch, and cellulose, store energy and provide structural support to an organism.

LIPIDS

Unlike the other macromolecules, lipids are not composed of repeated monomers and therefore are not true polymers. Lipids exhibit a range of structural diversity, but all are nonpolar and therefore insoluble in water. Lipids include the following molecules:

- **Fats:** store energy for future use in biological functions

- **Phospholipids:** make up the cell membrane

- **Steroids:** act as chemical messengers in an organism

FUNCTIONAL GROUPS

The atoms within a molecule that are most likely to be involved in chemical reactions are categorized by their **functional group.** All substances in a particular functional group share the same chemical properties. Although many functional groups exist, six functional groups are most commonly found in biological macromolecules. These six groups are described in the following table.

Functional Group	Structural Formula	Class of Compounds	Types of Macromolecules
Amino	$-N\begin{smallmatrix}H\\\\H\end{smallmatrix}$	Amines	Proteins
Carbonyl	$-C\begin{smallmatrix}=O\\\\\end{smallmatrix}$	Ketones & aldehydes	Lipids
Carboxyl	$-C\begin{smallmatrix}=O\\\\OH\end{smallmatrix}$	Carboxylic acids	Proteins
Hydroxyl	$-OH$	Alcohols	Carbo-hydrates
Phosphate	$-O-P(=O)(O^-)-O^-$	Phosphates	Nucleic acids
Sulfhydryl	$-SH$	Thiols	Proteins

Summary

Atoms

- **Atoms** are the smallest unit of an **element** that retains the properties of that element. Atoms contain **protons, neutrons,** and **electrons.**

- An atom's electrons are arranged in orbitals located in different energy levels, called electron shells.

- **Valence electrons** occupy the farthest, highest-energy electron shell.

- An **ion** is an atom that has acquired a net electric charge by gaining or losing electrons.

- **Isotopes** are atoms of the same element that have different numbers of neutrons.

Chemical Bonds

- **Electronegativity** describes the strength of the attraction an atom has for electrons.

- **Covalent bonds** form when atoms share electrons. Covalent bonds may be polar or nonpolar.

- **Ionic bonds** form when electrons are transferred between atoms.

- **Hydrogen bonds** are weak attractions between molecules that form when a hydrogen atom covalently bonded to an electronegative atom (usually O or N) is attracted to another electronegative atom.

Water

- **Solvents** dissolve molecules, called **solutes,** to form **solutions.** Water can form solutions with **hydrophilic** molecules but not with **hydrophobic** molecules.

- The **pH scale** measures the relative concentrations of H^+ and OH^- in a solution. Acidic solutions have greater H^+ and a pH of less than 7. Basic solutions have greater OH^- and a pH of greater than 7.

Biological Macromolecules

• Biological molecules fall into one of the following groups: nucleic acids, proteins, carbohydrates, or lipids.

• **Nucleic acids** are polymers of **nucleotides** composed of a phosphate group, a five-carbon sugar, and a nitrogenous base. Nucleic acids include DNA and RNA.

• **Proteins** consist of one or more **polypeptides,** which are polymers of **amino acids**. Proteins have up to four levels of structure: primary, secondary, tertiary, and quaternary.

• **Carbohydrates** include **monosaccharides** and **polysaccharides.**

• **Lipids** are nonpolar and include fats, phospholipids, and steroids.

• **Functional groups** are those atoms in a molecule most likely to be involved in chemical reactions. The six functional groups most commonly found in biological macromolecules are amino, carbonyl, carboxyl, hydroxyl, phosphate, and sulfhydryl.

Sample Test Questions

1. A thick, waxy coating composed of lipids covers the leaves of many desert plants and reduces water loss to the environment. What characteristic makes lipids, of all the biological macromolecules, uniquely suited to this purpose?

2. A scientist trying to determine the age of a fossilized leaf measures the ratio of carbon-14 to carbon-12. She finds that the ratio is $\frac{1}{16}$ of that which would be found in a living leaf $\frac{1}{6}x$. Based on this information, what determination could the scientist make about the age of the fossil?

3. Explain why water can form solutions with polar substances but not with nonpolar substances.

CHAPTER 1
The Chemistry of Life

4. A magnesium atom has twelve electrons. How many valence electrons does this atom have?

    A. One
    B. Two
    C. Six
    D. Seven
    E. Twelve

5. Which type of bond is formed between two chlorine atoms that share an electron?

    A. Ionic
    B. Covalent
    C. Hydrogen
    D. Metallic
    E. Network

6. Which type of molecule is formed by polymers composed of amino acids?

    A. Protein
    B. Lipid
    C. Nucleic acid
    D. Carbohydrate
    E. Polysaccharide

7. Which of the following functional groups is the main component of alcohols?

    A. Hydroxyl group
    B. Carboxyl group
    C. Amino group
    D. Phosphate group
    E. Carbonyl group

8. What is the pH of an acidic solution?

    A. 0
    B. 7
    C. 14
    D. Less than 7
    E. Greater than 7

9. Which of the following terms describes the different energy levels that can exist in an atom?

   A. Ions
   B. Electron shells
   C. Orbitals
   D. Valence electrons
   E. Atomic number

10. Is the pH of a substance that has $[H^+] = 10^{-4}$ acidic, basic, or neutral?

    A. Acidic
    B. Neutral
    C. Basic
    D. Both acidic and neutral
    E. Both basic and neutral

11. Monosaccharides are monomers of which of the following macromolecules?

    A. Nucleic acids
    B. Lipids
    C. Carbohydrates
    D. Proteins
    E. Steroids

12. Which level of protein structure forms α helices?

    A. Primary structure
    B. Secondary structure
    C. Tertiary structure
    D. Quaternary structure
    E. Hectonary structure

13. Which of the following structures is composed of a phosphate group, five-carbon sugar, and nitrogenous base?

    A. Carboxyl
    B. Amino acid
    C. Steroid
    D. Monosaccharide
    E. Nucleotide

14. Amino groups are essential components of which of the following macromolecules?

    A. Carbohydrates
    B. Proteins
    C. Lipids
    D. Nucleic acids
    E. Thiols

15. What role does salt play in a solution of salt water?

    A. Solvent
    B. Solute
    C. Acid
    D. Base
    E. Buffer

ANSWERS

1. Unlike the other biological macromolecules, lipids are nonpolar and are insoluble in water. Because they are hydrophobic, lipids do not dissolve in water and therefore make excellent barriers to water flow.

2. The fossil is approximately 22,920 years old. The fossil has a $\frac{1}{16}$ ratio of carbon-14 to carbon-12. Therefore, approximately four half-lives have likely passed: $\left(\frac{1}{2}\right)^4 = \frac{1}{16}$. Because the half-life of carbon-14 is 5,730 years, and $5,730 \cdot 4 = 22,400$, the fossil is approximately 22,920 years old.

3. Atoms in a water molecule are linked by hydrogen bonds, which are created by the polarity in each individual water molecule as hydrogen atoms with a partial positive charge are attracted to the slight negative charge in oxygen atoms. Polar molecules are pulled toward each other by attractions between oppositely charged atoms in each molecule. Water molecules form hydrogen bonds with other polar molecules, preventing nonpolar molecules from bonding. Therefore, a nonpolar molecule will not dissolve in water.

4. B The magnesium atom has two electrons in the innermost first shell, eight electrons in the second shell, and two valence electrons in the outermost third shell.

5. B   Covalent bonds form between atoms sharing valence electrons.

6. A   Proteins consist of one or more polypeptides, which are polymers of amino acids folded into complex three-dimensional shapes.

7. A   Hydroxyl groups are a component of alcohols.

8. D   Acidic solutions have a pH of less than 7.

9. B   The different energy levels that can exist in an atom are called electron shells.

10. A   The solution, which has a pH of 4, is acidic.

11. C   Monosaccharides are carbohydrate monomers.

12. B   Helices are an example of secondary protein structure.

13. E   A nucleotide consists of a phosphate group, a five-carbon sugar, and a nitrogenous base.

14. B   Amino groups are essential components of proteins.

15. B   Salt is the solute in salt water.

THE CELL

2

Cell Theory

Cells are the structural and functional building blocks of all organisms. They were first described by the English physicist Robert Hooke in his 1665 work *Micrographia*, which ushered in the science of microscopy and confirmed the microscope's importance to biology. While examining a thin slice of cork plant, Hooke observed the rectangular units that composed its structure. Because of their resemblance to a monk's chambers, Hooke called these units **cells,** from *cella*, the Latin word for "small room."

There are three main tenets of the Cell Theory:

1. Cells are the smallest structures that perform the processes essential to life, including food consumption, waste production, and the creation of new cells through reproduction.

2. All cells arise from the division of existing cells.

3. Every organism is made up of one or more cells.

Cells are made up of four key macromolecules: lipids, carbo-hydrates, proteins, and nucleic acids. Each of these macromolecules performs specific tasks in cell function. Many smaller molecules, such as water molecules, are also essential to cell function.

In a multicellular organism, different types of cells must work together to keep the organism alive and functioning. Although each cell performs a specific function, all cells carry the necessary hardware, contained in their DNA, for every other task being performed inside of the organism. We each start out as one cell: a **zygote** formed from the fusion of sperm and egg. This one cell divides into billions—trillions, by some estimates—of other cells with DNA identical to the original zygote cell. Each cell eventually takes on specific tasks, as in the case of brain cells, muscle cells, skin cells, and blood cells, to name a few. This process of **cell differentiation** is an intense area of study in cell biology.

> *EXAMPLE:* A stem cell is a cell that has not yet gone through the process of cell differentiation. Many scientists believe that stem cell research could hold the key to curing spinal cord injuries, among other disorders. In theory, if a stem cell can be manipulated, it can be made to grow into new nerve tissue, for example, and used to repair injuries.

MICROSCOPES

Most cells are too small to be seen with the naked eye. Biologists use several different kinds of microscopes to view and analyze them.

- **Light microscopes** bend and magnify visible light that passes through the specimen.

- **Scanning electron microscopes (SEMs)** use electrons to scan the surface of a specimen that has been coated with metal.

- **Transmission electron microscopes (TEMs)** use electrons to scan the internal structure of a thinly sliced specimen.

Light microscopes and electron microscopes have different powers of magnification and resolving powers. Magnification refers to the increase in the visual size of an object, while resolution refers to the clarity of the magnified image. Many details of cell structure can be viewed only with electron microscopes, which have much higher resolution and greater powers of magnification than light microscopes do.

EXAMPLE: Squid axons, which look like thin white strings, can be more than a meter (3.28 feet) in length. Most cells, however, are much smaller and can only be seen with a microscope.

Types of Cells

Every cell falls into one of two broad categories: prokaryotic cells and eukaryotic cells.

- **Prokaryotic cells** are small, simple cells that lack membrane-bound organelles and have very little internal division of labor.

- **Eukaryotic cells** are more complex cells with membrane-bound **organelles,** units that perform specific tasks.

Despite these differences, both types of cells share three basic components:

1. A **cell membrane** that surrounds the entire cell and separates it from the environment.

2. **Cytoplasm,** the substance between the cell membrane and the region of DNA. It is made up of **cytosol,** which is the

fluid-like material that fills the cell and can include cell components such as organelles, macromolecules, ions, and filaments.

3. **DNA,** the genetic material of the organism.

PROKARYOTIC CELLS

Prokaryotic cells are generally smaller and much simpler than eukaryotic cells. They have no membrane-bound organelles or central nucleus containing their DNA. Instead of a nucleus, their DNA is concentrated in a region called a nucleoid.

Bacteria and the archaea are the only two domains of organisms composed of prokaryotic cells. They are the most primitive organisms and are almost always single-celled, although some species form colonies in which labor is divided among specialized cells. Prokaryotic organisms were the earliest life forms on Earth and are highly adaptable, found in even the harshest environments today.

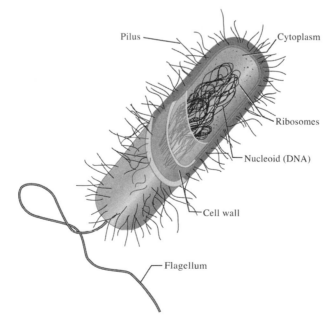

Prokaryotic Cell

EUKARYOTIC CELLS

Eukaryotic cells have a membrane-bound **nucleus** and membrane-bound organelles. The membrane-bound nucleus of each eukaryote contains **chromosomes,** pieces of DNA that are tightly folded up by proteins. The remaining substance of each eukaryotic cell consists of cytoplasm, which includes the cytosol and those organelles and materials suspended in it. Eukaryotic cells make up all organisms apart from bacteria and archaea. Although this encompasses all living organisms from the unicellular protists to the multicellular organisms including plants, fungi, and animals, prokaryotes still dominate life on Earth in terms of number and pervasiveness.

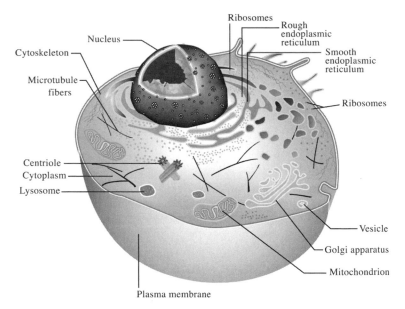

Eukaryotic Cells: Animal Cell

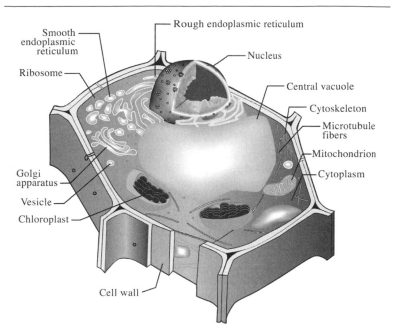

Eukaryotic Cells: Plant Cell

Organelles

Organelles perform functions in both prokaryotic and eukaryotic cells, although the organelles of prokaryotic cells are not membrane-bound. The following table describes the major organelles and their function and the type of cell in which they are found.

Organelles and Their Functions	Found in Prokaryotic Cell?	Found in Eukaryotic Cell?
Cell walls surround the cell membrane and give cells extra support and protection. Plant, fungi, and bacterial cells are supported by cell walls, but animal cells are not.	✓	In plants and fungi
The **nucleoid** is the region where the prokaryotic cell's DNA is located. It is not enclosed by a membrane.	✓	X

CONTINUED

Organelles and Their Functions	Found in Prokaryotic Cell?	Found in Eukaryotic Cell?
The **nucleus** is the command center for eukaryotic cells. It houses the cell's DNA and is differentiated from the prokaryotic nucleoid by the presence of a membrane.	X	✓
Ribosomes are structures that manufacture proteins. Prokaryotic ribosomes are scattered through the cytoplasm. In eukaryotic cells, ribosomes also attach to the endoplasmic reticulum.	✓	✓
The **endoplasmic reticulum,** or **ER** for short, is a system of membranes that performs many functions, such as the creation of different compartments within the cell called vesicles. Some portions of the ER have ribosomes attached to them and appear bumpy when viewed under an electron microscope. This so-called **rough ER** manufactures new membranes and transports proteins made by the attached ribosomes to the cell's Golgi bodies for further processing. **Smooth ER** is contiguous with rough ER but has no ribosomes attached to it. Smooth ER makes lipids, stores calcium ions, and contains enzymes that detoxify drugs and poisons.	X	✓
Vesicles are membranes within eukaryotic cells used for tasks such as transporting materials between the cell and the outside environment.	X	✓
Golgi bodies are stacks of flattened membranous sacs that receive proteins and lipids from ER, process them, and send them to their final destination. Golgi bodies are typically clustered together in a group known as a Golgi complex or a **Golgi apparatus.**	X	✓

CHAPTER 2
The Cell

CONTINUED

Organelles and Their Functions	Found in Prokaryotic Cell?	Found in Eukaryotic Cell?
Mitochondria (singular: mitochondrion) produce **ATP,** the energy currency of the cell, in a process called cellular respiration. For this reason, biologists refer to mitochondria as the cell's "powerhouses." Each mitochondrion has both an outer membrane and an inner folded membrane, which creates a separate inner compartment within the organelle, called the matrix.	X	✓
Chloroplasts convert light energy into ATP in a process called photosynthesis. Like mitochondria, chloroplasts are surrounded by two membranes. A third group of inner membranes form stacks, called grana. The grana are composed of disc-shaped compartments called **thylakoids,** where photosynthesis actually takes place. Chloroplasts are found in plant cells but not in animal cells.	X	In plants and some protists
Vacuoles store excess food, water, minerals, cellular fluid, and other matter for future use. Animal cells may contain several small vacuoles, whereas plant cells have a single large central vacuole. The pressure of the fluids in the central vacuole, called **turgor pressure,** helps maintain the shape of plants cells.	X	✓
Lysosomes contain digestive enzymes used to break down food molecules or damaged parts of the cell. Lysosomes are made by the Golgi apparatus.	X	✓
Centrioles are barrel-shaped organelles that produce **microtubules,** cylindrical protein fibers. These microtubule fibers give the cell its shape and assist in cell division. Centrioles exist in pairs. While microtubules form part of the plant and fungi cell cytoskeleton, plant and fungi cells do not contain centrioles.	X	In all but plants and fungi

CHAPTER 2
The Cell

CONTINUED

Organelles and Their Functions	Found in Prokaryotic Cell?	Found in Eukaryotic Cell?
Microtubule fibers form a network, called the cytoskeleton, which gives the cell its shape and prevents the cell from collapsing. They also assist in the movement of materials within the cell.	X	✓
Flagella and **cilia** are appendages that protrude from some cells and either provide motion or allow the movement of substances across the cell or tissue surface. Many unicellular organisms use flagella or cilia for movement in water or other fluids.	✓	✓

Key organelles and their functions can be remembered by using a city metaphor:

Organelle	Function
Nucleus	The mayor
Ribosomes	The protein factories
Endoplasmic reticulum (ER)	The highway
Golgi apparatus	The post office
Lysosomes	The trash collectors
Vacuoles	The storage warehouses
Mitochondria	The power plants

Membrane Structure

Cells collectively perform all functions needed to keep an organism alive. To do so, each cell must maintain itself while working with other cells. The **cell membrane,** also called a plasma membrane, plays a key role in maintaining the internal environment of the cell by regulating the entrance and exit of materials.

MEMBRANE PROTEINS

In eukaryotic cells, selective entry and exit of other molecules occurs through proteins embedded in the **phospholipid bilayer** of the cell membrane. The phospholipid bilayer is composed of two layers of phospholipids arranged so that their nonpolar "tails" form the interior of the membrane and the polar "heads" face outward. The phospholipid bilayer forms an effective barrier to all but the smallest nonpolar molecules, such as oxygen and carbon dioxide.

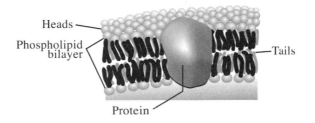

Heads
Phospholipid bilayer
Tails
Protein

Cell Membrane Structure

A number of specialized proteins within the phospholipid bilayer perform the functions necessary for regulation in the cell membrane:

• **Transport,** or **channel, proteins** transport molecules into or out of the cell.

• **Receptor proteins** receive or relay chemical messages between cells.

• Proteins known as **enzymes** facilitate chemical reactions within the cell.

• **Marker proteins** facilitate cell-to-cell recognition by identifying a cell's function or origin to other cells. Marker proteins are particularly crucial in sorting cells in an embryo into tissues and organs and rejecting or accepting foreign cells in transplanted organs.

• **Membrane proteins** interact with the cytoskeleton of the cell to stabilize the cell membrane.

• Membrane proteins form gap junctions and tight junctions that help attach cells to each other.

Membrane Permeability

Permeability describes the degree to which substances can pass through a cell's membrane. Cell membranes are **semipermeable,** allowing only some molecules to pass through the membrane. Because cell membranes are semipermeable, a **concentration gradient** exists such that some substances are present on one side of the membrane in higher concentrations than on the other side. All substances naturally move down their concentration gradient, from areas of high concentration to areas of low concentration, without the expenditure of energy. Movement up a concentration gradient is possible only with the expenditure of energy.

Both prokaryotic and eukaryotic cells engage in two activities to move materials across the cell membrane:

• **Passive transport** is movement down the concentration gradient. No energy is expended in this activity.

• **Active transport** is movement up the concentration gradient. This activity is powered by the cell's energy.

PASSIVE TRANSPORT

Passive transport refers to the movement of substances across a membrane without the expenditure of energy. Substances can be passively transported via diffusion, facilitated diffusion, or osmosis.

• **Diffusion:** Particles move spontaneously from an area of higher concentration to an area of lower concentration. Only small, uncharged particles, such as oxygen, carbon dioxide, and some ions, move across the cell membrane via diffusion.

- **Facilitated diffusion:** Particles move spontaneously across the cell membrane from an area of higher concentration to an area of lower concentration with the help of special transport proteins. Facilitated diffusion permits the passage of both large and charged molecules, such as glucose, through the cell membrane.

- **Osmosis:** Osmosis allows water to diffuse across the cell membrane from an area of higher water concentration to an area of lower water concentration.

Tonicity

The tonicity, or relative solute concentration, of a cell determines the direction in which osmosis occurs. Water diffuses from an area of low tonicity (high water, low solutes) to an area of high tonicity (low water, high solutes). There are three ways to describe relative levels of solute concentration:

- **Hypertonic:** higher solute concentration

- **Isotonic:** equal solute concentration

- **Hypotonic:** lower solute concentration

ACTIVE TRANSPORT

Cells must actively transport substances being moved up the concentration gradient. This movement requires the cell to expend energy and is enabled through the use of **carrier proteins,** which bind to a substance and shuttle it through pores in the cell membrane. Carrier proteins can perform two methods of active transport:

- **Endocytosis** occurs when proteins bring substances into the cell through the vacuoles or vesicles. Consumption of solid matter, such as cholesterol, through endocytosis is called **phagocytosis.** Consumption of extracellular liquid matter is called **pinocytosis.**

CHAPTER 2
The Cell

- **Exocytosis** occurs when carrier proteins remove substances from the cell through the vacuoles or vesicles. Insulin is delivered from the cell to the bloodstream by exocytosis.

Cellular Connection and Communication

The cells of multicellular organisms must stay bound together to keep the organism intact. These cells must also be able to communicate and cooperate with each other to function as a single organism.

The cell membrane is extremely important in the regulation of intercellular communication and cooperation. Eukaryotic cells, which constitute most multicellular organisms, have special structures that make cell-to-cell adhesion and interaction possible. (The cells of multicellular prokaryotes interact using different methods from those of eukaryotic cells. These methods are not yet fully understood by scientists.)

The following structures connect different eukaryotic cells to each other:

- **Extracellular matrices** are materials secreted by animal cells that surround and bind them together. A particularly thick and strong extracellular matrix surrounds bone cells.

- **Tight junctions, or occluding junctions,** bind cells so tightly together that molecules cannot pass between them. Tight junctions join the cells of the small intestine to prevent substances from leaking into other parts of the body.

- **Anchoring junctions** connect the cytoskeletons of two or more cells. Anchoring junctions allow greater mobility than other connective materials and are common in tissue that needs to be flexible, such as skin and muscle.

In addition to these connective materials, eukaryotic cells possess structures that facilitate cell-to-cell communication. These structures differ in plant and animal cells.

- **Plasmodesmata** are molecular channels that exist between adjacent plant cells.

- **Gap junctions** are molecular channels that exist between animal cells.

The biology of cell communication is still an area of intense investigation. There is much that is unknown about how organisms work together to maintain a functional organism.

Summary

Cell Theory

- Cells are the smallest structures that perform the processes essential to life, including consuming food, producing waste, and reproducing by making new cells.

- All cells arise from the division of existing cells.

- Every organism is made up of one or more cells.

Types of Cells

- **Prokaryotic cells** have no membrane-bound nuclei or membrane-bound **organelles.**

- Prokaryotic cells have ribosomes, cell walls, and sometimes flagella.

- Bacteria and archaea are made up of prokaryotic cells.

- **Eukaryotic cells** have membrane-bound nuclei and membrane-bound organelles.

- Eukaryotic organelles include the **nucleus, ribosomes, endoplasmic reticulum, Golgi bodies, mitochondria, lysosomes, vacuoles, chloroplasts, cytoskeleton, centrioles,** flagella, and cilia.

- Eukaryotic cells make up the protists, plants, fungi, and animals.

Membrane Structure

- The **cell membrane** consists of a **phospholipid bilayer** that is embedded with proteins. The cell membrane maintains the internal environment of the cell by regulating the movement of molecules into and out of the cell.

- **Membrane proteins** transport molecules into and out of the cell, receive or relay chemical messages between cells, facilitate chemical reactions, identify the cell to other cells in an organism, stabilize the membrane, and attach cells together.

Membrane Permeability

- Cell membranes are **semipermeable,** meaning some substances can pass through the membrane and others cannot.

- All substances naturally move down their **concentration gradient,** from areas of high concentration to areas of low concentration. Movement up a concentration gradient is possible only with the expenditure of energy.

- **Passive transport** is the movement of substances across a membrane without the expenditure of energy. Passive transport only occurs down a concentration gradient. Substances can be passively transported via **diffusion, facilitated diffusion,** or **osmosis.**

- **Active transport** occurs when cells expend energy to transport substances up a concentration gradient. Mechanisms of active transport include the use of **carrier proteins, endocytosis,** and **exocytosis.**

Cellular Connection and Communication

- Structures that connect eukaryotic cells together include **extracellular matrices, tight junctions,** and **anchoring junctions.**

- Structures that facilitate communication between eukaryotic cells include **plasmodesmata** and **gap junctions.**

Sample Test Questions

1. Shortly after someone eats, she has glucose in higher concentrations outside her body's cells than inside her body's cells. Explain the mechanism by which glucose enters the cell.

2. A scientist examines an unknown cell under a light microscope. Because the light microscope has poor magnification, the scientist can only identify three features of the cell: a nucleus, a cell wall, and a large vacuole that fills more than half the cell. What type of cell is this? What other features might be visible if the scientist examined the cell with an electron microscope?

3. Explain how cell membrane proteins might play a role in the immune system.

4. Which of the following would be most appropriate for studying the surface of a cell?

    A. TEM
    B. SEM
    C. Light microscope
    D. Dissecting scope
    E. Magnifying lens

5. The organelle best known as the "packaging and shipping" center of the cell is the

    A. Nucleus
    B. Mitochondria
    C. Golgi apparatus
    D. Peroxisome
    E. Vacuole

6. Which of the following is considered the most significant difference between prokaryotic and eukaryotic cells?

    A. Prokaryotic cells have a cell membrane, whereas eukaryotic cells do not.

    B. Prokaryotic cells possess ribosomes, whereas eukaryotic cells do not.

    C. Prokaryotic cells are structurally complex, whereas eukaryotic cells are not.

    D. Eukaryotic cells have DNA, whereas prokaryotic cells do not.

    E. Eukaryotic cells have membrane-bound organelles, whereas prokaryotic cells do not.

7. Protein synthesis occurs at which of the following organelles?

    A. Nucleus

    B. Mitochondria

    C. Ribosomes

    D. Golgi apparatus

    E. Chloroplasts

8. Which of the following organelles are NOT found in plant cells?

    A. Centrioles

    B. Cytoskeleton

    C. Mitochondria

    D. Golgi apparatus

    E. Ribosomes

9. Which of the following organelles are NOT found in animal cells?

    A. Gap junctions

    B. Tight junctions

    C. Anchoring junctions

    D. Plasmodesmata

    E. Extracellular matrices

10. Which of the following molecules may diffuse across a cell membrane without the expenditure of energy or the use of membrane proteins?

A. CO_2
B. Insulin
C. Na^+
D. Glucose
E. Cl^-

11. Which of the following organelles synthesizes lipids?

A. Golgi apparatus
B. Ribosomes
C. Smooth ER
D. Rough ER
E. Lysosomes

12. Suppose two compartments within a cell are separated by a semipermeable membrane. Compartment A has a higher solute concentration than compartment B. Compartment A is _____ to compartment B, and water will move via osmosis into _____ .

A. hypertonic; compartment A
B. hypertonic; compartment B
C. hypotonic; compartment A
D. hypotonic; compartment B
E. isotonic; both compartments

13. In which of the following processes is energy expended to transport a substance out of a cell?

A. Pinocytosis
B. Phagocytosis
C. Exocytosis
D. Facilitated diffusion
E. Osmosis

14. Which of the following structures are found in both prokaryotic and eukaryotic cells?

    A. DNA
    B. Ribosomes
    C. Cell membrane
    D. B and C only
    E. All of the above

15. Membrane-bound sacs that perform metabolic functions like breaking down fatty acids, detoxifying alcohol, and converting toxic hydrogen peroxide into water and oxygen are known as which of the following?

    A. Vacuoles
    B. Lysosomes
    C. Peroxisomes
    D. Golgi bodies
    E. Mitochondria

ANSWERS

1. Glucose will enter the cells through the process of facilitated diffusion. All molecules move spontaneously down their concentration gradients. If glucose concentrations are higher outside the cell than inside the cell, passive transport allows glucose to enter the cell without any expenditure of energy. However, since glucose is a relatively large molecule, it cannot pass directly through the cell membrane. Instead, a transport protein must assist it through the membrane.

2. The scientist is examining a plant cell. The presence of a nucleus indicates that the cell is eukaryotic, while the presence of a cell wall and large central vacuole indicates that the cell belongs to a plant, rather than an animal, fungus, or protist. Using an electron microscope, the scientist would also be able to see ribosomes, endoplasmic reticulum, Golgi bodies, mitochondria, peroxisomes, chloroplasts, and parts of the cytoskeleton.

3. One of the primary jobs of the immune system is to recognize and destroy foreign or damaged cells. Cell membrane proteins aid in this task because they are involved in cell-cell recognition. Marker proteins relay information about the cell's origin and function to other cells, allowing them to determine if the cell presents a threat to the organism.

4. B   SEM, or scanning electron microscopes, would be most appropriate for studying the surface of a cell.

5. C   The Golgi apparatus is best known as the "packaging and shipping" center of the cell.

6. E   The most significant difference between prokaryotic and eukaryotic cells is that eukaryotic cells have membrane-bound organelles, including a nucleus, whereas prokaryotic cells do not.

7. C   Protein synthesis occurs at the ribosomes.

8. A   Centrioles are not found in plant cells.

9. D   Plasmodesmata are not found in animal cells.

10. A CO_2 may diffuse across a cell membrane without energy or the use of membrane proteins.

11. C   The smooth ER synthesizes lipids.

12. A   Compartment A is hypertonic to compartment B. Water will move via osmosis into compartment A.

13. C   In exocytosis, energy is used to transport a substance out of a cell.

14. E   DNA, ribosomes, and a cell membrane are all found in both prokaryotic and eukaryotic cells.

15. C   Membrane-bound sacs that perform metabolic functions like breaking down fatty acids, detoxifying alcohol, and converting toxic hydrogen peroxide into water and oxygen are known as peroxisomes.

CELLULAR METABOLISM

Energy

Cellular Respiration

Photosynthesis

3

Energy

All of life's processes require **energy,** the capacity to do work. Energy comes in many different forms, including chemical, thermal, light, and mechanical. Despite its many forms, all energy can be classified as one of two types: kinetic energy and potential energy. **Kinetic energy** is the energy of moving objects; it is energy in use. **Potential energy** is stored energy; it is energy that has the potential to do work. A ball rolling down a hill has kinetic energy, whereas a ball at the top of a hill has potential energy.

A molecule stores potential energy until it is released in the kinetic form of chemical or thermal (heat) energy. Within a particular system, the sum of potential and kinetic energy is called the system's **free energy.** Free energy is the amount of energy that could be used to power other chemical reactions. The free energy in molecules can be transferred, stored, or released during chemical reactions.

Oxidation-reduction reactions transfer energy between molecules in the form of electrons. The molecule that loses an electron is **oxidized,** while the molecule that gains an electron is reduced.

> *EXAMPLE:* NADH is a common energy carrier within cells. In the equation below, through a chemical reaction with hydrogen (2H), NAD^+ is oxidized to NADH, while hydrogen is reduced to hydrogen ion (H^+). In the process, energy is transferred from hydrogen to NADH. The reaction can also occur in the reverse direction, with NADH being reduced to NAD^+ and hydrogen ion becoming oxidized to hydrogen.
>
> $$NAD^+ + 2H \rightleftarrows NADH + H^+$$

Endergonic reactions store energy within a molecule because the reactants have less free energy than the products. These reactions require energy input.

> *EXAMPLE:* The production of glucose from carbon dioxide (CO_2) and water (H_2O) is an endergonic reaction because it requires energy input. Because energy is expended in the process, this reaction cannot occur in the reverse direction.
>
> $$6CO_2 + 6H_2O + energy \longrightarrow C_6H_{12}O_6 \text{ (glucose)} + 6O_2$$

Exergonic reactions release energy, leaving the reactants with more free energy than the products.

> *EXAMPLE:* The breakdown of glucose into carbon dioxide (CO_2) and water (H_2O) is an exergonic reaction because it results in the release of energy. Because this reaction releases energy, it cannot occur in the reverse direction.
>
> $$C_6H_{12}O_6 \text{ (glucose)} + 6O_2 \longrightarrow 6CO_2 + 6H_2O + energy$$

Cells often use the energy released from exergonic reactions to power endergonic reactions; these are called coupled reactions.

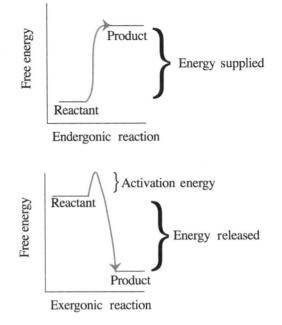

Endergonic and Exergonic Reactions

ATP

ATP, adenosine triphosphate, is referred to as "the energy currency of the cell" because it powers most of the reactions that take place in a cell. ATP consists of a ribose sugar, an adenine (a type of nucleotide), and a chain of three phosphate groups. The bonds that link the second and third phosphate group can be broken down to produce ADP (adenosine diphosphate), a free phosphate group (P), and a substantial amount of energy used for endergonic reactions.

$$ATP \rightleftarrows ADP + P + energy$$

EXAMPLE: The human body uses, on average, one kilogram of ATP every hour.

ENZYMES

Exergonic reactions require a small initial input of energy, called **activation energy,** before the reaction can proceed. **Enzymes** are proteins that lower the activation energy of a reaction. The **active site** of an enzyme binds with the reactants (also called substrates) and either changes them in some way or simply brings them in closer proximity to one another. Chemical reactions do not use up or change the enzyme. Once the reaction has taken place, the product is released, and the enzyme is free to catalyze other reactions.

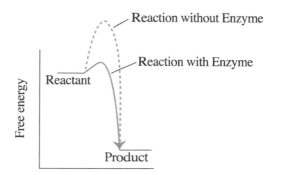

Enzyme Catalyzed Reaction

Enzyme Inhibitors

The presence of other molecules may inhibit an enzyme, or prevent it from functioning. Inhibition can occur in two ways:

- **Competitive inhibition** occurs when the inhibitor binds with the active site of an enzyme. With the active site already occupied, the enzyme cannot bind with the substrate.

- **Noncompetitive inhibition** occurs when the inhibitor binds with an **allosteric site** (any site other than the active site) and changes the shape of the enzyme so that it no longer bonds with the substrate.

EXAMPLE: Inhibitors are often used as drugs, in many cases to prevent detrimental reactions in an organism. Aspirin, for instance, inhibits the enzymes that cause pain and inflammation. However, inhibitors can also be poisonous. Cyanide is a lethal toxin because it competitively inhibits cytochrome coxidase, an enzyme involved with cellular respiration.

Cellular Respiration

Since energy cannot be created or destroyed, organisms must obtain their energy from the environment, usually in the form of food or solar radiation. The process of converting energy into a form that can be used by cells is called cellular metabolism. The two methods of cellular metabolism used by most organisms are **cellular respiration** and **photosynthesis.**

Cellular respiration converts the energy found in food molecules, especially glucose, to the more useable form of ATP. A single glucose molecule results in the production of approximately 36 ATP. Cellular respiration is a lengthy process that involves many reactions occurring in different parts of the cell. The entire process, however, can be summed up in one chemical formula:

$$C_6H_{12}O_6 \text{ (glucose)} + 6O_2 + ADP + P \longrightarrow 6CO_2 + 6H_2O + ATP$$

No transfer or use of energy is 100 percent efficient. Some energy is always lost to the environment in the form of heat. Cellular respiration is remarkably efficient for a biological process, but even so, only 40 percent of the energy in glucose is converted to ATP.

Cellular respiration occurs in four stages:

1. Glycolysis

2. The oxidation of pyruvate

3. The Krebs cycle

4. The electron transport chain

GLYCOLYSIS

The first stage of cellular respiration, **glycolysis,** takes place in the cytosol. Glycolysis converts glucose to two molecules of pyruvate, the compound from which energy will be extracted in the Krebs cycle. Glycolysis produces a net gain of 2 ATP and 2 NADH, an energy-carrying molecule. Water is also released in this reaction.

The figure below shows the conversion of glucose into two molecules of pyruvate. Other reactants and products, such as ATP, ADP, NAD^+, NADH, and CO_2 are also produced.

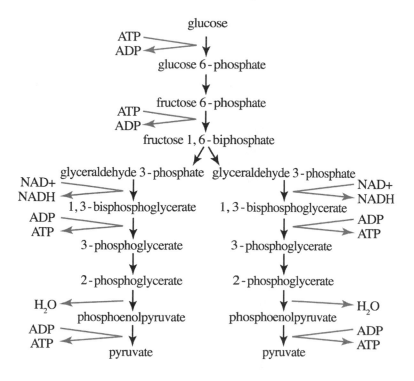

Glycolysis

OXIDATION OF PYRUVATE

In the second stage of cellular respiration, oxidation of pyruvate, the two molecules of pyruvate are oxidized and transformed into molecules of **acetyl CoA.** This reaction takes place across the mitochondrial membrane and produces 1 NADH per pyruvate, for a total of 2 NADH per glucose. Carbon dioxide is also released in this reaction.

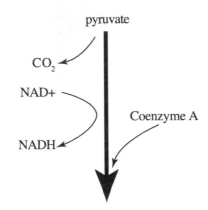

Oxidation of Pyruvate

KREBS CYCLE

The **Krebs cycle** is the third stage of cellular respiration and it takes place within the matrix of the mitochondria. The Krebs cycle processes each acetyl CoA to produce 3 NADH, 1 $FADH_2$, and 1 ATP, for a total of 6 NADH, 2 $FADH_2$, and 2 ATP per glucose. Carbon dioxide is also released in this reaction.

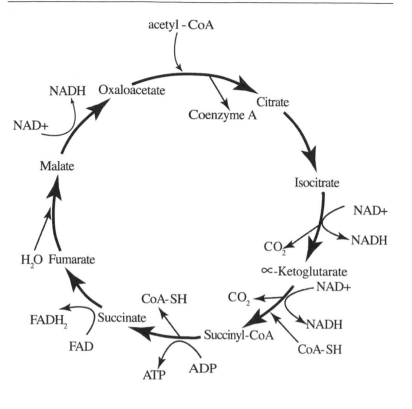

Krebs Cycle

ELECTRON TRANSPORT CHAIN

The **electron transport chain** is a series of molecules embedded in the inner membrane of the mitochondria. In this final stage of cellular respiration, the 10 NADH and 2 FADH$_2$ molecules produced in the three earlier stages power the production of the remaining 32 ATP. Oxygen is also consumed in this reaction to produce water.

CHAPTER 3
Cellular Metabolism

The production of 32 ATP from NADH and $FADH_2$ in the electron transport chain requires the following steps:

1. The electron carriers NADH and $FADH_2$ shuttle electrons to the inner mitochondrial membrane.

2. NADH and $FADH_2$ donate their electrons to the first in a series of membrane proteins. Each protein uses the energy in the electron to pump H^+ into the intermembrane space of the mitochondrion before passing the electron to the next carrier. The final electron receptor is O_2, which combines with two protons, H^+, to form water.

3. By pumping H^+ into the intermembrane space, the electron transport chain sets up a high concentration gradient. H^+ flow down gradient through the ATP synthase, a membrane protein that catalyzes the production of ATP from ADP.

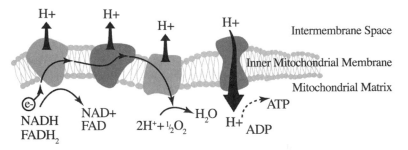

The Electron Transport Chain

Summary of Cellular Respiration

The four stages of cellular respiration are summarized in the table below.

	Stage	Location	Reaction
1.	**Glycolysis**	Cytosol	Converts 1 molecule of glucose to 2 molecules of pyruvate. 2 ATP and 2 NADH molecules are produced and water is released.
2.	**Oxidation of pyruvate**	Mitochondria	Converts 2 molecules of pyruvate to 2 molecules of acetyl CoA. 2 NADH molecules are produced and carbon dioxide is released.
3.	**Krebs cycle**	Mitochondrial matrix	Converts 2 molecules of acetyl CoA to 6 molecules of NADH, 2 molecules of $FADH_2$, and 2 molecules of ATP. Carbon dioxide is released.
4.	**Electron transport chain**	Mitochondria	10 NADH molecules and 2 $FADH_2$ are converted to 32 ATP molecules. Oxygen is consumed and water is produced.

Food molecules other than glucose can also be converted into ATP. Proteins and fats first undergo modification before entering the process of cellular respiration at different stages.

FERMENTATION

Eukaryotic cells can also produce ATP through **fermentation.** Fermentation is much less efficient than the four stages of cellular respiration described above, but it enables the cells to continue producing ATP when oxygen is unavailable. Fermentation begins with glycolysis, much the same way that cellular respiration begins. Following glycolysis, there are two general ways in which fermentation can proceed:

CHAPTER 3
Cellular Metabolism

- **Alcoholic fermentation:** Pyruvic acid is converted to ethanol. Alcoholic fermentation is used by fungi and some plants.

- **Lactic acid fermentation:** Pyruvic acid is converted to lactate. Lactic acid fermentation is used by animals and bacteria.

EXAMPLE: The sour taste of sourdough bread comes from the lactic acid produced by the fermentation of bacteria.

Photosynthesis

Plants, as well as some protists and bacteria, create food molecules (sugars) from carbon dioxide and solar energy through the process of photosynthesis. Photosynthesis is a complex process that can be summed up in the following equation:

Solar energy + $6CO_2$ $\longrightarrow$ $C_6H_{12}O_6$ (glucose) + $6H_2O$ + $6O_2$ + $12H_2O$

Photosynthesis in plants takes place in the chloroplasts and takes place in two stages:

1. The light-dependent reactions

2. The Calvin cycle (also called the light-independent reactions)

LIGHT-DEPENDENT REACTIONS

Light-dependent reactions convert solar energy into ATP and NADPH, the reduced form of the electron receptor, $NADP^+$. During these reactions, water is split, leaving oxygen (O_2) as a waste product. These reactions take place in **photosystems** in the chloroplasts.

Photosystems comprise clusters of molecules composed of light-absorbing pigments and a reaction center, which includes a primary electron acceptor and two chlorophyll *a* pigment molecules. Photosystems are found in the thylakoid membrane of the plant's chloroplast. There are two types of photosystems present in plants, each with small differences in structure. The two photosystems work sequentially, with light first being absorbed by **photosystem II** and later by **photosystem I.**

The light-dependent reactions following these steps in creating NADPH:

1. Photosystem II absorbs solar energy in the form of light.

2. The solar energy excites electrons in the reaction center of photosystem II, which then enter an electron transport chain. These electrons originate from the splitting of water, which produces free electrons and O_2.

3. As electrons pass down the electron transport chain, protons are pumped into the thylakoid membrane space of the chloroplast. Protons diffuse out of the thylakoid membrane space through an ATP synthase protein, creating ATP.

4. Photosystem I accepts electrons from the electron transport chain and uses light energy to excite the electrons further.

The electrons travel through a second electron transport chain before being passed to $NADP^+$ to create NADPH.

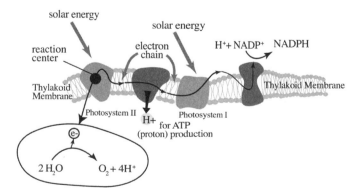

Light-Dependent Reactions

Cellular Respiration and Light-Dependent Reactions

Cellular respiration and the light-dependent reactions of photo-synthesis use similar processes to produce ATP. In both cases, an electron transport chain pumps H^+ across a membrane, creating a chemical gradient. The H^+ protons flow down their chemical gradients through ATP synthase proteins, resulting in the production of ATP. Scientists believe that the electron transport chain used in cellular respiration may have evolved from the transport system used in photosynthesis.

THE CALVIN CYCLE

The **Calvin cycle** uses ATP and NADPH from the light-dependent reactions to convert CO_2 into sugar that the plant can use. CO_2 is obtained from the outside environment though gas-exchanging organs on the plant's surface known as **stomata.** The process of **carbon fixation** incorporates the CO_2 into organic molecules. The incorporation of CO_2 is possible because of the energy-rich enzyme **rubisco** (ribulose biphos-phate carboxylase, or RuBP), a protein made during the light-dependent reactions of photosynthesis and abundant in plant leaves. A CO_2 molecule binds to RuBP. The molecule then splits into two 3-carbon molecules of PGA (3-phosphoglycerate). A series of reactions occur to convert the PGA into the 3-carbon sugar molecule glyceraldehyde 3-phosphate. This 3-carbon sugar molecule can then be used to make other sugars, includ-ing glucose and sucrose. The production of a single 3-carbon sugar molecule requires 3 CO_2, 9 ATP, and 6 NADPH.

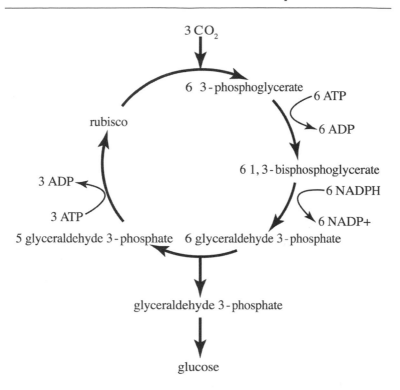

3 CO_2

6 3-phosphoglycerate

6 ATP

6 ADP

rubisco

6 1,3-bisphosphoglycerate

3 ADP

6 NADPH

3 ATP

6 NADP+

5 glyceraldehyde 3-phosphate   6 glyceraldehyde 3-phosphate

glyceraldehyde 3-phosphate

glucose

The Calvin Cycle

PHOTORESPIRATION

When the enzyme rubisco incorporates oxygen, rather than CO_2, into organic molecules, plants create energy through the process of **photorespiration.** Photorespiration occurs most often in arid regions where plants must close their stomata to prevent water loss to the air. This results in a buildup of oxygen levels in the leaf, which makes rubisco more likely to bind with the oxygen. Photorespiration is inefficient and detrimental to plants because it consumes more ATP to produce each 3-carbon sugar molecule, leading to a lower metabolic rate than photosynthesis.

Plants are divided into three different categories depending on their method of carrying out photosynthesis: the C_3 pathway, the CAM pathway, and the C_4 pathway. These modes of photosynthesis differ in how they adapt to the problem of photorespiration.

C_3 Plants

Photorespiration presents a major problem for **C_3 plants** because they have no special adaptations to reduce the process. The problem is exacerbated in hot, arid climates, where the rate of photorespiration increases as the temperature goes up. Consequently, C_3 plants are rarely found in these climates. Most plants, including wheat, barley, and sugar beet, are C_3 plants.

CAM Plants

CAM (crassulacean acid metabolism) plants reduce photorespiration and conserve water by opening their stomata only at night. CO_2 enters through the stomata and is fixed into organic acids, which are then stored in the cell's vacuole. During the day, the acids break down to yield high levels of CO_2 for use in the Calvin cycle. Through the periodic opening and closing of the stomata, CAM plants maintain a high CO_2 to O_2 ratio, minimizing the rate of photorespiration. CAM plants, such as cacti and pineapple, are most common in dry environments.

C_4 Plants

C_4 plants use the enzyme PEP carboxylase to fix CO_2 in the mesophyll cells of their chloroplasts. The fixed CO_2 is then shuttled to specialized structures known as the bundle-sheath cells, where it is released and incorporated into the Calvin cycle. This process is energetically expensive, but it limits photorespiration by allowing high concentrations of CO_2 to build up in the bundle-sheath cells. C_4 plants, such as corn and sugar cane, are common in warm environments.

Summary

Energy

- **Endergonic reactions** require energy input. **Exergonic reactions** release energy. Coupled reactions use the energy from exergonic reactions to power endergonic reactions.

- **Activation energy** is the small input of energy required before a reaction can proceed. **Enzymes** are proteins that bind with reactants to lower the activation energy of a reaction.

- **Oxidation** occurs when a molecule loses an electron, whereas reduction occurs when a molecule gains an electron.

- **ATP** is used to power most of the reactions in a cell.

Cellular Respiration

- **Cellular respiration** is the process that converts glucose into ATP.

- **Glycolysis** is the first stage of cellular respiration. In glycolysis, glucose is converted into two molecules of pyruvate.

- The second stage of cellular respiration is the oxidation of pyruvate and the formation of **acetyl CoA.**

- The third stage of cellular respiration is the **Krebs cycle.** In the Krebs cycle, acetyl CoA is broken down to produce NADH, $FADH_2$, and 2 ATP.

- The final stage of cellular respiration is the **electron transport chain.** In this stage, the NADH and $FADH_2$ molecules that were produced in the three earlier stages are used to produce 32 ATP.

- **Alcoholic fermentation** and **lactic acid fermentation** are methods of producing ATP in oxygen-poor environments.

Photosynthesis

- **Photosynthesis** is the process that converts CO_2 into sugars using solar energy.

- The **light-dependent reactions** convert solar energy into ATP and NADPH, and occur within **photosystems** in the chloroplasts.

- The **Calvin cycle** uses ATP and NADPH to convert CO_2 into sugar. An important step in the Calvin cycle is the **fixation of carbon** using the enzyme **rubisco.**

- **Photorespiration** occurs when the enzyme rubisco incorporates oxygen, rather than CO_2, into organic molecules.

- The three modes of photosynthesis, **C₃**, **CAM**, and **C₄**, each have a unique method for reducing photorespiration.

Sample Test Questions

1. How is energy stored in ATP? Why is ATP called the "energy currency of the cell"?

2. Why do C₄ plants have an advantage over C₃ plants in tropical environments? Why are C₄ plants not usually found in cool environments?

3. Why is cellular respiration less efficient when no oxygen is available?

4. Which of the following is a waste product of photosynthesis?

  A. CO_2
  B. O_2
  C. $C_6H_{12}O_6$
  D. ATP
  E. Solar energy

5. Before entering the Krebs cycle, pyruvate must be converted to which of the following?

  A. Lactic acid
  B. Alcohol
  C. Acetyl CoA
  D. $FADH_2$
  E. Rubisco

6. NADPH loses an electron to become NAD^+. In this process, NADPH is said to be which of the following?

    A. Oxidized
    B. Reduced
    C. Metabolized
    D. Synthesized
    E. Fermented

7. Plants that keep their stomata closed during the day and only absorb CO_2 at night use which mode of photosynthesis?

    A. Alcoholic fermentation
    B. Lactic acid fermentation
    C. C_3
    D. C_4
    E. CAM

8. The burning of wood is an example of which of the following?

    A. Coupled reaction
    B. Endergonic reaction
    C. Exergonic reaction
    D. Enzyme-catalyzed reaction
    E. Reduced reaction

9. NADH and $FADH_2$ are converted to ATP during which of the following?

    A. Glycolysis
    B. Oxidation of pyruvate
    C. Krebs cycle
    D. Electron transport chain
    E. Calvin cycle

10. The conversion of solar energy to ATP and NADPH occurs during which of the following?

    A. Light-dependent reactions
    B. Light-independent reactions
    C. Calvin cycle
    D. Krebs cycle
    E. Electron transport chain

11. Carbon fixation occurs during which of the following?

    A. Light-dependent reactions
    B. Calvin cycle
    C. Glycolysis
    D. Krebs cycle
    E. Electron transport chain

12. Substances that lower the activation energy of a reaction are which of the following?

    A. Electron receptors
    B. Electron donors
    C. ATP
    D. Enzymes
    E. Inhibitors

13. In photosynthesis, light energy is first absorbed by which of the following?

    A. Photosystem I
    B. Photosystem II
    C. O_2
    D. Bundle-sheath cells
    E. Mesophyll cells

14. Fungi, such as yeast, make ATP in anaerobic conditions using which of the following processes?

    A. Calvin cycle
    B. Krebs cycle
    C. Alcoholic fermentation
    D. Lactic acid fermentation
    E. Oxidation of pyruvate

15. In which of the following processes is the shape of an enzyme changed after binding with a substance at an allosteric site?

    A. Catalysis
    B. Oxidation
    C. Reduction
    D. Competitive inhibition
    E. Noncompetitive inhibition

ANSWERS

1. ATP stores energy in the bond between the second and third phosphate group. When a reaction breaks this bond, a substantial amount of energy is released. ATP is known as the "energy currency of the cell" because cells use the energy released in the breakdown of ATP to power endergonic reactions in the cell.

2. C_3 plants do not perform well in the tropics because their productivity is limited by photorespiration, an "error" of the Calvin cycle that occurs when the enzyme rubisco incorporates oxygen, rather than carbon dioxide, into organic molecules. Photorespiration is inefficient as it wastes both carbon and energy. As temperatures rise, plants close their stomata to prevent the loss of water, increasing the oxygen buildup that leads to photorespiration. C_4 plants fair better in the tropics because they have a mechanism for reducing photorespiration: CO_2 is concentrated in the bundle sheath cells, where the Calvin cycle occurs, thus reducing the likelihood that rubisco will bond with oxygen instead of carbon dioxide. However, this adaptation comes at a price: C_4 plants must expend extra ATP to produce glucose. Thus, C_4 plants outperform C_3 plants in the warm tropics, but not in cooler environments, where photorespiration is not a significant problem.

3. Of the maximum 36 ATP that can be produced in cellular respiration, 32 come from the electron transport chain. Oxygen is an important electron receptor, and without it the electron transport chain cannot function. Without the presence of oxygen, the metabolism of glucose is much less efficient at producing ATP.

4. B O_2 is a waste product of photosynthesis.

5. C   Before entering the Krebs cycle, pyruvate must be converted to acetyl CoA.

6. A   NADPH is oxidized when it loses an electron to become NAD^+.

7. E   Plants that keep their stomata closed during the day and only perform the Calvin cycle at night use CAM photosynthesis.

8. C   The burning of wood is an exergonic reaction because it releases energy.

9. **D** NADH and FADH$_2$ are converted to ATP in the electron transport chain.

10. **A** The conversion of solar energy to ATP and NADPH occurs during the light-dependent reactions of photosynthesis.

11. **B** Carbon fixation occurs during the Calvin cycle of photosynthesis.

12. **D** Enzymes are substances that lower the activation energy of a reaction.

13. **B** In photosynthesis, light energy is first absorbed by photosystem II.

14. **C** Alcoholic fermentation is the processes by which fungi make ATP in anaerobic conditions.

15. **E** Noncompetitive inhibition occurs when an inhibitor binds with an enzyme at an allosteric (nonactive) site, thus changing the shape of the enzyme.

CELL DIVISION

Cell Cycle

Chromosomes

Mitosis

Meiosis

4

Cell Cycle

The **cell cycle** encompasses the time between the creation of a new cell and that cell's division. **Cell division,** the splitting of one cell into two, is the process that makes growth and reproduction possible for any organism. Cell division follows a specific progression that depends on whether the cell is prokaryotic or eukaryotic.

The eukaryotic cell cycle is broken into two major phases: **interphase,** when the cell is not dividing, and **mitotic phase,** when it is. Subphases of these two phases are shown in the table on the following page.

Phase	Subphase	Description
Interphase	G_1 (Growth 1)	The main development period of cell growth, during which new organelles form within the cell.
	S (Synthesis)	The cell duplicates its DNA. Cells emerge from the S phase with two identical copies of their DNA.
	G_2 (Growth 2)	The second period of cell growth, during which the cell prepares for the division that will take place during the mitotic phase.
Mitotic phase, or M phase	Mitosis	The cell's chromosomes, or gene-carrying structures, divide.
	Cytokinesis	The cell's cytoplasm and cell membrane divide, completing cell division.

EXAMPLE: Some cells, including many nerve cells, are programmed never to divide. These cells are said to be in G_0, or resting, phase.

CELL CYCLE CONTROL

As they divide, cells must proceed through the various stages of the cell cycle, including the G_1, G_2, and M phases. All stages of the cell cycle are controlled by checkpoints. Triggers at each checkpoint assess the cell's readiness to proceed to the next stage. Regulation also ensures that each cell obtains the proper number and type of chromosomes (packages of DNA) and organelles. Without the controlled timing of cell division, an organism would be a shapeless blob of uncoordinated cells.

The table on the following page shows the three main checkpoints that control the cell cycle.

Checkpoint	Occurs at	Details
G$_1$ checkpoint	the end of G$_1$ phase	If conditions are not suitable for replication, the cell will not proceed to S phase but will instead enter a resting phase, G$_0$.
G$_2$ checkpoint	the end of G$_2$ phase	If conditions are not suitable, transition to the M phase will be delayed. If DNA is damaged, cell division will be delayed to allow time for DNA repair.
M checkpoint	between metaphase and anaphase stages of mitosis	If the chromosomes are aligned properly and ready for division, the cell will proceed from metaphase to anaphase, during which it will divide. If the chromosomes are not aligned properly, the anaphase stage will be delayed.

EXAMPLE: Malignant cancer cells are deadly, in part, because they undergo unregulated cell division, which enables them to spread rapidly throughout the body. Scientists have discovered one reason behind this uncontrolled growth: a defective *p53* gene. Proteins produced by the *p53* gene assess the cell's DNA for damage at the G$_1$ checkpoint. If the DNA is intact, cell division proceeds. If the DNA is damaged, however, the *p53* proteins halt cell division until the DNA is repaired or the cell is destroyed. If the *p53* gene itself has been damaged, as in the case of cells that are cancerous, the G$_1$ checkpoint will fail and a malignant cancer cell may develop.

Chromosomes

In eukaryotic cells, DNA and associated proteins are wrapped together in packages called chromosomes. Eukaryotic cell division can take many forms depending on the chromosome structure of the cell involved. DNA in eukaryotic cells is wrapped around the proteins to form a complex called a **chromatin.** Throughout most of a cell's life cycle, the chromatin is loosely packed within the nucleus. During cell division, however, the chromatin becomes highly condensed and folds up to form condensed **chromosomes.** DNA is always replicated, or copied, before becoming condensed. Therefore, the characteristic *x* shape associated with chromosomes actually represents a replicated chromosome consisting of two identical sister **chromatids** joined at the **centromere.**

> *EXAMPLE:* Unlike eukaryotes, prokaryotes do not have chromosomes. Prokaryotic DNA exists in a single loop and is not extensively folded the way eukaryotic DNA is.

CHROMOSOME NUMBER

Chromosome number refers to the number of chromosomes within each cell of an organism. Most animals possess two nonidentical versions of every chromosome. These pairs are known as **homologous chromosomes.** Homologous chromosomes have the same size, shape, and function but may have slightly different versions of most **genes,** the basic unit of hereditary information. A pair of homologous chromosomes may both have genes for eye color, but one may have genes encoding blue eyes, while the other has genes encoding brown eyes. Cells with two sets of every chromosome between their homologous chromosomes are **diploid** ($2n$), while cells with one set of every chromosome are **haploid** ($1n$).

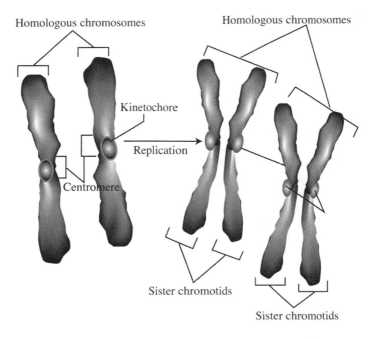

Chromosome Structure

Humans' Chromosome Number

The chromosome number of nearly every cell in the human body can be written as $2n = 46$. This notation indicates that human cells are diploid, possessing 2 sets each of 23 chromosomes. The exceptions are human egg and sperm cells. These cells have a chromosome number of $1n = 23$, indicating that egg and sperm are haploid and possess one set each of 23 chromosomes. The union of sperm and egg that occurs during **fertilization** restores the chromosome number of the resulting embryo to $2n = 46$.

Mitosis

Mitosis is the method of eukaryotic cell division that produces two genetically identical cells. All cells in an organism, except sperm and eggs, are produced via mitosis. Mitosis progresses along five stages: prophase, metaphase, anaphase, telophase, and cytokinesis.

PROPHASE

During prophase, the duplicated chromosomes condense and become visible as distinct sister chromatids. The nuclear envelope, the membrane surrounding the nucleus, breaks down and the centrosomes move toward the poles of the cell. The mitotic spindle, which is made of microtubules, attaches to a specialized structure called the **kinetochore,** located at the centromere of each replicated chromosome.

METAPHASE

During metaphase, the replicated chromosomes align at the equator, or metaphase plate, of the cell.

ANAPHASE

During anaphase, the sister chromatids separate and are moved toward opposite poles of the cell by the spindle. As this happens, the cell begins to elongate toward the poles.

TELOPHASE

The cell continues to elongate throughout telophase, and the mitotic spindle breaks down. A new nuclear envelope forms at each end of the cell, and the chromosomes within begin to unfold into chromatin.

CYTOKINESIS

The cytoplasm and organelles are evenly divided between the two new cells during **cytokinesis,** completing the process of cell division. Cytokinesis differs slightly in animals and plants.

- In animals, a ring of microfilaments contracts in the center of the elongated cell, producing a **cleavage furrow** that eventually pinches off the two cells.

- In plants, a **cell plate** is formed as vesicles containing cell membrane materials fuse together along the equator of the cell. Once the cell plate has fused with the plasma membrane and the two cells are completely divided, cellulose is secreted to form the cell wall.

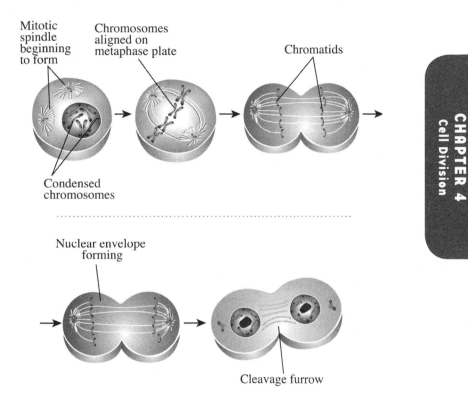

The Five Steps of Mitosis

Binary Fission

Cell division in prokaryotes occurs through a process called **binary fission.** Because prokaryotic cells have just a single double-stranded loop of DNA, instead of the multiple chromosomes of eukaryotic cells, the steps in binary fission are much simpler than those of mitosis. Binary fission occurs in four steps:

1. The DNA is replicated.

2. The cell doubles in size.

3. The cell membrane grows into the center of the cell, between the two circles of DNA, dividing the cell in two.

4. The two cells separate, and a cell wall forms around each new cell.

Meiosis

Meiosis is the method of cell division that takes place in sexually reproducing organisms specifically for the creation of **gametes**—sperm and egg cells. Meiosis results in the production of four haploid cells, each of which is genetically different. To produce four cells, meiosis requires two rounds of cell division:

1. **Meiosis I:** Homologous pairs of each chromosome join and might exchange genetic material. The homologous chromosomes are pulled to opposite poles in the cell, at which point the cell separates, resulting in two cells. Each cell contains half of the chromosome number of the original diploid cell. Each chromosome remains in the duplicated state and is made up of two sister chromatids.

2. **Meiosis II:** The second stage of meiosis follows similar steps as mitosis in the creation of two more cells. Chromosomes do not replicate between Meiosis I and Meiosis II. The result is four haploid cells genetically different from one another.

MEIOSIS I

Meiosis I occurs in five stages, which are similar to the steps of mitosis in other eukaryotic cells: prophase I, metaphase I, anaphase I, telophase I, and cytokinesis I. During the S phase of interphase, prior to the start of meiosis I, each chromosome replicates to produce two sister chromatids. The two genetically identical (homologous) sister chromatids remain attached at their centromeres.

Prophase I

The most important events in prophase I are synapsis and crossing over. **Synapsis** occurs when the two homologous chromosomes condense and combine to form complexes called **tetrads. Crossing over** is the exchange of genetic material that then takes place between these homologous chromosomes along several junctions known as **chiasmata.** As a result of crossing over, the cells produced through meiosis are genetically variable. The other events in prophase I are similar to those occurring during prophase in mitosis: The centrosomes move to the poles of the cell, the spindle forms and attaches to the kinetochores of the chromosomes, and the nuclear envelope breaks down.

Metaphase I

The tetrads align along the metaphase plate of the cell.

Anaphase I

The homologous chromosomes of each tetrad separate and are pulled toward opposite poles of the cell by the spindle. The side of the cell toward which a homologous chromosome is pulled is random, depending only on the orientation of the tetrad. The independent assortment of chromosomes for each cell is a result of this random mix of chromosomes derived from that organism's parents.

> *EXAMPLE:* Crossing over and the independent assortment of chromosomes during meiosis are two forces that help to produce genetic variation. By independent assortment alone, a single human can produce more than 8 million genetically different gametes. When crossing over is also considered, the possible number of genetically different gametes is nearly limitless.

Telophase I

Telophase I is essentially identical to telophase in mitosis. The cell continues to elongate, and the mitotic spindle breaks down. A new nuclear envelope forms at each end of the cell and the chromosomes within unfold into chromatin.

Cytokinesis I

Cytokinesis I is identical to cytokinesis in mitosis: The cytoplasm and organelles are divided between the two cells, completing the process of cell division. By the end of this stage, two genetically different haploid cells have been produced. Each chromosome is still in the duplicated state and is made up of two sister chromatids. However, because of crossing over during prophase I, the sister chromatids are no longer identical.

MEIOSIS II

Meiosis II can be divided into stages that follow a similar sequence as meiosis I: prophase II, metaphase II, anaphase II, telophase II, and cytokinesis II. These stages are identical in each of the haploid cells produced during meiosis I.

Prophase II

The chromosomes within the haploid cell condense, and the spindle attaches to the kinetochore of each chromosome. The nuclear envelope breaks down and the centrosomes move toward the poles of the cell.

Metaphase II

The chromosomes align along the center of the metaphase plate.

Anaphase II

The sister chromatids separate and are moved toward opposite poles of the cell by the spindle. The cell begins to elongate toward the poles.

Telophase II

The cell continues to elongate and the mitotic spindle breaks down. A new nuclear envelope forms at each end of the cell and the chromosomes within may unfold into chromatin.

Cytokinesis II

The cytoplasm and organelles are divided between the two cells, completing the process of cell division. By the end of this stage, four genetically different haploid cells have been produced.

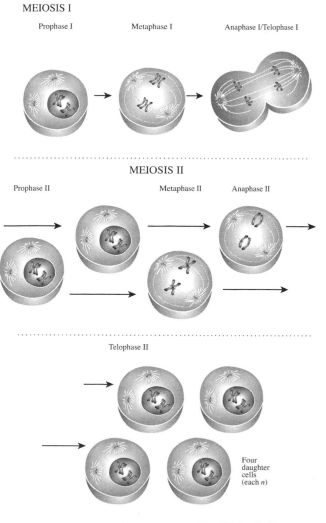

The Stages of Meiosis I and II

EXAMPLE: The process of meiosis results in four genetically different haploid cells. In animals, these haploid cells develop into gametes, a sperm in males and an egg in females. Fertilization is the process by which a sperm and an egg fuse together. The resulting zygote is diploid, with half the chromosomes coming from the mother and half coming from the father. The processes of meiosis and fertilization both account for the genetic variation found in animals of the same species. Meiosis is responsible for creating gametes whose genetic material varies from that of the parent. Fertilization then combines the genetic material of the two parents to produce the genetic material of the offspring.

Summary

Cell Cycle

- The **cell cycle** comprises the events occurring between the origin of a new cell and that cell's division. The cell cycle takes place over five phases: G_1, S, G_2, **mitosis,** and **cytokinesis.**

- The transitions between phases of the cell cycle are regulated by three checkpoints: the G_1 checkpoint, the G_2 checkpoint, and the M checkpoint.

Chromosomes

- **Chromosomes** are pieces of DNA that have been tightly folded and wrapped around proteins.

- Replicated chromosomes consist of two sister **chromatids** joined at the **centromere.**

- **Homologous chromosomes** are different versions of the same chromosome. Homologous chromosomes have the same size, shape, and function but may have slightly different versions of most genes.

- **Diploid** ($2n$) cells have two versions of every chromosome, whereas **haploid** ($1n$) cells have one version of every chromosome.

Mitosis

- **Mitosis** is the method of cell division that produces two genetically identical cells. Mitosis takes place over five stages: prophase, metaphase, anaphase, telophase, and cytokinesis.

- In prophase, the chromosomes condense. In metaphase, the chromosomes line up along the metaphase plate. In anaphase, the sister chromatids separate. In telophase, one nucleus forms at each end of the cell. Finally, in cytokinesis, the cytoplasm is divided and cell division is completed.

Meiosis

- **Meiosis** is the process of cell division used to create **gametes**—sperm and egg cells. Meiosis results in the production of four haploid cells, each of which is genetically different.

- In meiosis I, homologous chromosomes are separated and the chromosome number is reduced by half. In prophase I, **synapsis** and **crossing over** takes place. In metaphase I, the **tetrads** line up along the metaphase plate. In anaphase I, the homologous chromosomes separate. In telophase I, one nucleus forms at each end of the cell. Finally, in cytokinesis I, the cytoplasm is divided and cell division is completed.

- In meiosis II, the sister chromatids are separated. Meiosis II follows the same process as mitosis.

Sample Test Questions

1. Compare and contrast both the processes and the results of mitosis and meiosis.

2. Children always get 50% of their genes from their mother and 50% from their father, but the percentage of genes derived from each grandparent varies. Why?

CHAPTER 4
Cell Division

3. Many chromosomal diseases in humans, including Down syndrome, are the result of errors in meiosis. Down syndrome occurs when a person possesses three copies of chromosome 21. Explain how an error in meiosis might cause a person to have Down syndrome.

4. What is the name for chromosomes that are similar in size, shape, and function but may have different versions of each gene?

   A. Sister chromosomes
   B. Sister chromatids
   C. Replicated chromosomes
   D. Homologous chromosomes
   E. Homologous chromatids

5. Sister chromatids separate during which phase of mitosis?

   A. Prophase
   B. Metaphase
   C. Anaphase
   D. Telophase
   E. Cytokinesis

6. Sister chromatids separate during which phase of meiosis?

   A. Prophase I
   B. Prophase II
   C. Anaphase I
   D. Anaphase II
   E. Cytokinesis

7. Which checkpoint determines whether a cell will proceed to the S phase or enter a resting phase (G_0)?

   A. G_0 checkpoint
   B. G_1 checkpoint
   C. G_2 checkpoint
   D. M checkpoint
   E. S checkpoint

8. Crossing over occurs during which phase of meiosis?

    A. Prophase I
    B. Prophase II
    C. Metaphase I
    D. Metaphase II
    E. Cytokinesis

9. Bacteria divide using which of the following methods?

    A. Binary fission
    B. Mitosis
    C. Meiosis I
    D. Meiosis II
    E. Recombination

10. In which type of cell does a cell plate form during cytokinesis?

    A. Animal cells
    B. Plant cells
    C. Fungal cells
    D. Protist cells
    E. Bacterial cells

11. Sister chromatids align along the metaphase plate during which phase of meiosis?

    A. Metaphase I
    B. Metaphase II
    C. Anaphase I
    D. Telophase I
    E. Telophase II

12. What is the name for a cell with one copy of every chromosome?

    A. Replicated
    B. Prokaryotic
    C. Eukaryotic
    D. Diploid
    E. Haploid

CHAPTER 4
Cell Division

13. In which phase do most cells spend the majority of their life cycle?

    A. G_1
    B. G_2
    C. G_3
    D. S
    E. M

14. Sister chromatids are joined at which of the following?

    A. Centrosome
    B. Centromere
    C. Kinetochore
    D. Spindle
    E. Cleavage furrow

15. DNA synthesis occurs during which of the following phases?

    A. Prophase
    B. Metaphase
    C. G_1 phase
    D. G_2 phase
    E. S phase

ANSWERS

1. Mitosis and meiosis are the two processes of eukaryotic cell division. Mitosis has one round of cell division resulting in two genetically identical diploid cells, whereas meiosis has two rounds of cell division resulting in four genetically different haploid cells. Meiosis I is similar to mitosis in most respects, except the following:

- In prophase I of meiosis I, homologous chromosomes pair up to form tetrads, and crossing over occurs.

- In metaphase I of meiosis I, tetrads, rather than sister chromatids, line up along the metaphase plate.

- In anaphase I of meiosis I, homologous chromosomes, not sister chromatids, are separated.

- By the end of meiosis I, the chromosome number has been halved, and the result is two genetically different cells. Meiosis II follows essentially the same process as mitosis in the production of two more haploid cells.

2. Children get one set of chromosomes from their mother's egg and one set of chromosomes from their father's sperm. As a result, 50% of their genes come from one parent, and 50% from the other. The number of genes ultimately derived from their grandparents varies, however, because the selection of chromosomes during meiosis is entirely random. By chance, some eggs or sperm cells will contain a higher percentage of chromosomes derived from one grandparent than the other. On average, children get 25% of the chromosomes from each grandparent, but the actual percentage could vary anywhere from 0% to 50%.

3. If chromosomes fail to separate properly during anaphase of meiosis, the egg or sperm that results may have too few or too many chromosomes. If, as a result of such an error, an egg or sperm has two copies of chromosome 21 instead of one, the child that is produced from that egg or sperm will have three copies of chromosome 21 and will develop Down syndrome.

4. **D** Homologous chromosomes are similar in size, shape, and function but may have different versions of each gene.

5. **C** Sister chromatids separate during anaphase of mitosis.

6. **D** Sister chromatids separate during anaphase II of meiosis.

7. **B** The G_1 checkpoint determines whether a cell will proceed to the S phase or enter a resting phase, G_0.

8. **A** Crossing over occurs during prophase I of meiosis.

9. **A** Bacteria divide using binary fission.

10. **B** In plant cells, a cell plate is produced during cytokinesis.

11. **B** Sister chromatids align during metaphase II of meiosis.

12. **E** A haploid cell has one copy of every chromosome.

13. A Most cells spend the majority of their life cycle in phase G_1.

14. B Sister chromatids are joined at the centromere.

15. E DNA synthesis occurs during the S phase of the cell cycle.

CHAPTER 4
Cell Division

MENDELIAN GENETICS

Heritability

Phenotype vs. Genotype

Punnett Squares

Other Mechanisms of Inheritance

5

Heritability

Biologists studying genetics and evolution seek to answer questions about **heritability,** which is the transmission of traits from one generation to the next. Early theories of inheritance were often crude and inaccurate. The ancient Greeks, for example, thought characteristics were passed directly from parent to child in a material known as *gonos*, or seed, with each trait existing independent of the others. The twentieth century saw major leaps forward in the study of inheritance through the modern study of DNA. Scientists figured out that characteristics are not passed directly from parent to child as independent packages. Characteristics are interconnected, and many of the traits passed from generation to generation are also heavily influenced by environmental factors.

MENDEL'S EXPERIMENTS

Important experimental work took place in the nineteenth century that laid the foundation for the modern genetic revolution. In the mid-nineteenth century, Austrian monk Gregor Mendel set out to determine patterns of inheritance. Mendel performed a series of breeding experiments on the pea plant, *Pisum sativum,* in which he demonstrated predictable patterns for the inheritance of physical characteristics across several generations.

Mendel found the pea plant to be the ideal subject in his study of inheritance for several reasons:

- Short generation time

- Many offspring

- Small and easy to grow

- Many varieties with easily visible characteristics (such as flower color and seed shape)

- Many characteristics already known to be heritable

- Eggs can be manually fertilized with sperm, guaranteeing known parentage

MENDEL'S EXPERIMENTAL DESIGN

Mendel sought to find out how heritable traits pass from parent to offspring. His pea plant breeding experiments consisted of three major stages:

1. Each variety of plant was allowed to self-fertilize over the course of many generations. By self-breeding, Mendel created a **true-breeding** line for each variety, meaning all potential varieties were bred out. For example, one line would produce only wrinkled peas, and no other varieties. The end result was a true-breeding P generation, or parental generation.

2. P generation plants of alternative varieties were then crossed together. For example, Mendel would cross a wrinkled pea variety and a smooth pea variety. Mendel called the offspring of these two opposing parental varieties the F_1 **generation.**

3. Finally, the F_1 generation was allowed to self-fertilize, creating a F_2 **generation.** Mendel then counted the offspring.

MENDEL'S RULES OF INHERITANCE

Based on the results of his experiments, Mendel formulated five important rules of inheritance. In the time since his experiments, new technologies and methods of research have allowed scientists to refine and expand on Mendel's work; however, the essence of his basic rules of inheritance are still correct.

1. The units that determine characteristics, or **traits,** come in more than one form. For example, the gene that produces seed color in pea plants comes in two forms: one that codes for green seeds and another that codes for yellow seeds. In modern genetics, the different forms of genes are called **alleles.** The characteristics that Mendel bred for had only two allele forms, but many genes have more than two alleles for a given trait.

2. Each **gamete** (a sperm or egg cell) contains one copy of a particular gene. When gametes fuse during fertilization, the resulting offspring contain two copies of each gene. This rule forms the basis of Mendel's principle of segregation, which states that pairs of genes separate during meiosis and form new pairings as gametes fuse during fertilization.

3. Each organism possesses two alleles for each gene, one from each parent. An organism is said to be **homozygous** for a specific trait if it inherits two identical alleles for the corresponding gene. An organism is **heterozygous** if the two alleles are different.

4. In a heterozygous organism, one trait may be fully expressed, or **dominant,** while the opposing trait is

not expressed at all, or **recessive.** Recessive alleles are only expressed if the organism is homozygous for that allele.

5. The alleles of different genes are assorted randomly and are independent of each other. This rule forms the basis of Mendel's principles of independent assortment.

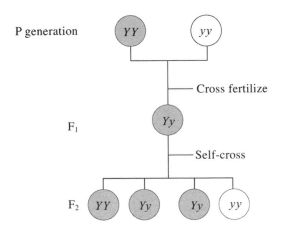

Distribution of Alleles in the Cross-breeding of Pea Pods

Phenotype vs. Genotype

An organism's outward appearance (purple flowers instead of white, for example) is its **phenotype.** An organism's **genotype** refers to its actual allelic (genetic) composition—for example, two dominant alleles, two recessive alleles, or one of each allele for flower color.

> *EXAMPLE:* Pea plant A possesses two alleles for yellow color (Y) and is therefore homozygous for seed color. Because it possesses two dominant yellow alleles, the phenotype is yellow and the genotype is YY. Pea plant B possesses one allele for yellow seeds (Y) and one allele for green seeds (y) and is therefore heterozygous for seed color. The phenotype is yellow, because the dominant yellow allele is expressed. Although the phenotype is the same as that of pea plant A, pea plant B's genotype is Yy for seed color.

Mendel studied several pea plant traits and found the following phenotypic ratios over successive generations. The letter in parentheses denotes the label for that particular allele. Note that alleles are typically labeled with the letter corresponding to the dominant allele capitalized; that same letter is lowercase to indicate the recessive allele.

Charac-teristic	Dominant Trait	Recessive Trait	Ratio of Dominant: Recessive Phenotype in F_1 Generation	Ratio of Dominant: Recessive Phenotype in F_2 Generation
Seed color	Yellow (Y)	Green (y)	1 : 0	2.96 : 1
Seed shape	Smooth (W)	Wrinkled (w)	1 : 0	3.01 : 1
Pod color	Green (G)	Yellow (g)	1 : 0	2.95 : 1
Pod shape	Inflated (I)	Wrinkled (i)	1 : 0	2.82 : 1
Flower color	Purple (P)	White (p)	1 : 0	3:14 : 1
Flower position	On the stem (S)	At the tip of the stem (s)	1 : 0	3.14 : 1
Stem length	Tall plant (T)	Short plant (t)	1 : 0	2.84 : 1
		Average Ratio	1 : 0	3 : 1

Punnett Squares

Punnett squares represent a cross of two individuals and their potential offspring and are used to illustrate the ratio of dominant to recessive phenotypes for a specific trait. In each square, the alleles of one individual are laid individually across the top, representing the potential gametes carrying that allele. The alleles of the second individual are laid vertically along the left. The potential offspring are placed in individual boxes of the square, represented by a combination of the two potential alleles indicating the genotype for that trait. Punnett squares can be used to project the distribution of alleles for any given cross with any number of traits.

	P	P
p	Pp	Pp
p	Pp	Pp

Cross of PP vs. pp: The purple flower can only produce gametes with the *P* allele, and the white flower can only produce gametes with the *p* allele. When gametes from these two plants fuse, all potential offspring will have the genotype *Pp*. Since the purple flower allele is dominant, all of these flowers will have a purple flower phenotype. A cross of these two parental types will yield purple offspring 100% of the time.

	P	p
P	PP	Pp
p	Pp	pp

Cross of Pp vs. Pp: The two parents crossed in this Punnett square have a 75% chance of having a purple-flowered offspring and a 25% chance of having a white-flowered offspring. If these parents produce one thousand offspring, the expected offspring phenotype would include 750 offspring with purple flowers and 250 offspring with white flowers. The expected genotypes should be 50% *Pp*, or heterozygous dominant, 25% *PP*, or homozygous dominant, and 25% *pp*, or recessive.

A cross with one specific characteristic is known as a **monohybrid cross.** Punnett squares can also be used to determine multiple characteristics of offspring. A cross with two specific characteristics is known as a **dihybrid cross.**

	RP	Rp	rP	rp
RP	RR PP	RR Pp	Rr PP	Rr Pp
Rp	RR Pp	RR pp	Rr Pp	Rr pp
rP	Rr PP	Rr Pp	rr PP	rr Pp
rp	Rr Pp	Rr pp	rr Pp	rr pp

Cross of *Rr Pp* vs. *Rr Pp*: Two plants are crossed using flower color (purple or white) *and* seed shape (round or wrinkled) as characteristics. The gametes possess a single allele for both characteristics, creating far more alleles than if just one trait were in question. For this particular cross, offspring possessing a dominant allele (*R* or *P*) will display the dominant trait. Only offspring possessing two recessive alleles (either *rr* or *pp*) will display a recessive trait.

Test Cross

Geneticists use Punnett squares to calculate the probability that offspring will display a specific characteristic type in any simple cross. This calculation is only possible when the parental genotypes are known. For example, if an organism is phenotypically recessive, it is known that it is homozygous for that allele. If the organism is phenotypically dominant, it could be either homozygous or heterozygous. To illustrate that either genotype is possible, the individual is typically labeled *P_*. To determine a dominant genotype, a test cross can be performed in which an unknown dominant individual is crossed several times with a recessive individual. If the cross yields any recessive offspring, the unknown dominant is heterozygous.

Other Mechanisms of Inheritance

Not all alleles are strictly dominant or recessive, although the characteristics Mendel studied were. The following table describes the different ways alleles can interact to create a given phenotype.

Type of Inheritance	How It Works	Example
Incomplete dominance	Allele traits create an intermediate phenotype when an organism is heterozygous.	Snapdragons can have red, white, or pink flowers. When a plant is heterozygous with a red and white allele, it will produce pink flowers. A cross between two pink flowers will produce red, pink, and white-flowered plants in a 1:2:1 ratio.
Codominance	Both alleles are expressed fully in heterozygous individuals.	Humans have four blood types: O, A, B, and AB. Blood type is determined by three alleles symbolized by i, I^A, and I^B. The i allele produces no cell-surface carbohydrate, the I^A allele produces a "type A" cell-surface carbohydrate, and I^B produces a "type B" cell-surface carbohydrate. People with blood type O have the genotype ii, people with type A have either $I^A I^A$ or $I^A i$, people with type B have either $I^B I^B$ or $I^B i$, and people with type AB have $I^A I^B$. In those with blood type AB, both the type A carbohydrate and the type B carbohydrate are produced.
Polygenic inheritance	The interactions of many genes are responsible for a single characteristic.	Skin color is a result of the interactions of multiple genes to create a range of color.

CONTINUED

Type of Inheritance	How It Works	Example
Epistasis	The interaction of one gene with another will alter the phenotypic outcome.	Albinism, or lack of pigment, occurs when one gene prevents the production of pigment by another gene. In the absence of the albinism gene, an organism will have perfectly normal color.
Pleiotropy	One gene influences many different characteristics.	Individuals with sickle-cell anemia are homozygous for the sickle-cell allele. The sickle-cell allele creates an abnormal hemoglobin protein, which crystallizes, leading to the breakdown of red blood cells, the accumulation of sickle cells in the spleen, and the clogging of smaller blood vessels. All of these characteristics in turn cause the multiple symptoms of sickle-cell anemia.

Summary

Mendel's Experiments

- Austrian monk Gregor Mendel used the pea plant, *Pisum sativum*, to study rules of inheritance across generations.

- Pea plants proved to be an ideal organism for Mendel's experiments because they grow quickly, have large numbers of offspring, are small, and have many easily observable characteristics.

- Mendel's experimental design consisted of three stages:

  1. Creating **true-breeding** lines for all characteristics, or **traits**, of interest.

  2. Crossing true-breeding plants of alternative characteristics to create a first generation of offspring.

  3. Self-fertilizing these offspring to create a second generation of offspring.

CHAPTER 5
Mendelian Genetics

- Mendel's experiments led him to settle on five basic rules of inheritance:

  1. The units that determine characteristics come in more than one form. The different forms are called **alleles.**

  2. The principle of segregation states that pairs of genes separate during meiosis and therefore each **gamete** contains one allele. When egg and sperm fuse they create an offspring that carries the alleles from both parents on one gene.

  3. If an organism has two identical alleles on a gene, it is **homozygous** for that characteristic. If the two alleles are different, the organism is **heterozygous** for that characteristic.

  4. A trait that is fully expressed in a heterozygous organism is the **dominant allele.** The trait that is masked when the organism is heterozygous is the **recessive allele.**

  5. The principle of independent assortment states that the alleles of different genes are assorted randomly and independently.

Phenotype vs. Genotype

- A **genotype** is the combination of alleles an organism possesses for a given trait.

- An organism's **phenotype** is its outward appearance based on genotype.

- A given allele is generally represented by a letter. A capitalized letter denotes the dominant allele and a lowercase letter represents the recessive allele.

Punnett Squares

- **Punnett squares** are used to represent a given cross between two organisms for any number of characteristics for which the parental genotypes are known.

- Punnett squares predict the probability of offspring characteristics with respect to genotype and phenotype.

- To determine the genotype of an unknown dominant phenotype, a test-cross can be conducted.

Other Mechanisms of Inheritance

- In incomplete dominance, two different alleles combine to create an intermediate phenotype of a given trait.

- Codominant traits occur when both alleles are expressed fully in one individual.

- Polygenic traits are controlled by multiple genes.

- Epistasis occurs when the interactions of two genes alters the phenotypic outcome.

- Pleiotropy occurs when many characteristics are controlled by one gene.

Sample Test Questions

1. A white rat mates with a black rat, producing gray offspring. Describe the type of inheritance demonstrated by the color of their offspring.

2. Give three reasons why pea plants were a good model organism for Mendel to use.

3. Explain how a Punnett square for a dihybrid cross is set up differently from a Punnett square for a monohybrid cross.

4. Which of the following describes a cross between two varieties of pea plants that differ in two characteristics?

    A. Monohybrid cross
    B. Dihybrid cross
    C. Trihybrid cross
    D. Test cross
    E. Codominant cross

5. Which of the following is an example of polygenic inheritance?

    A. Flower color in snapdragons
    B. Human blood typing
    C. Human skin color
    D. Sickle-cell anemia
    E. Colorblindness

6. What is the name for a cross used to determine the unknown allele of a phenotypically dominant individual?

    A. Test cross
    B. Law of segregation
    C. Trihybrid cross
    D. Dihybrid cross
    E. Monohybrid cross

7. Which of the following best describes Mendel's principle of segregation?

    A. Each pair of alleles segregates independently during gamete formation.
    B. The units that determine traits come in more than one form.
    C. Alleles can be either dominant or recessive.
    D. An organism may possess two different alleles, and one may be fully expressed while the other is not expressed at all.
    E. Pairs of genes separate during the formation of gametes, and the fusion of gametes restores gene pairings during fertilization.

8. What can be said about an individual organism that has the genotype *AA* for a certain characteristic?

    A. It is heterozygous for a dominant allele.
    B. It is homozygous for a dominant allele.
    C. It is heterozygous for a recessive allele.
    D. It is homozygous for a recessive allele.
    E. It could be either heterozygous or homozygous.

9. Which of the following terms describes the expression of both alleles of a trait in one individual?

    A. Incomplete dominance
    B. Codominance
    C. Inversion
    D. Polyploidy
    E. Crossing-over

10. The condition of having blue eyes is a display of which aspect of an organism?

    A. Genotype
    B. Stenotype
    C. Alleles
    D. Autosomes
    E. Phenotype

11. What ratio describes the offspring produced in a cross between two individuals heterozygotic for one trait?

    A. 3:1 phenotypic ratio
    B. 3:1 genotypic ratio
    C. 9:3:3:1 phenotypic ratio
    D. 9:3:3:1 genotypic ratio
    E. 1:1 phenotypic ratio

12. What is the name of the hybrid offspring produced by the cross-fertilization of a P generation?

    A. Parental generation
    B. F_1 generation
    C. F_2 generation
    D. Autosomic generation
    E. Pleiotropic generation

13. Albinism is an example of which mechanism of inheritance?

    A. Polygenic inheritance
    B. Codominance
    C. Incomplete dominance
    D. Pleiotropy
    E. Epistasis

14. Which of the following statements is NOT true of alleles?

    A. There are typically four alleles for every gene.
    B. Alleles can be alike or different.
    C. There are many allele possibilities for a given trait.
    D. It is possible to have only one allele for a given trait.
    E. Both alleles can be expressed by one individual.

15. Human blood typing is one example of which of the following mechanisms of inheritance?

    A. Pleiotropy
    B. Multiple genes
    C. Polygenic inheritance
    D. Incomplete dominance
    E. Codominance

ANSWERS

1. The gray-only offspring of a white rat and a black rat is the result of incomplete dominance. In incomplete dominance, hybrids have an appearance that is an intermediate phenotype of two parental phenotypes. If either gene had been completely dominant, the offspring color would correspond to the dominant allele.

2. Pea plants were a suitable model organism for Mendel to use because a) they are small enough to be easily manipulated in the lab setting, b) they have a relatively short lifespan, unlike mice or other animals, c) they have a few easily manipulated and visible traits, like flower color and pod shape, and d) they're easy to cross-fertilize—unlike animals, whose fertilization would be harder to control.

3. The two alleles in a monohybrid cross are split into potential gametes. If the organism is homozygous, there is only one potential gamete type; if the organism is heterozygous, there are two potential gamete types. In a dihybrid cross, there is the potential for more than two gamete types.

4. B A dihybrid cross is a cross between two varieties of a species that differ in two characteristics rather than one, as is the case if a pea plant with yellow, smooth seeds (*YYSS*) is crossed with a pea plant that has green, wrinkled seeds (*yyss*).

5. C Human skin color is an example of a polygenetically inherited trait. In polygenic inheritance, the actions of many genes are responsible for a single characteristic.

6. A A test cross is used to determine if a phenotypically dominant individual is homozygous or heterozygous by crossing it with a recessive individual and observing the traits of its offspring.

7. E Mendel's principle of segregation states that pairs of genes separate during the formation of gametes. During fertilization, two gametes fuse, and the gene pairings are restored.

8. B This individual has the genotype *AA* for a certain characteristic. Since both letters are capitalized, we know that it is homozygous for this dominant allele.

9. B With codominance, both alleles' traits are expressed in an individual.

10. E The condition of being blue eyed corresponds with an organism's phenotype. An organism's phenotype corresponds with its appearance, whereas an organism's genotype corresponds with its allelic composition.

11. A If both parents are heterozygous for the trait in question, their genotype is *Xx*. Using a Punnett square to cross the two results in one offspring with genotype *XX*, which expresses the dominant phenotype; two offspring with genotype *Xx*, which also express the dominant genotype; and one offspring with the genotype *xx*, which expresses the recessive genotype. So the phenotypic ratio of this cross is 3:1.

12. **B** In a cross, the hybrid offspring produced by the cross-fertilization of the P generation produces the F_1 generation.

13. **E** Epistasis explains albinism.

14. **A** Most organisms only have two alleles for every one gene. Some organisms have more, but it is rare.

15. **E** Human blood type is an example of codominance. Codominance is the expression of two different alleles of a gene.

DNA, GENES, AND PROTEINS

DNA Structure

DNA Replication

From DNA to Protein

Control of Gene Expression

DNA Structure

Genes are the basic unit of heredity, and they are housed in the DNA of every organism. DNA, short for deoxyribonucleic acid, is composed of two chains, or strands, of nucleotides wrapped around each other in a helical (spiral staircase) shape. Each individual nucleotide is composed of three parts:

1. a phosphate group

2. a deoxyribose sugar

3. nitrogenous base

A covalent bond forms between the phosphate group of one nucleotide and the sugar group of an adjacent nucleotide, resulting in a strong sugar-phosphate frame. Hydrogen bonds form between the nitrogenous bases of two nucleotide chains to form the unique double helix of the DNA molecule. Because hydrogen bonds are much weaker than covalent bonds, the two strands of DNA can unzip relatively easily.

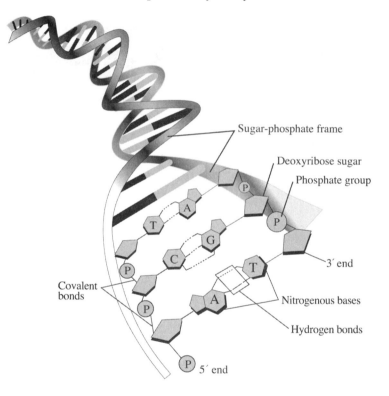

Sugar-phosphate frame

Deoxyribose sugar

Phosphate group

3′ end

Nitrogenous bases

Hydrogen bonds

Covalent bonds

5′ end

DNA

COMPLEMENTARY BASE PAIRING

A single DNA chain contains an arrangement of four nitrogenous bases: adenine (A), cytosine (C), guanine (G), and thymine (T). The bases on one chain form complementary base pairings with the bases on the opposite chain in a single DNA molecule, with specific rules governing those pairings. For example, adenine pairs with thymine, while cytosine pairs only with guanine. As a result, the order of nitrogenous bases on one strand of DNA determines the order of bases on the opposite chain.

> **EXAMPLE:** If the order of nitrogenous bases on one chain of DNA is ATCGGTC, then the opposite chain must be TAGCCAG.

Owing to the nature of complementary base pairing, a relationship exists between the amounts of nitrogenous bases in DNA: The percentage of adenine equals the percentage of thymine, and the percentage of cytosine equals the percentage of guanine. This relationship holds true for all species and is known as **Chargaff's rule,** after Austrian biochemist Erwin Chargaff.

ANTIPARALLEL CHAINS

The two nucleotide chains that make up a single molecule of DNA are **antiparallel,** meaning that the chains run in opposite directions. One end of the chain, known as the five prime end (usually written 5'), is composed of a phosphate group that has not bonded with a sugar unit. The other end of the chain, the three prime end (usually written 3'), is composed of a sugar unit whose hydroxyl group has not bonded with a phosphate group. From top to bottom, one chain will run from 5' to 3', while the other will run from 3' to 5'.

The names for the ends of a nucleotide chain refer to the number of the carbon atom in the sugar molecule to which each group attaches. The carbon atoms in the sugar of the nucleotide are numbered from 1' to 5'. In a five prime end, the phosphate group attaches to the sugar at the five prime (5') carbon. In a three prime end, the hydroxyl group attaches to the sugar at the three prime (3') carbon.

COMPARISON OF DNA AND RNA

DNA requires the assistance of a ribonucleic acid **(RNA)** to perform a number of critical functions, such as replication

CHAPTER 6
DNA, Genes, and Proteins

and the production of protein. An RNA molecule has a similar structure to DNA. Both molecules consist of chains of nucleotides. Unlike DNA, RNA is usually single stranded. RNA is also composed of a ribose sugar group, instead of deoxyribose, and a uracil base (U), instead of thymine (T). Adenine can pair with either base.

DIFFERENCES BETWEEN DNA AND RNA		
Characteristic	DNA	RNA
Number of chains	2	1
Type of sugar group	Deoxyribose	Ribose
Type of nitrogenous bases	A, T, C, G	A, U, C, G

DNA Replication

DNA houses an organism's **genome,** which is that organism's complete genetic material. When a cell divides, its DNA must be duplicated in a process called **DNA replication,** so that both resulting cells receive a complete copy of the genome. During replication, the two strands of DNA unwind, or unzip. Each strand then serves as a template for the formation of a new opposite strand, built with near-perfect accuracy according to rules of complementary base pairing. If a template strand has an A in one spot, then a T is placed in the opposite strand. Because each new DNA molecule consists of one old strand and one new strand, DNA replication is said to be semiconservative.

THE STEPS IN DNA REPLICATION

The process of DNA replication can be described in six basic steps:

1. The enzyme helicase unwinds DNA at specific sites along the chain, known as origins of replication.

2. **Primers,** small sections of RNA, are laid down along the template strand by the enzyme primase. The RNA acts as the starting point for DNA to build on.

3. Elongation of the new strands begins. During elongation, the enzyme **DNA polymerase** adds nucleotides to the 3' end of the growing strand. Because the two strands of DNA are antiparallel, elongation proceeds in a different manner on each strand. Nucleotides are added continuously to the leading strand, the strand that is opened from 5' to 3'. On the opposite strand, known as the lagging strand, replication occurs in pieces, called Okazaki fragments.

4. The RNA primers are removed and replaced with DNA by the DNA polymerase.

5. The enzyme **DNA ligase** binds together the Okazaki fragments of the lagging strand.

6. Two chemical processes occur throughout elongation to prevent errors from creeping in. In a process known as proofreading, DNA polymerase enzymes recognize mistakes in DNA synthesis as they elongate the strand. At points where mistakes occur, the DNA polymerases reverse direction to remove the nucleotide they have added and replace it with the correct one. Following DNA synthesis, enzymes conduct what is known as mismatch repair by scanning the newly formed DNA strand and repairing any errors that are found.

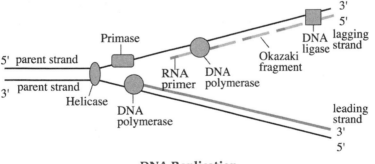

DNA Replication

EXAMPLE: Mutations in mismatch repair genes, which encode the enzymes that proofread and correct errors in DNA, are known to be responsible for certain hereditary forms of cancer.

From DNA to Protein

Specific proteins perform a variety of important functions in the body. They may act as catalysts to control reactions, act as antibodies in the immune system, transport materials around the body, form the physical structures that allow organisms to move, or control the movement of substances into and out of cells. The instructions for making specific proteins are controlled by genes housed in DNA. Since genes control the production of proteins, they have enormous influence on the physical characteristics of an organism.

Francis Crick, one of the discoverers of the structure of DNA, is also known for formulating the central dogma of molecular biology. This dogma states that information flows from DNA to RNA to protein, but not from the protein back to RNA or DNA. The pathway of protein information runs one way:

$$DNA \longrightarrow RNA \longrightarrow protein$$

Protein production occurs in two phases:

1. **Transcription:** During transcription, information in DNA is transcribed, or rewritten, onto a strand of **messenger RNA (mRNA).**

2. **Translation:** During translation, the information written onto the mRNA is used to assemble a protein.

TRANSCRIPTION

Transcription, the process by which information from DNA is transferred to mRNA, occurs in the nucleus of the cell. DNA is unwound at a specific gene, and mRNA is formed via complementary base pairing at one strand of the DNA.

Transcription can be described in three steps:

1. **Initiation:** The enzyme **RNA polymerase** binds to the **promoter** region of DNA at the start of a gene. RNA polymerase unwinds the DNA and starts to assemble the mRNA.

2. **Elongation:** The RNA polymerase moves down the DNA strand, assembling the mRNA from 5' to 3'. Through complementary base pairings, the mRNA assembles along one strand of DNA, known as the **template strand.**

3. **Termination:** The RNA polymerase reaches a stop signal, a certain series of bases within the DNA. Transcription ceases, and the mRNA dissociates from the template strand of DNA.

EXAMPLE: As a result of complementary base pairings, the template strand 3' TACTTGGCGATT 5' would be transcribed into the following mRNA: 5'AUGAACCGCUAA 3'.

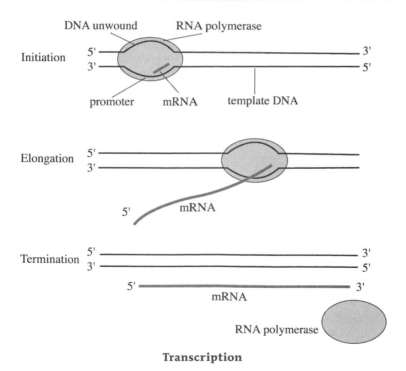

Transcription

mRNA Modification

After transcription, the mRNA of eukaryotic cells undergoes two modifications to protect it as the mRNA moves from the nucleus to the cytoplasm. (Since transcription occurs in the cytoplasm for prokaryotic cells, these modifications do not occur in prokaryotic cells.)

Extra bases are added at the beginning and end of the mRNA strand, forming a cap and tail. The cap and tail protect the mRNA from degradation caused by enzymes as the mRNA travels from the nucleus to the cytoplasm for translation.

Introns, noncoding regions of the mRNA strand, are removed in a process called **RNA splicing.** The coding regions of mRNA, called **exons,** are attached together.

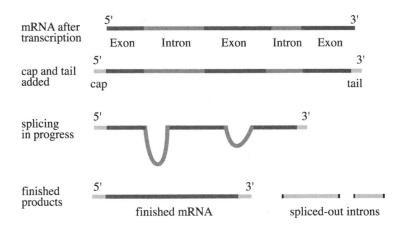

Posttranscriptional RNA Processing

EXAMPLE: Alternative splicing refers to the process in which the splicing of the mRNA strand leads to different RNA strands, which in turn produce different proteins. Alternative RNA splicing is one reason humans appear to have fewer genes than scientists originally estimated. Scientists once thought that humans might have as many as 100,000 genes, but currently it is believed that humans have 20,000 to 25,000 genes.

TRANSLATION

During translation, proteins assemble according to the information encoded in mRNA. Each series of three mRNA bases, called **codons,** corresponds to a specific amino acid. The association of codons with specific amino acids is known as the genetic code. As the mRNA molecule is read from 5' to 3', each codon indicates which specific amino acid should be added to the growing protein. Once a stop codon is reached, protein synthesis is completed. There are three stop codons, named by their three bases: UAG, UGA, and UAA.

THE GENETIC CODE					
First Letter	**Second Letter**				**Third Letter**
	A	**G**	**C**	**U**	
A	Lysine	Arginine	Threonine	Methionine	A
	Lysine	Arginine	Threonine	Methionine / START	G
	Asparagine	Serine	Threonine	Isoleucine	C
	Asparagine	Serine	Threonine	Isoleucine	U
G	Glutamate	Glycine	Alanine	Valine	A
	Glutamate	Glycine	Alanine	Valine	G
	Aspartate	Glycine	Alanine	Valine	C
	Aspartate	Glycine	Alanine	Valine	U
C	Glutamine	Arginine	Proline	Leucine	A
	Glutamine	Arginine	Proline	Leucine	G
	Histidine	Arginine	Proline	Leucine	C
	Histidine	Arginine	Proline	Leucine	U
U	STOP	STOP	Serine	Leucine	A
	STOP	Tryptophan	Serine	Leucine	G
	Tyrosine	Cysteine	Serine	Phenylalanine	C
	Tyrosine	Cysteine	Serine	Phenylalanine	U

CHAPTER 6
DNA, Genes, and Proteins

Caption: The three letter columns of the table on the previous page show the possible combinations of nitrogen bases that form the various mRNA codons. The central column provides the name for all possible amino acids depending on each second-letter combination. For example, if the first, second, and third letters are all A, lysine is coded. If the first letter is A, the second letter is G, and the third letter is A, arginine is coded. UAA, UAG, and UGG are all stop codons.

EXAMPLE: The mRNA strand 5' AUG AAC CGC UAA 3' translates into the following protein: Methionine – Asparagine – Arginine, with the stop codon UAA.

Transfer RNA and Ribosomes

Translation is facilitated by two key molecules: transfer RNA and ribosomes.

- **Transfer RNA (tRNA)** molecules transport amino acids to the growing protein chain. Each tRNA carries an amino acid at one end and a three-base pair region, called the **anticodon,** at the other end. The anticodon bonds with the codon on the protein chain via base pair matching. For example, if the codon on the protein chain is GCC, it will pair with a CGG anticodon. By binding with the codon of the mRNA, the anticodon is able to place the amino acid in the correct position in the growing protein strand.

- **Ribosomes** bind to the mRNA and facilitate protein synthesis by acting as docking sites for tRNA. Each ribosome is composed of a large and small subunit, both made of **ribosomal RNA (rRNA)** and proteins. The ribosome has three docking sites for tRNA: the A site, P site, and E site.

STEPS IN TRANSLATION	
Step	**Details**
Initiation	Ribosome binds to the mRNA and the first tRNA enters the P site of the ribosome. The first tRNA always carries a methionine amino acid and is able to bind with the start codon, 5'AUG 3', of the mRNA.
Elongation	Each new tRNA enters the ribosome at the A site and bonds with the codon of the mRNA. Amino acids attached to the tRNA at the P site are then transferred to the tRNA at the A site. The tRNA at the P site moves to the E site and leaves the ribosome. The tRNA at the A site, which now carries a chain of amino acids, moves to the P site. This process is repeated as the ribosome moves down the mRNA strand.
Termination	When the ribosome reaches a stop codon, a release factor binds to the A site of the ribosome. The release factor ends translation and releases the protein.

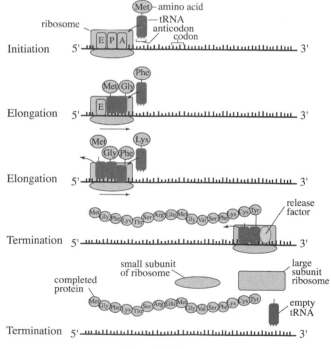

Translation

Control of Gene Expression

Gene expression refers to the synthesis, or production, of proteins according to information encoded in DNA. Gene expression is the phenotypic characteristics, or physical and functional display, of a specific genotype. Since proteins determine the structure and function of the cell, gene expression, in effect, controls the cell. Genes are not expressed randomly, or even at equal frequency, within the cell. Instead, the timing and frequency of gene expression is controlled by the individual cell.

EXAMPLE: Biologists believe that the differences between humans and chimpanzees arise mainly from differences in gene expression, rather than differences in the genes themselves. The genes of humans and chimps are 99% similar, yet the patterns of where and when each gene is expressed in the body are vastly different between the two species.

REGULATION IN PROKARYOTES

Gene expression in prokaryotes is regulated through a process known as **transcriptional control.** Many prokaryotic genes have one or more binding sites near a promoter region of the DNA template strand. Regulatory proteins, enzymes and molecules that influence enzymes, either allow or block the initiation of transcription as they bind to these sites. Regulatory proteins alter gene expression by improving or reducing the ability of RNA polymerase to bind to the promoter.

Two classes of transcription regulating proteins are present in prokaryotes: activators and repressors.

•   **Activators** increase rates of transcription by improving the ability of RNA polymerase to bind to the promoter.

- **Repressors** decrease rates of transcription by blocking the ability of RNA polymerase to bind to the promoter, usually by binding to the DNA near the promoter region.

REGULATION IN EUKARYOTES

The regulation of gene expression in eukaryotes occurs at every stage of protein synthesis. Scientists traditionally divide gene regulation into those events occurring before transcription **(transcriptional control)** and those occurring after transcription **(posttranscriptional control).**

Transcriptional Control

Most gene regulation in eukaryotic cells takes place in the transcriptional control phase. Two main proteins affect transcriptional control.

- Sets of proteins called **transcription factors** are required for RNA polymerase to bind to the promoter of eukaryotic genes. If these transcription factors are absent, transcription will not occur.

- Proteins called **enhancers** activate transcription by binding to the DNA at a site far from the transcription site. Enhancers cause the DNA to loop so that the promoter comes into contact with the transcription site.

Posttranscriptional Control

Regulation of gene expression can also occur at any point following transcription:

- In the process of alternate splicing, mRNAs can be spliced in a variety of ways to create different proteins; each affects gene expression in specific ways.

- Following transcription, the mRNA is shuttled out of the nucleus to the ribosome for translation.

- Selective destruction can occur as mRNA is degraded by enzymes in the cytoplasm.

- Selective destruction of proteins can occur, in which proteins are destroyed, or denatured, by enzymes in the cytoplasm.

- Selective translation of mRNA can occur when the translation of certain mRNAs are slowed or blocked.

Summary

DNA Structure

- DNA comprises four nitrogenous bases: adenine (A), cytosine (C), guanine (G), and thymine (T). These bases bind with each other in specific ways. For example, adenine binds only with thymine, while cytosine binds only with guanine.

- **Chargaff's rule** describes the relationship between the amounts of nitrogenous bases in DNA.

- RNA also comprises four nitrogenous bases: adenine (A), cytosine (C), guanine (G), and uracil (U). Adenine binds only with uracil, whereas cytosine binds only with guanine.

DNA Replication

- **DNA replication** is the process in which the two strands of DNA are separated, each strand serving as the template for the formation of a new strand.

From DNA to Protein

- Genes are regions of DNA that carry instructions for making proteins. During **transcription,** the genetic information in DNA is transcribed, or rewritten, onto a strand of **messenger RNA (mRNA).** During **translation,** the information on the mRNA is used to assemble a new protein.

- Transcription occurs when DNA is unwound and one strand serves as a template for the formation of mRNA.

- In eukaryotes, the mRNA is processed after transcription. **Introns** are spliced out, and a cap and tail are added to the mRNA.

- During translation, the three base sequences on the mRNA called **codons** specify the sequence of amino acids to be assembled in the creation of a specific protein. **Ribosomes** assist in protein synthesis by serving as docking points for **tRNAs,** amino acid–carrying molecules that bond to the mRNA codons.

Control of Gene Expression

- **Gene expression** refers to the synthesis of proteins according to information encoded in DNA. The timing and frequency of gene expression is controlled by many factors in the cell.

- Gene regulation in prokaryotes occurs via **activators** and **repressors** that promote or reduce transcription.

- Gene regulation in eukaryotes occurs via both **transcriptional** and **posttranscriptional control.** In eukaryotes, proteins called **transcription factors** are required for transcription to begin, and proteins called **enhancers** activate transcription.

- Posttranscriptional controls in eukaryotes include RNA splicing, transporting the mRNA out of the nucleus, selective destruction of mRNA, selective destruction of proteins, and selective translation of mRNA.

CHAPTER 6
DNA, Genes, and Proteins

Sample Test Questions

1. Explain how the structure of DNA facilitates the replication and storage of information.

2. What would be the order of bases on the opposite strand of the DNA template for the strand of a gene shown below?

3' TAC ATG CCG GCT CAG ATT 5'

3. Scientists studying gene expression of an enzyme in a culture of the bacteria *E. coli* find that they can alter the production of the enzyme by adding different combinations of two substances, A and B, to the culture. Use the information in the chart below to determine which substance is acting like an activator and which substance is acting like a repressor. What will happen if both A and B are present? What if both are absent?

Substance A	Substance B	Enzyme Produced
Present	Absent	No
Absent	Present	Yes

4. Which of the following characteristics describes RNA?

   A. Double stranded, ribose sugar, thymine instead of uracil
   B. Single stranded, ribose sugar, thymine instead of uracil
   C. Single stranded, deoxyribose sugar, thymine instead of uracil
   D. Single stranded, deoxyribose sugar, uracil instead of thymine
   E. Single stranded, ribose sugar, uracil instead of thymine

5. Which of the following enzymes binds the breaks in a single strand of DNA?

   A. DNA polymerase
   B. RNA polymerase
   C. DNA ligase
   D. Helicase
   E. Primase

6. What is the name for the noncoding regions of mRNA removed during RNA splicing?

   A. Introns
   B. Exons
   C. Codons
   D. Okazaki fragments
   E. Caps

7. Which of the following is an example of posttranscriptional control?

   A. Activators
   B. Enhancers
   C. Repressors
   D. Alternative splicing
   E. Transcription factors

8. Which of the following molecules facilitates translation by acting as a docking site for tRNA?

   A. Okazaki fragments
   B. Ribosomes
   C. mRNA
   D. rRNA
   E. Promoters

9. Okazaki fragments form on the _____ during replication because DNA can be synthesized only in the _____ direction.

   A. leading strand, 5' to 3'
   B. leading strand, 3' to 5'
   C. lagging strand, 5' to 3'
   D. lagging strand, 3' to 5'
   E. template strand, 5' to 3'

10. Which of the following describes a method of prokaryotic regulation of gene expression?

    A. Activators
    B. Enhancers
    C. Transcription factors
    D. Selective destruction of mRNA
    E. Selective destruction of proteins

11. Which type of RNA carries genetic information from the nucleus to the ribosome?

    A. mRNA
    B. rRNA
    C. sRNA
    D. tRNA
    E. uRNA

12. Which type of RNA carries amino acids to the ribosome?
    A. mRNA
    B. rRNA
    C. sRNA
    D. tRNA
    E. uRNA

13. Which type of RNA is found in the ribosome?

    A. mRNA
    B. rRNA
    C. sRNA
    D. tRNA
    E. uRNA

14. Which of the following is the entrance point for tRNA in the ribosome?

    A. A site
    B. E site
    C. P site
    D. S site
    E. T site

15. Which of the following enzymes adds nucleotides to the growing mRNA strand during transcription?

    A. Primase
    B. Ligase
    C. Helicase
    D. DNA polymerase
    E. RNA polymerase

ANSWERS

1. Easy, accurate replication of DNA is possible due to the complementary base-pairing nature of its structure: Each strand of DNA separates and the opposite strand is assembled based on these pairings. Information storage is made possible through the long sequence of nitrogenous bases on each strand. Each sequence of three bases forms a codon that specifies a particular amino acid. The bonds within DNA also facilitate both replication and transcription: Covalent bonds in each strand provide structural stability, while the weaker hydrogen bonds between strands allow each strand to be unwound for replication and transcription.

2. Because the first strand reads from 3' to 5', the second strand reads from 5' to 3'. Each base will be matched with its pair as follows:

    3' TAC ATG CCG GCT CAG ATT 5'

    5' ATG TAC GGC CGA GTC TAA 3'

3. Since the enzyme is produced in the presence of substance B and in the absence of substance A, substance B must be the activator, while substance A must be the repressor. If this is correct, then the enzyme will not be produced if both substances are present (A will repress expression). Neither will the enzyme be produced if both substances are absent (B is needed to activate expression).

4. **E**   Unlike DNA, RNA is single stranded, has a ribose sugar, and uses the base uracil instead of thymine.

5. **C**   DNA ligase is the enzyme that binds together breaks in a single strand of DNA.

6. **A**   Introns are noncoding regions of mRNA removed during RNA splicing.

7. **D**   Alternative splicing is an example of posttranscriptional control.

8. **B**   Ribosomes facilitate translation by acting as docking sites for tRNA.

9. **C**   Okazaki fragments form on the lagging strand during replication because DNA can be synthesized only in the 5' to 3' direction.

10. **A**   Activators are a method of prokaryotic regulation of gene expression.

11. **A**   mRNA carry genetic information from the nucleus to the ribosome.

12. **D**   tRNA carry amino acids to the ribosome.

13. **B**   rRNA is found in the ribosome.

14. **A**   The A site is the entrance point for tRNA in the ribosome.

15. **E**   RNA polymerase adds nucleotides to the growing mRNA strand during transcription.

THE GENE REVOLUTION

Transgenic Biotechnology

Medicine and Biotechnology

Cloning Research

Genomics

Transgenic Biotechnology

An organism's complete genetic makeup is represented by its **genome,** its complete DNA sequence. As scientists have explored the properties of DNA and the genomes of whole species, they have discovered many of the functions and processes by which organisms survive and reproduce. **Genetically modified organisms** are those whose genome has been scientifically altered, either by the insertion of foreign genes or by the manipulation of genes already present in the organism. Genetic modification is an area of intense study in a range of fields including medicine, agriculture, and zoology. For example, farmers around the world currently grow produce that has been modified with genes from other organisms to improve specific traits. In some cases, genes are manipulated to alter the taste of the crop or to increase its resistance to droughts or diseases. As the genome of an organism is altered, changes are passed down to subsequent generations.

Biotechnology describes technology based on biology and biological processes. **Transgenic biotechnology** is a field of research in which scientists transfer genes from one genetically engineered organism, usually a bacterium, to another organism. The organisms that result from this transfer of genes are called transgenic organisms. Gene transfer occurs naturally among bacteria, where one bacterium can insert its genes into a bacterial cell of a different species.

VIRUS

A **virus** is a microscopic parasitic molecule whose genome is wrapped in a protective protein coat. A virus lacks cytoplasm and contains no organelles; therefore, viruses can be considered nonliving organisms. Viruses differ from other organisms in that they are unable to perform any function, including reproduction, unless they have infected a host organism. Once a virus has infected the cells of another organism, it can begin to use the host's cellular faculties for its own reproductive and living requirements.

Outside of a host cell, a virus particle is inert, or unable to perform any metabolic function, and is called a **virion.** A virion infects an organism by first inserting its genome into the host cell and then reprogramming the cell to divert resources to the reproduction of the virus genome. This process generally damages the host. New viruses emerge from the host cell and can go on to infect new cells, where the entire process is repeated. Viruses infect virtually every type of organism, including plants, animals, protists, fungi, and bacteria. However, each type of virus is designed to infect only a limited number of cell types.

VIRUS INFECTION IN BACTERIA

Viruses that infect bacteria are called **bacteriophages,** or phages. Phages can be used by humans as **vectors** to transfer genes from cell to cell. Phages undergo one of two cycles when infecting bacteria.

1. **The lytic cycle:** The virus injects its genetic material into a bacteria cell. The cell replicates the virus's material and packages up new viral particles. The bacteria cell bursts, releasing the new viruses. These viruses move on to find new bacteria cells to infect.

2. **The lysogenic cycle:** The virus injects its genetic material into a bacteria cell. The virus's genetic material combines with the bacteria's DNA. When the host cell reproduces, the virus material is also reproduced and packaged into new host cells. The end product is more infected bacteria cells.

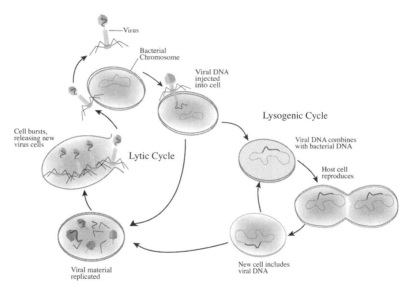

Lytic Cycle and Lysogenic Cycle in Viruses

BACTERIA

Bacteria are single-celled organisms that contain their genome in a single double-stranded, circular piece of DNA called a **chromosome.** Bacteria reproduce through the process of **binary fission,** in which the chromosome is replicated and packaged into a new cell. Binary fission is asexual, and variations occur only when errors in chromosome replication, or **mutations,** appear.

GENE TRANSFER IN BACTERIA

In addition to errors in binary fission, genetic variation in bacteria can also be caused by the transfer of genetic material in a process known as **recombination.** Recombination in bacteria occurs in three different ways: transformation, transduction, and conjugation.

Recombination Method	Process
Transformation	Bacteria take up genetic material from their surrounding environment. Transduction is possible if a live bacterial cell exists in the same medium as a dead bacterial cell. The live bacterial cell absorbs DNA from the dead bacterial cell and incorporates it into its own genome.
Transduction	DNA is transferred between bacterial cells via viruses called bacteriophages. There are two ways bacteriophages undergo transduction. In the first method, a bacteriophage in the lytic cycle accidentally packages a virus containing DNA from the host bacterial cell. When the virus infects another bacterial cell, it injects the DNA from the first bacterial cell. The new cell can then incorporate the DNA into its genome. In the second method, after genetic material is combined from the bacterium and virus in the lysogenic cycle, the virus infects a new bacterial cell carrying some of the first bacterial's DNA with it. This DNA can then be incorporated into the genome of the newly infected bacterial cell.
Conjugation	A direct transfer of DNA occurs between two bacterial cells. The transfer occurs in a one-way direction when one bacterial cell extends an appendage, called a pilus, to another bacterial cell, creating a bridge between the two. The first cell replicates the desired DNA and sends the copy along the bridge to the second cell.

The conjugation method of recombination involves the transfer of portions of DNA. This transfer occurs with the aid of small, circular DNA molecules called **plasmids.** A plasmid contains only a small portion of the bacterium's genome, usually only a portion that would produce advantageous traits in the other bacterial cell. Geneticists often use plasmids when transferring genes between different organisms.

GENE TRANSFER BY HUMANS

Gene transfer allows scientists to engineer organisms with improved traits through the addition of new genes into the organism's genome. Scientists incorporate the methods of gene transfer in viruses and bacteria by inserting a specific gene fragment into rapidly reproducing bacterial colonies to create multiple copies of that gene fragment.

To manipulate gene transfer, scientists first create the DNA molecules that they want to transfer. Scientists manufacture recombinant DNA by using enzymes to cut up and splice together strands of DNA. Recombinant DNA is transferred to the host cell through the use of **vectors,** viruses and bacteria that transport genes from organism to organism.

The following steps describe the production of recombinant DNA:

1. Restriction enzymes cut specific sites on the vector DNA. The site of the cut depends on the particular restriction enzyme used. The cutting of DNA results in two double-stranded DNA pieces, both of which have single-stranded ends known as sticky ends.

2. The DNA containing the gene of interest is cut with the same restriction enzyme as the vector DNA, resulting in complementary sticky ends on both DNA strands.

3. The two strands are mixed and the complementary sticky ends of the strands pair up by forming hydrogen bonds.

4. The DNA fragments are bonded together permanently by the addition of an enzyme called **DNA ligase.** The end result is a strand of DNA containing the gene to be transferred attached to the vector.

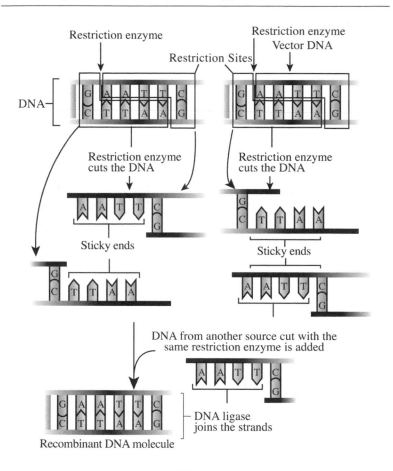

Recombinant DNA

Vectors

Scientists use bacterial plasmids and phages (viruses that infect bacteria) as vectors to transfer genetic material into host cells. In both cases, fragments of the DNA to be transferred are inserted into the plasmid or phage. A section of the vector's genetic material is sliced out, the required gene fragment is inserted, and the DNA is bonded back together again. The restriction enzymes used to slice the vector material are the same used in creating the DNA fragment to be inserted. Once the DNA is inserted, a new genome is formed in the host cell. Both bacterial and virus vectors clone numerous identical copies. Scientists can then create multiple copies of this new genome by allowing the vectors to reproduce in the host.

> *EXAMPLE:* Vectors introduce a new gene into the host genome by infecting the host's cells. By using vectors to manipulate an organism's genome, scientists can alter species for their benefit. For example, plasmids containing genes resistant to certain disease can be inserted into crops, resulting in a decrease in crop loss caused by a particular disease. Vectors are also used to insert genes that code for medically useful proteins into bacterial cells. The bacteria then proliferate rapidly to produce a large store of proteins used in medical treatments.

Marker Proteins

Scientists screen clones to ensure that they contain the vector and/or new genetic fragment. During screening, a marker is added to the vector. A marker is an innocuous, or harmless gene that attaches to the DNA fragment inserted into the vector. The marker gene is detectable when scientists conduct tests on the newly created clones. For example, scientists often use marker genes that create resistance to a particular antibiotic in a bacterial cell. Clones that have the vector will be resistant to that antibiotic and will survive when the antibiotic is applied to a colony of clones. Through this testing method, scientists can effectively kill off all clones that did not replicate with the vector, leaving the vector-containing clones unharmed.

Medicine and Biotechnology

Science has been greatly advanced by progressions in genetic research. Biotechnology, in which scientists utilize genetically engineered bacteria to produce human products (such as hormones), has led to the development of many new tools, both practical and life-saving. The field of medicine has also benefited greatly, as many new techniques, such as gene therapy and stem cell research, have led to advances.

The table below summarizes some of the most prominent areas of study and the advances made through work in genetic research.

Area of Study	Advances	Example
Biotechnology	Genes for the production of medically useful substances can be encoded into rapidly reproducing bacterial colonies, resulting in large-scale production of useful substances.	Biotechnology has been used to produce insulin.
Gene therapy	Genes found to cause particular genetic diseases in humans can be fixed when genes that either prevent or fix the disease are placed into a person's genome.	Gene therapy has been used to cure sickle-cell anemia in mice.
Stem cell research	Stem cells, which are cells that form into new tissues, can be harvested and reproduced. Embryonic stem cells are cells harvested from human embryos and have the potential to become the cell of any body tissue. Adult stem cells can only become one type of body tissue cell, such as a skin cell, a bone cell, or a nerve cell.	Stem cell research presents the potential to regrow damaged tissue, which could lead to effective treatments for spinal cord injuries, Parkinson's disease, and other conditions. Stem cell research could potentially be used to grow entire organs for transplant.

Cloning Research

A **clone** is a genetic copy of an original. Biologists can clone an entire organism or segments of DNA. Contrary to popular belief, cloning, especially in higher organisms, rarely yields an exact replica of the clone's parent. While cloning reproduces an exact *copy* of an organism's genotype, the physical expression of the genotype, the phenotype, is influenced by environmental characteristics. A clone would likely be subjected to different environmental stimuli as it develops into adulthood and therefore would most likely look and act very differently from the original organism.

Cloning has the potential to be used in the mass production of important bacteria, the preservation of endangered species, or even the recreation of deceased pets. Because of the dramatic possibilities it offers in the duplication of life, the field of cloning is quite possibly the most controversial field in gene research today.

> EXAMPLE: A cloned tree contains the same genotype as the original tree. However, the clone may be exposed to less light or might be infected by a growth-stunting disease at some point in development. These environmental factors would affect the physical appearance of the cloned tree, making it different from the original.

CHAPTER 7
The Gene Revolution

REPRODUCTIVE CLONING

The process of reproductive cloning is similar to in-vitro fertilization. Many organisms, from bacteria to sheep to horses, have been successfully cloned in the laboratory via reproductive cloning. Dolly the sheep, the first and perhaps most famous animal to be cloned, was created using the following process of reproductive cloning:

1. A somatic cell, a cell from the body of a multicellular organism, was taken from the udder of one sheep parent and incubated in a lab.

2. An egg cell from another sheep was harvested and the DNA was extracted from its nucleus.

3. The somatic cell was placed next to the DNA-empty egg cell and an electric current was applied to facilitate the fusion of the two cells, much like egg-sperm fusion. This process is called somatic cell nuclear transfer (SCNT).

4. The new cell, a zygote containing only the DNA material from the somatic donor, was incubated and allowed to divide into a larger embryo.

5. The larger embryo was placed in the uterus of a surrogate mother.

6. Dolly was born. She was genetically identical to the somatic sheep donor.

Since the birth of Dolly, many animals have been cloned using reproductive cloning. The efficiency of this method is still quite low, with only a small percentage (3–5%) of implanted embryos resulting in live birth. In addition, cloned offspring are subject to problems throughout their development into adulthood. Many offspring become oversized, a condition known as large offspring syndrome, leading to their early death. Other offspring develop liver failure, arthritis, and infections, among other potentially fatal health conditions.

Genomic Imprinting

One vital factor that scientists have recognized as part of the development process in all organisms is genomic imprinting. This process takes place naturally as sperm and egg cells mature. The DNA in the cell undergoes the process of **reprogramming,** in which the expression of genes is programmed to occur at specific times during development. Under natural circumstances, reprogramming can take months or even years in sperm and eggs; in donor cells used for cloning, by contrast, this process must occur within a few hours. Scientists performing clone research have determined that the cloning process does not allow sufficient time for genomic imprinting to occur.

Genomics

Genomics is a relatively new field of genetics that studies the genomes of different animals to determine such things as gene functionality, relatedness between species, and the genetic requirements for life. The field of genomics is divided into two smaller fields:

- **Comparative genomics,** which compares the genomes of various organisms.

- **Functional genomics,** which studies the function of genes and their products.

Scientists study genomics with several major goals in mind:

- To determine the minimal genome necessary to support a cell.

- To investigate the functional and behavioral processes of proteins, the major constituents of an organism's body.

- To determine the relatedness between various organisms through the comparison of genomes. Insights in this area can be used to answer major evolutionary questions, such as how closely related humans are to other primates.

- To map the human genome by identifying all of its genes. Geneticists believe that if genes and their purposes can be identified, biotechnology may be able to be used to manipulate the genome and improve conditions for life. For example, if the genes that cause genetic diseases such as cystic fibrosis can be identified, biotechnology can be developed to fix or switch off those genes.

EXAMPLE: Criminal investigators can use genomic information to identify criminals and their victims in a process known as **DNA fingerprinting.** Since every person, with the exception of identical twins, has a unique genome, investigators can use DNA sequencing technology to hone in on genetic markers, or segments of DNA known to differ from person to person. These genetic markers can be used to identify people linked to a crime scene if investigators have a sample of the suspect's or victim's DNA.

Summary

Transgenic Biotechnology

- **Biotechnology** is technology based on biology or biological processes.

- **Transgenic** organisms are the result of the transfer of genes from one organism to another.

- **Genetically modified** organisms are those whose **genome** has been modified.

- **Viruses** infect an organism by injecting their genetic material into host cells.

- **Bacteria** exchange genes with other bacteria via the processes of **transduction,** transformation, and conjugation.

- Because bacteria are naturally prone to gene swapping, scientists often use them in transgenic biotechnology. **Plasmids,** circular DNA, are often used as **vectors** to transfer given DNA. Viruses are also used as vectors during transgenic biotechnology.

- The term recombinant DNA refers to genes from different organisms that are spliced together in the laboratory. Restriction enzymes and **DNA ligase** are used in the process of gene splicing.

Medicine and Biotechnology

- The use of biotechnology in medicine is incredibly important and has been the basis for curing many diseases and disabilities.

- Biotechnology has been used to transfer genes into bacterial cells so they can be colonized to increase supplies of a specific substance, such as a protein or hormone for medicinal purposes.

- Gene therapy is the process of manipulating or removing genes known to cause disease in an attempt to cure that disease.

- Stem cell research has the potential to cure many diseases and debilitations by repairing or regrowing damaged tissue.

Cloning Research

- **Cloning** is the creation of an exact genetic copy of an original.

- Cloning has many positive uses, such as harvesting mass quantities of useful bacteria and saving endangered species from extinction. Cloning is also very controversial, as it raises ethical questions regarding the sanctity of life.

- Dolly the sheep was cloned via reproductive cloning, in which the DNA from one sheep was electrically fused with a DNA-less egg from another sheep. The resulting egg was then planted in a surrogate sheep's uterus.

- Cloning is still inefficient. Many clones die before birth, and many others develop debilitations as they grow into adulthood.

Genomics

- This new field of genetics focuses on studying the genomes of different animals for a range of potential uses. **Genomics** can be used to compare the relatedness between organisms, uncover the functionality of genes and proteins, and determine the minimal genetic requirements for life.

- Once geneticists identify the regions of the genome code that produce specific behaviors or products, biotechnology may be used to manipulate those genes to cure genetic disease and further improve other areas of life.

CHAPTER 7
The Gene Revolution

Sample Test Questions

1. Describe the method scientists use to create recombinant DNA.

2. Describe the process used to clone Dolly the sheep.

3. What questions of safety and ethics might arise from the use of biotechnology in gene transfer between organisms?

4. Organisms that contain genes from other species are called

    A. Recombinant species
    B. Recombinant organisms
    C. Clones
    D. Transgenic organisms
    E. Hybrids

5. What are the single-stranded ends created by the cutting of a piece of DNA by a restriction enzyme called?

    A. Sticky ends
    B. Base pairs
    C. Tacky ends
    D. Complementary ends
    E. Restriction zones

6. Which of the following is NOT a means by which bacteria can gain new genetic material?

    A Mutation
    B. Organization
    C. Conjugation
    D. Transduction
    E. Transformation

7. Which of the following presents a simple definition of cloning?

   A. Self-fertilization
   B. The creation of offspring with similar characteristics to the parent
   C. The creation of a genetically identical offspring via sexual reproduction
   D. The creation of a genetically identical copy of an original individual
   E. The creation of at least thousands of offspring from one individual

8. In which cycle does a virus inject its genetic material into a bacterial cell, replicate, and then release numerous cloned virus cells, resulting in the death of the bacterial cell?

   A. Lysogenic cycle
   B. Krebs cycle
   C. Menstruation cycle
   D. Photosynthetic cycle
   E. Lytic cycle

9. Why do embryonic stem cells have wider applications for use than adult stem cells?

   A. They never deteriorate.
   B. There are no arguable moral repercussions.
   C. They are easier to harvest.
   D. They can become a cell of one particular body tissue.
   E. They can become any cell in the body.

10. What are plasmids?

   A. Junk DNA
   B. Long stretches of repeating DNA sequences
   C. DNA enzymes
   D. Strands of genetic material made up of RNA
   E. Small, circular segments of DNA

CHAPTER 7
The Gene Revolution

11. What is the name for the enzyme used to glue together the strands of DNA in making recombinant DNA?

   A. DNA ligase
   B. DNA gyrase
   C. DNA vaccine
   D. DNA polymerase
   E. DNA splicer

12. Which of the following is the type of bacterial gene transfer that involves the construction of a bridge between two bacterial cells?

   A. Transformation
   B. Conjugation
   C. Transduction
   D. Cytoplasm
   E. Transference

13. Why might a clone show different physical characteristics than its parent?

   A. The clone receives different genes from another parent.
   B. The environment influences the clone's phenotype in a manner different from its parent.
   C. Clones are always physically different from their parent.
   D. Genes change each generation and thus the clone has different genetic material than its parent.
   E. Cloning involves the recombination of genetic material resulting in genetic variation.

14. What are the small circular pieces of bacterial DNA commonly used as vectors?

   A. Plasmids
   B. Cosmids
   C. Haploids
   D. Diploids
   E. Capsids

15. Which field of study explores the use of living organisms to perform operations useful to humans?

    A. DNA technology
    B. Biotechnology
    C. Genetics
    D. DNA manipulation
    E. DNA engineering

ANSWERS

1. There are four steps to creating recombinant DNA. First, restriction enzymes cut the DNA that is to be used as the vector at specific sites along the genome. This cutting results in two double-stranded DNA pieces with sticky ends. The DNA containing the gene of interest is also cut with the restriction enzyme so that each DNA strand has complimentary sticky ends. The two DNA strands are mixed, and the sticky ends bind together. The enzyme DNA ligase is then added to the mix to glue the DNA fragments together.

2. Dolly was cloned from the DNA of one sheep. This DNA was taken from the sheep's udder and placed next to another sheep's egg cell from which the DNA had been removed. An electric shock was applied to fuse the DNA with the egg cell. The new hybrid egg was then implanted into the uterus of a surrogate sheep to grow.

3. The future implications of genetic research are relatively unknown. As a result, any negative effects associated with transferring genes between organisms, such as what may occur from the consumption of genetically modified foods, are also unknown. In addition, there are many ethical issues associated with creating variations of organisms to suit human requirements. Harvesting embryos for their stem cells is particularly controversial because it requires the use of human embryos.

4. D Transgenic organisms contain genes from other species. Molecular biologists insert new genes into an organism's genome to create transgenic organisms.

5. A Sticky ends are created when a restriction enzyme cuts a piece of DNA.

6. B   The other answers are all ways in which bacteria can gain genetic variation in their DNA.

7. D   Cloning is the creation of a genetically identical individual from a given host or parent.

8. E   In the lytic cycle, a virus injects its genetic material into a bacterial cell, uses the cell to replicate its material, and then packages up numerous new virus cells. Once the cells are ready to be released, they burst the bacterila cell, which kills it.

9. E   Embryonic stem cells can become any cell, whereas adult stem cells can only become the cell of a specific body tissue.

10. E   Plasmids are small, circular strands of DNA most often found in bacteria.

11. A   DNA ligase glues together the strands of DNA in making recombinant DNA.

12. B   In conjugation, two bacterial cells mate. A cytoplasmic bridge, called a pilus, forms between them, and the cell donating the DNA replicates its plasmid as it passes a copy along to the recipient cell.

13. B   The genetic material of a clone is identical to its parent organism. However, environmental influences can influence the phenotype (the physical characteristics of an individual) in different ways.

14. A   Plasmids are common vectors.

15. B   Biotechnology explores the use of living things to perform operations that are useful to humans.

MECHANISMS OF EVOLUTION

Darwin's Theory of Evolution

Evolution of Populations

Speciation

Phylogeny and Systematics

8

Darwin's Theory of Evolution

Evolution, the development of species over time, is the single greatest unifying concept in biology. Natural scientists and philosophers have debated varying theories of evolution for hundreds of years. As early as the sixth century BCE, Greek philosopher Anaximander proposed that species change over time. Scientists currently believe that all life on Earth evolved from a single common ancestor about 3.8 billion years ago. Since that time, environmental changes have prompted organisms to evolve continually as they adapt to their environments, leading to the diversification of life as we know it.

Evolutionary thinking was invigorated in the eighteenth century as scholars studied the influence of heredity on evolutionary development. Charles Darwin (1809–1882) was the first to propose two essential mechanisms of evolution that are accepted by scientists today: natural selection and sexual selection.

NATURAL SELECTION

Darwin based his theory of evolution on four observations he made while working as a naturalist aboard the HMS *Beagle*:

1. Resources, such as food and shelter, are limited in the environment.

2. Not all individuals in a population survive long enough to reproduce.

3. Physical characteristics in a population vary from individual to individual.

4. Many physical characteristics are heritable from parents to offspring.

Darwin proposed that individuals who have characteristics that make them successful at acquiring resources, surviving, and reproducing will leave more offspring. If the successful characteristics are heritable, then they will increase in frequency over time. Darwin termed this process **natural selection.**

EXAMPLE: Darwin's theory of natural selection was supported by his observations of fourteen species of finches living on the Galápagos Islands, an island group 600 miles off the coast of Ecuador. In 1977, Peter and Rosemary Grant documented natural selection among Darwin's finches during a major drought on the islands. In normal years, seeds and the finches that eat them appear in a range of sizes. Larger finches with bigger beaks tend to eat the larger, harder seeds, while smaller finches eat the smaller seeds. During the drought, only large, hard seeds were available in the environment. The large finches survived and reproduced more successfully than the small finches, passing their traits on to their offspring. Large bodies and large beaks became more common in the population, while small bodies and small beaks became less common. Through their research, the Grants showed that natural selection can work at a rate rapid enough that scientists can observe changes as they happen.

SEXUAL SELECTION

Sexual selection describes the theory that traits are passed down at a higher frequency from individuals that are more successful at acquiring mates than other individuals. Attractive individuals tend to pass along traits to their offspring, leading to the dominance of these traits in a species over time. Sexual selection can explain the development of ornaments and weapons that characterize some species.

- **Ornaments:** Features that attract the opposite sex, such as peacocks' tails.

- **Weapons:** Features that help the animal compete with members of the same sex, such as deer's antlers.

EXAMPLE: Female guppies prefer mates with large, colorful spots. In comparison with dull males, brightly colored males attract and mate with more females and therefore produce more offspring. Furthermore, brightly colored males produce brightly colored sons, perpetuating the dominance of brightly colored males in future generations.

ADAPTATION AND FITNESS

Natural selection and sexual selection are the only mechanisms of evolution that produce **adaptations,** inherited characteristics that enhance an organism's ability to survive and reproduce. Examples of adaptations include color patterns and behaviors that keep animals hidden from predators and help the animals survive and reproduce.

An organism's evolutionary **fitness** describes its contribution to the gene pool of the next generation, relative to other members of its population. High fitness indicates that an organism is well suited to its environment and likely to survive and reproduce. An organism that is fit will pass its genes to more offspring, and therefore its traits will become more common.

CHAPTER 8
Mechanisms of Evolution

EXAMPLE: If the average number of surviving offspring in a population is two, then an individual who produces three surviving offspring has high fitness, while an individual who produces one surviving offspring has low fitness.

The more fit an organism is, the more it will contribute to the gene pool of the whole species. Therefore, traits of the fittest individuals will be present with more frequency and will influence the overall adaptation of the species. Over time, the pressure exerted by selective forces will influence adaptations, and the eventual result will be the evolution of the species.

EXAMPLE: A population of birds consists of individuals with large beaks that open large seeds, and individuals with small beaks that open small seeds. If there is a lack of small seeds, the birds with small beaks will be more likely to starve. The birds with large beaks will contribute to the gene pool of the species at a greater frequency than those with small beaks, causing the large beak trait to be passed on to more offspring. Eventually, the species could adapt to consist almost completely of individuals with large beaks.

Evolution of Populations

Evolutionary changes take place over a wide range of time scales, from days to millennia. **Macroevolution** refers to large-scale evolutionary changes that take place over long time periods, such as the splitting of one species into two. Although some processes of macroevolution are observable in human time frames, biologists must often make inferences about macroevolution based on fossil records or patterns in existing organisms.

Microevolution refers to small changes in the characteristics of a population, which is a group of individuals of the same species that live in the same geographic area and are therefore capable of interbreeding. Microevolution, which includes changes in a single gene, physical trait, or behavior, can occur over very short time periods, and is often observable to biologists. For example, biologists witness microevolution when they observe a bacteria that has developed resistance to a particular drug.

The study of evolution at the level of the population is known as population genetics. Population geneticists examine a population's **gene pool** (the total of all the different variations of genes that exist in a given population) and observe changes in the frequency of different alleles (single genes) and genotypes (the collection of alleles in an individual) over time.

HARDY-WEINBERG THEOREM

The **Hardy-Weinberg theorem** states that the genotype and allele frequencies of a population will remain constant from one generation to the next if the population has the following characteristics:

• **Large size**

• **No gene flow into or out of the population**

• **No mutations**

• **Random mating**

• **No natural selection**

A population that meets these conditions is said to be in Hardy-Weinberg equilibrium. Because the genotype and allele frequencies are not changing, a population in Hardy-Weinberg equilibrium is not evolving. Very few natural populations are ever in true Hardy-Weinberg equilibrium, since they rarely have all the characteristics listed above. However, the theorem presents an important reference point for population geneticists. By using the Hardy-Weinberg theorem, a geneticist can determine what a population should look like at genetic equilibrium (the state at which a population's gene pool is stable), and then compare her findings to that population as it exists in nature. Any deviation from the Hardy-Weinberg equilibrium suggests that the population is adapting and that the process of **micro-evolution** is taking place.

HARDY-WEINBERG EQUATION

Geneticists use the **Hardy-Weinberg equation** to estimate allele or genotype frequencies in a given population at the Hardy-Weinberg equilibrium. The two-allele Hardy-Weinberg equation is: $p^2 + 2pq + q^2 = 1$, where p equals the frequency of allele A

- q equals the frequency of allele B

- p^2 equals the frequency of genotype AA

- $2pq$ equals the frequency of the genotype AB

- q^2 equals the frequency of the genotype BB

Each organism contains two alleles, either a homogenous pair (for example, AA or BB) or a heterogenous pair (for example, AB). Geneticists can determine the frequency of one allele given the frequency of the other. According to the Hardy-Weinberg equation, the combined frequencies of the two alleles—the frequency of allele A (p) plus the frequency of allele B (q)—must equal 1. Because $p + q = 1$, if the value of p is known, the value of q can be found using the equation $q = 1 - p$.

Geneticists use the Hardy-Weinberg equation to convert allele frequencies into expected genotype frequencies and genotype frequencies into expected allele frequencies, allowing them to predict the frequency of a certain genotype or allele in a population.

EXAMPLE: A certain flower has two alleles for color, R and W. Flowers with the genotype RR are red, those with RW are pink, and those with WW are white. If we know that 25% of flowers in the population are red, the frequency of each allele can be calculated as follows:

- The frequency of red flowers $(RR)(p^2) = 0.25$ (25%)

- The frequency of $R = p = \sqrt{0.25} = 0.5$ (50%)

- The frequency of $W = q = 1 - p = 1 - 0.5 = 0.5$ (50%)

EXAMPLE: A certain disease is caused by an allele, d, which occurs in 1% of a population. Only people homozygous for that trait (genotype dd) express the disease. The frequency of people with the dd genotype can be calculated as follows:

- The frequency of $d = q = 0.01$ (1%)

- The frequency of dd genotype $= q^2$

- $q = \sqrt{0.01} = 0.1$

- The frequency of the recessive allele (d) in the population is 0.1 and the frequency of the dominant allele (D) is 0.9.

MECHANISMS OF MICROEVOLUTION

Geneticists reduce the many causes of microevolution (evolution in a given population) to five basic processes:

1. **Mutation**

2. **Genetic drift**

3. **Gene flow**

4. **Non-random mating**

5. **Selection**

Any of these processes may alter the genetic structure of a population, including its allele frequencies, genotype frequencies, and **genetic variation,** or the number of different alleles per gene.

Mutation

Mutation, which refers to a random change in the DNA sequence of a gene, is the only process that produces new alleles. Due to the constant production of new alleles in a population's gene pool, the allele frequencies of a population alter over time. As the number of alleles present in a gene pool increases, the likelihood of mutation and genetic variation in the population also increases.

A number of events in DNA replication can bring about a mutation: changes in nucleotide sequencing, the alteration of gene positioning, the loss or duplication of a gene, or the insertion of a foreign substance. Generally speaking, however, the rates of mutation are so low that they have a relatively minor effect compared with the other evolutionary processes. Even so, mutation is the key to evolution, and without it evolution would not take place.

EXAMPLE: Scientists estimate that one mutation occurs in every 100,000 genes per generation. Humans are thought to have about 25,000 genes. This means that approximately one in every four babies has a new mutation! Still, chances are slim that the mutation will have an effect on survival and will therefore be selected as a trait contributing to fitness.

Genetic Drift

Genetic drift refers to unpredictable, chance changes in allele frequencies that cause one allele to become more common in a population than another allele. Genetic drift can occur slowly over time, or it can be the result of a sudden decrease in population size. In most cases, the larger a population is, the more stable its allele frequencies are, and the less likely it is to experience genetic drift.

Two extreme cases of genetic drift are **bottlenecks** and **founder effects.**

• **Bottlenecks** occur when a population undergoes a sudden and drastic reduction in size. Natural disasters can cause bottlenecks, for example. The rapid reduction in population size may lead to the loss of alleles if they are not present in the surviving population. This reduction changes allele frequencies in that population. A reduction in genetic variation may also result, as there are fewer alleles available to be passed on to the next generation.

EXAMPLE: Suppose the majority of members of a small population of fish have the allele for yellow fins. An unusually large shark feeding frenzy results in the death of most of the fish population. The next season, a fishing boat catches the only yellow-finned fish left in the population. The surge of shark feeding coupled with the selective fishing causes the allele for yellow fins to decrease in the population, resulting in a shift in allele frequency. In a couple of generations, the yellow fin allele may disappear entirely from the population's gene pool.

• **Founder effects** are what happen when a few individuals are isolated from their original group and form a new population. For example, when two individuals colonize an island, they may not have all the alleles that were once present in the original population. The result is a change in allele frequency and a decrease in genetic variation in the new population.

Both bottlenecks and founder effects lead to changes in allele frequencies if the few surviving or colonizing individuals are not representative of the original population. Often rare alleles are lost during these events as individuals who possess them die off or leave, lowering the genetic variation of the entire population.

> *EXAMPLE:* Northern elephant seals were hunted to the brink of extinction during the late nineteenth century. As a result, many alleles were lost. Although the population has recovered since that time, the effects of the bottleneck are still present: Northern elephant seals today share many of the same alleles and possess extremely low genetic variation across the entire species.

Gene Flow

Gene flow occurs when organisms **migrate,** or move, from one population to another. A population might lose some alleles when individuals leave the population and might gain new alleles when individuals join the population.

Nonrandom Mating

Nonrandom mating describes mating that is not equally likely to occur between all members of the population. Mates will often match up due to choice, or selection, as members of a population favor certain characteristics possessed by some members over those possessed by others. Nonrandom mating is likely to alter both allele and genotype frequencies in a population, since the alleles for the preferred characteristics will increase in frequency in the population. Other types of nonrandom mating, such as inbreeding, may alter genotype frequencies as well.

Selection

Selection, both natural and sexual, occurs when some individuals leave more progeny than others, resulting in dramatic changes in allele frequencies in a population. Selection is often classified according to those individuals favored, or more fit, to leave more offspring.

- **Stabilizing selection:** Individuals with average phenotypes are favored, and those with phenotypic extremes are selected against.

- **Directional selection:** Individuals at one phenotypic extreme are favored, and those at the other extreme are selected against.

- **Disruptive selection:** Individuals at both phenotypic extremes are favored, and those with intermediate phenotypes are selected against.

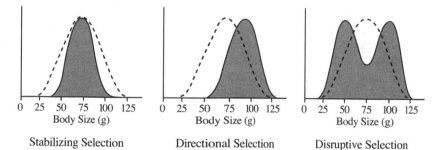

Stabilizing Selection   Directional Selection   Disruptive Selection

Modes of selection: The graphs above illustrate the different modes of selection for body size in a given population. The dotted peak in each graph represents the average body size. The shaded area represents the individuals selected for in each population.

EXAMPLE: Disruptive selection can be observed in populations of sticklebacks, a species of fish found in freshwater lakes in Canada. Sticklebacks often have two distinct phenotypes: a benthic morph that specializes in feeding at the bottom of the lake and a limnetic morph that specializes in feeding at the top of the lake. These morphs represent two phenotypic extremes. Individuals with intermediate phenotypes are selected against because they can't feed efficiently at either the top or the bottom of the lake.

Speciation

Speciation is the term biologists use to describe the process by which new species arise through evolution. According to the Biological Species Concept of Ernst Mayr, **species** are "groups of interbreeding natural populations reproductively isolated from other such groups." According to this definition, organisms are considered different species if they would not naturally mate and produce healthy, fertile offspring.

REPRODUCTIVE ISOLATING MECHANISMS

Speciation is essentially the process by which groups evolve **reproductive isolating mechanisms,** traits that prevent different groups from successfully interbreeding. Reproductive isolating mechanisms evolve when species differ in four possible ways:

- **Genetics:** The genetic makeup of the species.

- **Morphology:** The form and structure of most individuals in a species.

- **Behavior:** The actions exhibited by most individuals in a species.

- **Ecology:** Interactions between a species and its environment.

Prezygotic Isolating Mechanisms

Prezygotic isolating mechanisms refer to traits that specifically prevent mating and fertilization between different species. There are five basic types of prezygotic isolating mechanisms, any of which will prevent a zygote from being formed between two organisms:

- **Habitat isolation:** Organisms live or breed in different areas.

- **Temporal isolation:** Organisms breed at different times.

- **Behavioral isolation:** Courtship rituals and other mating cues differ among organisms.

- **Mechanical isolation:** The genitalia of organisms are incompatible.

- **Gametic isolation:** The gametes of organisms cannot fuse during fertilization.

EXAMPLE: Lake Victoria in Africa is home to hundreds of species of cichlid fish, many of which are similar in most respects, with the key exception of color pattern. Female cichlid fish use color patterns to identify mates of their own species, preventing them from mating with the "wrong" species. Because they exhibit this form of behavioral isolation, the many species of cichlid fish found in Lake Victoria will not mate with one another.

CHAPTER 8
Mechanisms of Evolution

Postzygotic Isolating Mechanisms

Postzygotic isolating mechanisms refer to traits that prevent **hybrids,** the offspring of two different species, from developing into healthy, fertile adults. There are two main types of postzygotic isolation:

- **Hybrid inviability:** Refers to hybrids that are not physically able to survive into adulthood.

- **Hybrid sterility:** Refers to hybrids that are themselves healthy but are sterile and therefore unable to produce offspring.

EXAMPLE: Mules are hybrids produced from the successful mating of a male donkey and a female horse. Mules are almost always sterile, but there are exceptions. In 1984, a female mule named Krause mated with a donkey and successfully gave birth. Since this was such an unusual event, the owners named the foal Blue Moon.

GENETIC DIVERGENCE

Genetic divergence of a population leads to **speciation,** the separation of one species into two. Speciation can occur quickly or it can take place over a long time, depending on the factors that lead to separation. The two methods by which speciation occur are **allopatric speciation** and **sympatric speciation.**

- **Allopatric speciation:** Two populations are geographically isolated and over time evolve into different species. Two separate species develop due to the lack of gene flow between the two populations.

- **Sympatric speciation:** One species splits in two when a small proportion of individuals within a population reproduce exclusively with each other. Sympatric speciation, which is more common in plants than animals, is rarer than allopatric separation and may be the result of non-random mating or a mutation that leads to reproductive isolation. Geographic isolation is not necessary for this division within a population.

EXAMPLE: Many species of flowering plants have originated from a process of sympatric separation known as polyploidization. An error during cell division produces a plant with an extra set of chromosomes. As a result of the error, new plants are reproductively isolated from the original species.

THE TEMPO OF SPECIATION

Speciation does not proceed at a single, constant tempo or rate: Evolutionary change occurs rapidly in some groups of organisms and slowly in others. Biologists employ two models to describe the rate of evolution, each reflective of the tempo of speciation in certain organisms:

- The **gradualist model** of evolution states that evolution is a slow, continuous process.

- The **punctuated equilibrium model** of evolution states that evolution proceeds in rapid spurts interrupted by long periods during which a population undergoes little evolutionary change.

These two models explain the differing patterns of evolution observed by scientists. Ancestral lines of some populations show many changes in short periods of time, while other lines show no changes over many millions of years.

Phylogeny and Systematics

Phylogeny is the study of evolutionary history across groups of species. Phylogeneticists investigate the evolutionary relationship between different species and seek to determine how long ago they diverged. **Systematics** is the classification of organisms into these different groups, usually on the basis of their evolutionary relatedness.

PHYLOGENETIC TREES

Phylogenetic trees describe evolutionary history by illustrating the patterns of relatedness among different groups of species. In some cases, these trees also depict the time and/or degree of divergence between species.

Interpretation of Phylogenetic Trees

Each branch on a phylogenetic tree represents the divergence of two groups from a common ancestor. Closely related groups thought to share a recent common ancestor will occupy branches in closer proximity to each other than distantly related groups. Biologists use phylogenetic trees to represent hypotheses regarding species relatedness that can be tested.

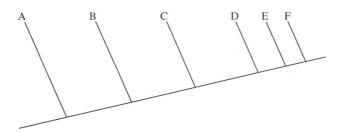

Sample Phylogenetic Tree: In the phylogenetic tree above, species D is more closely related to species E and F than it is to species C.

Biologists classify groups of species in a phylogenetic tree as monophyletic, paraphyletic, or polyphyletic, depending on the exact relationship they are studying.

- **Monophyletic groups** include all descendants of a single common ancestor. Monophyletic groups are also called clades.

- **Paraphyletic groups** include some, but not all, descendants of a single common ancestor.

- **Polyphyletic groups** include descendants, but not the common ancestor, of all species in the group.

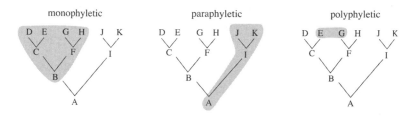

Three Types of Phylogenetic Groupings

Construction of Phylogenetic Trees

Biologists construct phylogenetic trees by grouping organisms according to **shared derived characters,** traits that are evolutionarily unique among a particular group of organisms. Shared derived characters may be physical traits, such as the presence of hair, or they may be individual nucleotides in a sequence of DNA. Biologists often use an **outgroup** to differentiate between shared derived characters and **primitive characters,** traits that were present in the ancestor of the species of interest. The outgroup must be more distantly related to the ingroup, the organisms of interest, than the members of the ingroup are to each other.

Similarity between physical traits may be the result of convergent evolution, rather than recent common ancestry. **Convergent evolution** occurs when distantly related organisms face similar environmental challenges and consequently evolve functionally similar traits. Traits that are similar owing to convergent evolution are **analogous,** while those traits that are similar owing to common ancestry are **homologous.** Because the presence of analogous traits can present a misleading case for relatedness, phylogenetic trees based on multiple genes are generally considered more accurate than those based on multiple physical traits.

EXAMPLE: A dolphin flipper, human hand, and bat wing all share striking similarities in bone structure, which suggests that they descended from a common ancestor. This is an example of homology. On the other hand, while the wings of an insect and the wings of a bird are functionally similar, insects and birds do not share a common ancestor. Insects and birds both happened to develop wings because of environmental influences that selected for an ability to fly. This is an example of analogy.

Homoplasy

Instances of **homoplasy,** shared character states not inherited from a common ancestor, create conflicts for biologists seeking to determine the relationship between different species. Once shared derived characters have been determined for multiple species, biologists construct a phylogenetic tree using one of two rules to address this conflict: maximum parsimony or maximum likelihood.

- **Maximum parsimony:** A phylogenetic tree is designed to demonstrate the fewest possible evolutionary events (changes in characters) and the fewest instances of homoplasy between branches. Trees constructed using the rule of maximum parsimony seek to simplify relationships by requiring the fewest assumptions about relatedness as possible.

- **Maximum likelihood:** Evolutionary events are weighted according to the likelihood of occurrence. As genomic research advances, biologists are finding that DNA characters evolve at different rates, which suggests that homoplasy may happen more often than the rule of maximum parsimony would allow.

EXAMPLE: Scientists believe that the loss of a complex trait, such as eyes, can occur more readily than the addition of a complex trait. A phylogenetic tree constructed according to maximum parsimony would give equal weight to the gain and loss of eyes, despite the difference in how readily each occurs. A tree constructed according to maximum likelihood would minimize the number of times eyes were gained, the least likely of the two events, even though that leads to a greater number of total changes in the tree.

HIERARCHICAL ORGANIZATION OF LIFE

Biologists categorize all living organisms according to the **binomial system of nomenclature** first developed by the botanist Carolus Linnaeus in the mid-1700s. The binomial system of nomenclature is a two-word system used to name each species. The first name identifies the **genus** and the two names together identify the **species.** When written, both words are italicized and the genus name is capitalized.

EXAMPLE: The binomial name for modern humans is *Homo sapiens.* The name indicates that modern humans are from the genus *Homo*, which includes both ancestors and cousins such as *Homo erectus, Homo ergaster, Homo heidelbergensis, Homo floresiensis,* and *Homo neanderthalensis.* While some of these other species of humans existed at the same time as *Homo sapiens*, most notably the *Homo neanderthalensis, Homo sapiens* are the only currently surviving member of the *Homo* genus.

Linnaeus grouped species into progressively broader categories, ranging from species to kingdom, based on similarities in traits. Present-day biologists continually update this classification scheme as they learn more about the evolutionary relatedness among organisms.

CHAPTER 8
Mechanisms of Evolution

The modern categories of the hierarchical classification of life, from broadest to smallest, are:

Category	Example
Domain	Eukarya
Kingdom	Animalia
Phylum	Chordata
Class	Mammalia
Order	Primates
Family	Hominidae
Genus	*Homo*
Species	*Homo sapiens*

Summary

Darwin's Theory of Evolution

- **Natural selection** is the mechanism by which factors in the environment influence whether an individual in a population will be more successful than others at surviving and reproducing. Natural selection is the primary force causing evolution.

- **Sexual selection** is the theory that traits are passed down at a higher frequency from individuals that are more successful at acquiring mates than other individuals.

- **Adaptations** are inherited characteristics that enhance an organism's ability to survive and reproduce.

- An organism's evolutionary **fitness** describes its contribution to the **gene pool** of the next generation relative to other members of its population.

Evolution of Populations

- **Macroevolution** refers to large-scale changes, such as the splitting of a single species into two.

- **Microevolution** refers to small changes in the characteristics of a population.

- The **Hardy-Weinberg theorem** states that a population will not evolve if it has a large size, no gene flow, no mutations, random mating, and no selection.

- The **Hardy-Weinberg equations** can be used to estimate genotype frequencies from allele frequencies, and vice versa.

- The five mechanisms of microevolution are **mutation, genetic drift, gene flow, nonrandom mating,** and **selection.**

Speciation

- **Speciation** is the process by which a new species arises through evolution.

- **Prezygotic isolating mechanisms** are traits that prevent mating and fertilization between different species. **Postzygotic isolating mechanisms** are traits that prevent **hybrids,** the offspring of different species, from developing into healthy, fertile adults.

- **Allopatric speciation** occurs when two geographically isolated populations split into different species. **Sympatric speciation** occurs when one species splits into two through means other than geographic isolation.

- The **gradualist model** of evolution states that evolution is a slow, continuous process. The **punctuated equilibrium model** states that evolution takes place in rapid spurts interrupted by long periods marked by little evolutionary change.

Phylogeny and Systematics

- **Phylogeny** is the study of the evolutionary history of groups of species.

- **Systematics** is the classification of organisms into different groups, usually on the basis of evolutionary relatedness.

- Phylogenetic trees describe the evolutionary relatedness of groups of organisms.

- Biologists use a multi-level hierarchical system to classify organisms. From broadest to smallest, the major levels are **domain, kingdom, phylum, class, order, family, genus,** and **species.**

CHAPTER 8
Mechanisms of Evolution

Sample Test Questions

1. Male long-tailed widowbirds, *Euplectes progne*, possess extremely long tails (up to 50 cm long), while females possess short tails. Outline a series of observations or experiments that would allow you to determine whether long tail length in the male widowbird is a trait influenced by sexual selection.

2. A scientist examines a population of birds with varying beak lengths. Birds with long beaks have the genotype *BB* or *Bb*; birds with short beaks have the genotype *bb*. The scientist finds that the frequency of allele *B* is 0.4 and the frequency of allele *b* is 0.6. She also finds that the frequency of birds with long beaks is 0.48 and the frequency of birds with short beaks is 0.52. What would be the expected frequency of beak length among this population? Why might the expected and actual frequencies of beak lengths differ?

3. Two scientists are designing a phylogenetic tree to explain the presence of hair in mammals and its absence in reptiles. What are the two methods they could use to design a phylogenetic tree, and how might these methods lead to different conclusions regarding species relatedness?

4. Species A breeds in March, while species B breeds in April. The difference in the breeding season is an example of which of the following?

   A. Spatial isolation
   B. Temporal isolation
   C. Behavioral isolation
   D. Gametic isolation
   E. Postzygotic isolation

5. Which rule applies to changes in characters that are weighted according to the probability of occurrence in a phylogenetic tree?

   A. Maximum likelihood
   B. Maximum parsimony
   C. Maximum homology
   D. Maximum convergence
   E. Hardy-Weinberg theorem

6. Which term describes inherited characteristics that enhance an organism's ability to survive and reproduce?

   A. Shared derived characters
   B. Homologies
   C. Analogies
   D. Evolved traits
   E. Adaptations

7. Which term describes groups that include some, but not all, of the descendants of a single common ancestor?

   A. Monophyletic
   B. Paraphyletic
   C. Polyphyletic
   D. Homologous
   E. Analogous

8. Which of the following classification levels falls between class and family in the hierarchical classification scheme?

   A. Phylum
   B. Kingdom
   C. Domain
   D. Order
   E. Genus

9. Hybrid sterility is an example of which of the following?

   A. Prezygotic isolation
   B. Postzygotic isolation
   C. Gametic isolation
   D. Mechanical isolation
   E. Behavioral isolation

CHAPTER 8
Mechanisms of Evolution

10. Which of the following processes is the most likely cause for the development of antlers in deer?

    A. Mutation
    B. Genetic drift
    C. Nonrandom mating
    D. Natural selection
    E. Sexual selection

11. Which term refers to the classification of organisms on the basis of evolutionary relatedness?

    A. Phylogenetics
    B. Systematics
    C. Genetics
    D. Speciation
    E. Homology

12. Which term refers to the idea that evolution occurs in rapid spurts interrupted by long periods of little evolutionary change?

    A. Homology
    B. Gradualist model
    C. Punctuated equilibrium
    D. Phylogenetics
    E. Systematics

13. Which mechanism is at work in populations that experience a sudden and drastic reduction in size?

    A. Bottleneck
    B. Founder effect
    C. Gene flow
    D. Migration
    E. Mutation

14. Two populations that begin to diverge after being separated by a mountain range are said to be undergoing which of the following forms of speciation?

A. Sympatric speciation
B. Parapatric speciation
C. Monophyletic speciation
D. Allopatric speciation
E. Polyphyletic speciation

15. Which mechanism of selection would most likely cause animals of medium size to be selected against in favor of animals of small or large size?

A. Stabilizing selection
B. Normalizing selection
C. Directional selection
D. Disruptive selection
E. Extreme selection

ANSWERS

1. If tail length is a trait maintained by sexual selection, then two things must be true: First, males with longer tails must acquire more mates, and therefore leave more offspring, than males with shorter tails; second, tail length must be heritable. To test the first point, you could manipulate the tail length of male widowbirds and observe the effect on mating success. During the experiment, a large number of male widowbirds would be collected and randomly assigned to one of three groups: shortened tail, control, or lengthened tail. After their tails had been shortened, lengthened, or not changed, each bird would be released. The mating success and offspring production of each male would then be observed. If the first point is true, males whose tails were artificially lengthened would have the highest number of mates and offspring, and males whose tails were shortened would have the smallest number of mates and offspring. To determine if tail length is heritable, one could randomly mate male and female widowbirds and collect the offspring. If heritable, the tail length of male offspring would correlate to the tail length of their fathers.

2. Based on allele frequencies and the Hardy-Weinberg equation, the expected frequency of the *BB* genotype is 0.16, because $0.4 \times 0.4 = 0.16$. The expected frequency of the *Bb* genotype is 0.48, because $2 \times 0.4 \times 0.6 = 0.48$. Since both *BB* and *Bb* genotypes produce big beaks, the expected frequency of big beaks is the sum of their two frequencies (0.16 + 0.48 = 0.64). The expected frequency of short beaks is 0.36, because 0.6 × 0.6 = 0.36. The actual frequency of beak sizes indicates fewer large beaks than expected. One possible explanation is non-random mating. If birds prefer mates with similar beak sizes, then fewer heterozygous genotypes (*Bb*) will be produced. This will lower the frequency of big beaks, despite the higher frequency of their genotype.

3. The scientists can construct phylogenetic trees based on either maximum parsimony or maximum likelihood. Because it is assumed under maximum parsimony that all events have an equal chance of occurring, hair is just as likely to have been lost in birds as it is to appear in mammals. This results in the minimum number of branches on the tree. Maximum likelihood suggests that one event is more likely to occur than another. In this scenario, scientists believe that the loss of a complex trait, such as hair, is more likely to occur than the gain of a complex trait. Therefore, a tree based on maximum likelihood would illustrate that animals evolved to lose hair rather than to develop hair, resulting in more branches on the tree.

4. **B** Difference in the timing of breeding between species is an example of temporal isolation.

5. **A** Maximum likelihood is a rule for constructing phylogenetic trees in which changes in characters are weighted according to the probability of occurrence.

6. **E** Inherited characteristics that enhance an organism's ability to survive and reproduce are known as adaptations.

7. **B** Groups that include some, but not all, of the descendants of a single common ancestor are paraphyletic.

8. **D** In the hierarchical classification scheme used by biologists, order falls between class and family.

9. **B** Hybrid sterility is an example of postzygotic isolation.

10. **E** Since they are primarily used to challenge other male deer in pursuit of a mate, antlers most likely evolved because of sexual selection.

11. B   The field of systematics classifies organisms into different groups, usually on the basis of evolutionary relatedness.

12. C   The idea that evolution occurs in rapid spurts interrupted by long periods with little evolutionary change is called punctuated equilibrium.

13. A   Populations that undergo a bottleneck experience a sudden and drastic reduction in size.

14. D   Two populations that begin to diverge after being separated by a mountain range may undergo allopatric speciation.

15. D   Disruptive selection would select against animals of medium size.

CHAPTER 8
Mechanisms of Evolution

THE ORIGIN OF LIFE

The Early Earth

Early Life Forms

9

The Early Earth

Physicists estimate the age of the universe at no less than 14 billion years and the age of Earth at around 4.5 billion years. The early Earth was composed of extremely hot liquid rock, even at the surface. No protective ozone layer existed to prevent the surface from being bombarded by UV radiation. The atmosphere was a mix of gases that would be toxic to all forms of life. The hostile conditions of the Earth at this state were unsuitable for life. After hundreds of millions of years the Earth began to cool and solidify. Water collected on the surface, creating vast oceans rich in chemicals. Around 3.8 billion years ago, ocean temperatures dropped and between 3.8 and 2.5 billion years ago, life first appeared.

THE REDUCING ATMOSPHERE

The atmosphere at the advent of life was most likely full of water vapor and noxious gases, such as hydrogen sulfide, methane, carbon dioxide, ammonia, and nitrogen, resulting from volcanic explosions. Without plants, the atmosphere was reducing, meaning it did not contain oxygen. The Earth's early reducing atmosphere allowed any small molecules that formed to remain stable and react with other small molecules to form larger molecules. This would not be possible in an oxidizing atmosphere, like the one present today, because the small molecules would be broken down in reactions with oxygen. Scientists theorize that energy provided by lightning and other atmospheric activity facilitated the formation of these smaller molecules and, eventually, early life formed as the molecules came together.

The Miller-Urey Experiment

In 1953, scientists Stanley Miller and Harold Urey re-created the conditions of an early reducing atmosphere to test whether organic molecules could be created. The experimental atmosphere yielded many organic molecules, including amino acids, providing compelling support for the hypothesis that an early reducing atmosphere yielded the building blocks for life. The experiment has been repeated many times since the success of Miller and Urey, with many trials also yielding organic molecules.

THE PATH OF CHEMICAL EVOLUTION

Although scientists agree that elements present on Earth in some way combined to form organic molecules billions of years ago, they are divided over which organic compound appeared first: the nucleic acid RNA, proteins, or a combination of the two. The reasoning behind each theory is outlined below:

- **RNA appeared first:** A hereditary molecule, such as RNA, is required for the propagation of more molecules.

- **Proteins appeared first:** Without the enzymes formed from proteins, replication cannot occur, regardless of heritability.

- **A combination of both RNA and proteins appeared first:** A molecule known as protein-nucleic acid (PNA), which is more simple than RNA but still able to replicate, may have existed prior to RNA.

WHERE LIFE FIRST APPEARED

Most scientists agree that the organic chemicals required for life formed spontaneously at a time when the Earth was still cooling. Although it is not currently known where on the Earth life first originated, scientists have developed several theories.

- **At the ocean's edge:** Life arose in bubbles around the ocean's edge.

- **Under frozen seas:** Life originated beneath a frozen ocean.

- **Deep in the earth's crust:** Life was a byproduct of volcanic activity beneath the earth's crust.

- **Within clay:** Chemical activity within clay promoted synthesis of the molecules required for life.

- **At deep-sea vents:** Chemical activity around deep-sea vents promoted synthesis of the molecules required for life.

THE BUBBLE MODEL

The bubble model presents a theory for how cells may have formed from the collection of large molecules in the early stages of life. The bubble model states that gases escaping from hydrothermal vents, fissures at the ocean floor where gases escape from beneath the surface of the Earth, become trapped inside protective bubbles. These gases interact within the enclosed environment to form larger, organic molecules. The bubbles pop at the ocean's surface, releasing their molecular contents into the atmosphere, where they then react with other such molecules. Fueled by lightning and UV radiation, these molecules may have formed more complex organic compounds. Finally, these molecules, regardless of whether they became more complex or not, fall back into the ocean, potentially to be encased in bubbles again, starting the whole process over.

As the complexity of the bubbles increased, their ability to incorporate more molecules and energy also increased, extending the span of existence for the most complex bubbles.

Furthermore, some bubbles may have been able to multiply themselves by cleaving in two in a process similar to cellular reproduction, in which a parent cell produces an identical offspring cell. Although bubbles in this state would not have been considered living, current theory suggests that if molecules with the ability to store and pass on genetic material existed in these bubbles, life may have formed from them.

Although much remains unknown about these early origins, one of the appeals of the bubble method is the protective barrier the bubble provides. A protective boundary the separates the inside of an organism from its environment is one characteristic of all living things. Bubbles may have provided the early membrane for potentially fragile early life forms, until a true cell membrane evolved.

Early Life Forms

Whatever the mechanism, organic molecules must have preceded the origin of life. Once these molecules multiplied to a sufficient quantity in an appropriate environment, life may have formed. The earliest life forms biologists have found in the fossil record date from at least 3.5 billion years ago. These early organisms were small, single-celled, aqueous (water-dwelling) prokaryotes lacking external appendages and possessing no nucleus and limited internal structure.

PROKARYOTES

There are two groups of prokaryotes existing on Earth today: Archaebacteria and bacteria. Biologists believe both groups also existed at the beginning of life.

- **Archaebacteria** usually inhabit extreme conditions, either habitats high in toxic gases but oxygen-free **(methanogens),** extremely salty habitats **(extreme halophiles),** or extremely hot habitats **(extreme thermophiles).**

- **Bacteria** make up most of the prokaryotes alive today. They have very strong cell walls, and a simpler gene architecture than Archaebacteria. Some bacteria, known as photosynthetic bacteria, evolved the ability to absorb energy from light and convert it to chemical bonds within the cell.

EUKARYOTES

Approximately 1.5 billion years ago, more complex cells appeared in the fossil record. These cells show key innovations, notably the presence of a nucleus, that differentiate them from prokaryotic cells.

Key innovations in the development of eukaryotic cells include:

- **Complex structures:** Simple structures present inside the cells of prokaryotic cells are thought to have developed into more sophisticated structures, such as the nucleus and endoplasmic reticulum in eukaryotes. Biologists also believe that early cells would engulf bacteria that possessed specific functions, such as photosynthetic bacteria. These bacteria became part of the cell structure, such as a chloroplast. This process eventually led to the formation of permanent structures in the eukaryotic cell.

- **Sexual reproduction:** Most eukaryotic cells reproduce sexually, resulting in the recombination of genetic material. This recombination generates variation, a key requirement for evolution that has led to the comparatively large diversity of eukaryotes.

- **Multicellularity:** Single eukaryotic cells formed colonies in which individual cells began performing specific tasks for the benefit of the colony. Eventually, these colonies became individual organisms, leading to even further diversity in the eukaryotes.

The majority of life's diversity can be attributed to the great variation present among the eukaryotes, and both sexual reproduction and multicellularity provided the mechanisms for this diversity.

CHAPTER 9
The Origin of Life

DIVERSITY OF LIFE

Scientists have developed a system of taxonomy to distinguish between the different organisms on Earth. Taxonomy is a branch of science devoted to categorizing organisms into three **domains:** the Archaea, Bacteria, and Eukarya. From here, organisms then fall into six **kingdoms:** Bacteria (in the domain Bacteria), Archaebacteria (in the domain Archaea), Protista, Fungi, Plantae, and Animalia (all in the domain Eukarya).

Summary

The Early Earth

- The Earth was created 4.5 billion years ago.

- The early Earth was hostile, constantly bombarded by meteorites and UV radiation, and possessing an atmosphere full of water vapor and noxious gases.

- As the Earth cooled, oceans began to cover the surface, creating an environment conducive to life.

- According to the fossil record, unicellular life appeared between 3.8 and 2.5 billion years ago.

- Life must have been preceded by the formation of organic molecules.

- A **reducing atmosphere,** with minimal amounts of oxygen, would have facilitated the formation of organic molecules such as amino acids and nucleic acids. These molecules could have aggregated to form early, simple life forms. The Miller-Urey experiment supports this hypothesis.

- The bubble model suggests that bubbles protected and promoted the formation of organic molecules at hydrothermal vents. These bubbles popped at the surface and released the molecules in the air, where the molecules may have grown in complexity, eventually creating an early life form.

Early Life Forms

- Early life was almost certainly prokaryotic, single-celled, and aqueous.

- Two types of prokaryotic cells that biologists believe were present when life began and are still around today: **Archaebacteria** and **bacteria.** Archaebacteria usually inhabit extreme environments. Bacteria are more diverse and make up most of the prokaryotic cells present today.

- Following a number of key innovations, prokaryotic cells increased in complexity and led to the evolution of eukaryotic cells. Eukaryotic cells first appeared approximately 1.5 billion years ago and contained sophisticated internal structures, such as a nucleus, and underwent sexual reproduction. They may have also aggregated into multicellular organisms.

- To categorize the great diversity of life on Earth, scientists employ a system of taxonomy, in which all organisms are divided into three **domains,** which are in turn separated into six **kingdoms.**

Sample Test Questions

1. What conditions made early Earth hostile to life?

2. What are some of the main barriers scientists face when seeking to explain how life began?

3. Explain the major innovations that distinguish eukaryotic cells from prokaryotic cells.

4. What disadvantage did the lack of an ozone layer present to early Earth?

    A. Heat could not be contained.
    B. The Earth's surface was bombarded with solar radiation.
    C. Gases could not be contained.
    D. Wind currents were unsustainable.
    E The Earth's surface didn't receive sufficient radiation.

5. Which of the following was an important component of the Earth's early atmosphere?

    A. Reducing atmosphere
    B. Oxidizing atmosphere
    C. Radiating atmosphere
    D. Absorbing atmosphere
    E. Regenerating atmosphere

6. How old is the earliest definitive fossil record of life?

    A. 14 billion years
    B. 4.6 billion years
    C. 3.5 billion years
    D. 2.5 billion years
    E. 10,000 years

7. Which characteristic describes a reducing atmosphere?

    A. It is highly difficult for amino acids to form.
    B. Oxygen is the predominant gas.
    C. Electrons are removed.
    D. Organic molecules may be readily formed.
    E. It never rains.

8. Which condition most likely preceded the origin of life?

    A. Pure hydrogen in the atmosphere
    B. The presence of hydrothermal vents
    C. The presence of viruses
    D. The presence of bacteria
    E. The formation of organic molecules

9. Which type of prokaryote converts energy from the sun into compounds within the cell?

    A. Photosynthetic bacteria
    B. Chemosynthetic bacteria
    C. Chemotrophic bacteria
    D. Photochemotrophic bacteria
    E. Phototrophic bacteria

CHAPTER 9
The Origin of Life

10. How might bubbles show behavior similar to sexual reproduction in eukaryotic cells?

    A. Two bubbles with identical gas compositions may join into one bubble.
    B. One bubble may break into two new bubbles with identical gas compositions.
    C. Molecules in a bubble may form a nucleus.
    D. Bubbles may undergo mitosis.
    E. Bubbles may undergo meiosis.

11. What two groups of organisms make up the prokaryotes?

    A. Archaebacteria and Plantae
    B. Fungi and Animalia
    C. Bacteria and Archaebacteria
    D. Plantae and Animalia
    E. Fungi and Bacteria

12. To which domain does the kingdom Animalia belong?

    A. Protista
    B. Fungi
    C. Bacteria
    D. Archaea
    E. Eukarya

13. What did the experiments of Miller and Urey show?

    A. Under the conditions existing on early Earth, organic molecules could have formed spontaneously from the materials present.
    B. Traits acquired during a lifespan are not passed on to offspring.
    C. Electrons exist around the nucleus of an atom in areas of probability known as orbitals.
    D. Evolution occurs in spurts of activity, followed by long periods during which relatively little activity occurs.
    E. DNA replication occurs via a replication fork.

CHAPTER 9
The Origin of Life

14. What two key innovations have led to the large diversity in eukaryotic life forms?

    A. Chemosynthesis and autotrophism
    B. Multicellularity and sexual reproduction
    C. Asexual reproduction and transcription
    D. DNA replication and photosynthesis
    E. Complex structures and translation

15. Which group of Archaebacteria inhabits environments rich in toxic gases and lacking oxygen?

    A. Thermophiles
    B. Halophiles
    C. Methanogens
    D. Pathogens
    E. Hydrophiles

ANSWERS

1. Early Earth was hot with a molten lava surface. There was no standing water and the surface was constantly being bombarded by meteorites and other massive objects. The UV radiation from the sun was very intense and there was no ozone layer.

2. The fossil record indicates that life on Earth began as far back as 3.5 billion years ago. However, scientists can only speculate about what life looked like in its early stages. In addition, the Earth and its atmosphere were very different. Experiments to reconstruct past conditions provide only inconclusive and speculative evidence, and debate over the composition of the atmosphere continues. A better understanding of the Earth's early environment is necessary to understand how life evolved.

3. Eukaryotic cells developed more sophisticated internal structures, such as a nucleus and endoplasmic reticulum, that the relatively simple prokaryotes do not possess. By engulfing bacteria that possessed special functions, they were also able to use those special functions for their own purposes. Eukaryotes underwent sexual reproduction, unlike prokaryotes, which allowed greater genetic variation and increased diversity. Lastly, eukaryotic cells began to congregate and form multicellular organisms, which led to even greater diversity.

CHAPTER 9
The Origin of Life

4. B The ozone is a protective layer that prevents the Earth from being bombarded with solar radiation.

5. A Early Earth had a reducing atmosphere, which allowed organic molecules to remain stable.

6. C The earliest life forms discovered in the fossil record are unicellular organisms that date back 3.5 billion years, although scientists believe that life began as far back as 3.8 billion years ago. The first multicellular organisms appeared approximately 1.5 billion years ago.

7. D In a reducing atmosphere, less energy is required for the formation of complex organic molecules.

8. E Organic molecules are vital for life to exist.

9. A Photosynthetic bacteria convert the sun's energy into stored molecules in their cells.

10. B Bubbles that cleave and form two new identical bubbles undergo a process similar to reproducing cells, in which parents split and leave two daughter cells.

11. C Bacteria and Archaebacteria are the two groups of prokaryotes. All other organisms are eukaryotic.

12. E Animalia is a kingdom in the Eukarya domain.

13. A Stanley Miller showed that organic molecules could have formed spontaneously from the materials present on early Earth.

14. B Sexual reproduction in eukaryotes led to the start of genetic variation necessary for evolution to occur. Multicellularity led to the development of varying organisms composed of multiple cells, further increasing diversity among organisms.

15. C Methanogens are Archaebacteria that are able to survive in environments toxic to other organisms. However, they cannot survive in the presence of oxygen.

THE EVOLUTION AND DIVERSITY OF LAND PLANTS

Kingdom Plantae

Plant Adaptations

Nonvascular Plants

Seedless Vascular Plants

Seed Vascular Plants

Kingdom Plantae

Organisms that belong to the kingdom Plantae share, with very few exceptions, certain common characteristics, described in the table below.

Characteristic	Details
Multicellularity	Almost all plants are multicellular organisms.
Eukaryotic cells	All plant cells are eukaryotic. They contain a membrane-bound nucleus and membrane-bound organelles.
Chloroplasts	Almost all plant cells contain chloroplasts, organelles that contain chlorophyll for the task of photosynthesis.
Cell walls	All plant cells have cell walls.
Nonmotility	Plants are incapable of independent movement.

Plant Adaptations

Water-bound green algae, called **charophyceans,** existed in abundance in oceans 500 million years ago. Green algae created their own food via photosynthesis, though they were not able to survive outside of water. Biologists believe that the first organisms to colonize land were plants that evolved from green algae.

The earliest land plants were simple compared with many of the plants that exist today. They depended on water for reproduction and therefore grew in close proximity to moist swampy environments. After the evolution of the seed and new methods of pollination, plants moved farther inland and occupied drier and harsher habitats. Over millions of years, plants have developed adaptations that allow them to grow over most of the Earth's surface.

MAJOR EVOLUTIONARY BENCHMARKS

Biologists point to four major evolutionary benchmarks that have led to the development and diversification of plants. Each successive adaptation allowed plants to continue to radiate across the land and survive in previously uninhabitable environments.

Evolutionary Benchmarks	Significance
Alternation of generations is a reproductive cycle during which plants move from a haploid phase, where a multicellular gametophyte forms, to a diploid phase, where a multicellular sporophyte forms.	Though algae have a similar reproductive cycle, the much longer diploid phase that plants undergo provides more protection to the developing embryo. Additional protection allowed plant embryos to disperse farther and survive the dry conditions on land. The longer diploid stage also provided a longer growth time, allowing further development and greater range.
Vascular tissue, including the root system, allows fluids and nutrients to be transported throughout the plant body.	Plants with vascular tissues were able to grow much taller than their primitive plant counterparts and to exploit more resources from their environment.

CONTINUED

Evolutionary Benchmarks	Significance
Seeds form a protective coating around the plant embryo and provide the seedling with nourishment.	The development of seeds allowed plant embryos to survive longer, gave seedlings a greater chance of survival, and allowed plant embryos to be dispersed farther from the parent plants. All of these advantages enabled seed-bearing plants to inhabit farther reaches of land.
Flowers and fruits attract animals to the plant, where the animals can inadvertently assist in the plant's reproductive cycle.	Fruit protects the embryo as an animal disperses it far from the parent plant. Flowers attract animals that pollinate the plant, again allowing for greater dispersal.

Reproductive Adaptations

The reproductive processes of the earliest plants to evolve from green algae still required an aquatic environment. Consequently, these primitive land plants could live only in highly moist, swampy areas. As plants evolved, they acquired the ability to reproduce without water and began to populate drier environments. The seed in particular allowed safe dispersal of plant embryos by supplying it with nutrition and protection against hostile conditions. The alternation of generations in the reproductive cycle also allowed a longer diploid stage that better equipped plants to survive in harsh environments.

Adaptations to Reduce Water Loss

Strategies to catch and retain water were vital for the survival of land plants. Once removed from moist, swampy areas, plants risk **desiccation,** or drying out. An adaptation common to many plants to prevent desiccation is the **cuticle,** a thick, waxy, watertight barrier that covers the plant and prevents loss of moisture to the air. Molecules pass through the cuticle barrier via **stomata,** highly specialized openings in the cuticle that allow carbon dioxide to enter and water and oxygen to exit as the plant performs photosynthesis.

CHAPTER 10
Land Plants

Mineral Absorption

Land plants also evolved methods to gain vital minerals from the soil. Green algae absorb the nutrients they need directly from the water they live in. Land plants have to live without being suspended in water. **Roots** developed as a structure that allowed plants to obtain minerals and water from the soil. Some plants also developed **vascular tissue,** vessels that act as a transport system to bring water and other substances such as minerals up the plant body, allowing for increased plant size.

Nonvascular Plants

The first land plants arose between 505 and 435 million years ago. These highly simplistic plants possessed neither vascular tissue nor leaves. Nonvascular plants are limited in size and must grow close to the ground, where they can obtain nutrients through the processes of osmosis and diffusion.

Reproduction in nonvascular plants occurs via the production of flagellated sperm, which swim to eggs. This means that water must be present for fertilization to take place. The haploid stage of these organisms is dominant, producing **spores,** single haploid reproductive cells that grow into a new generation of plants via mitosis.

Early nonvascular plants evolved from green algae to include three divisions, whose descendents are still seen today:

- **Division Bryophyta:** The mosses

- **Division Hepatophyta:** The liverworts

- **Division Anthocerophyta:** The hornworts

Seedless Vascular Plants

The evolution of the vascular system followed the initial land colonization by nonvascular plants around 435 million years ago. The earliest vascular plants lacked a seed and reproduced in the same manner as their primitive ancestors, requiring water and the production of spores.

The development of vascular tissue, in particular a root system, provided a means to transport water and nutrients throughout the plant body. As a result, plants no longer gained water and nutrients via osmosis and diffusion alone. In addition, the diploid phase dominated in the life cycle of early vascular plants. As a result, vascular plants could grow higher off the ground during a longer development period. With larger plant life on the land, more food existed for other organisms that also began to move onto land.

Seedless vascular plants include two divisions:

1. **Division Lycophyta:** The club mosses

2. **Division Pterophyta:** The ferns and horsetails

Seed Vascular Plants

Seed-bearing plants followed the evolution of vascular plants and first appeared around 360 million years ago, signifying a crucial step in land colonization. Seeds provided a protective coating along with nutritional substances for plant embryos. This protection allowed embryos to survive in harsher environments. Seeds also allowed plants to break their dependence on water for reproduction. Sperm can travel by wind or by animals such as insects. Seed plants quickly radiated away from swampy areas and started to colonize all areas of the Earth.

CHAPTER 10
Land Plants

There are two types of seed plants: **gymnosperms** and **angiosperms.** Gymnosperms evolved around 300 million years ago and produce seeds that lack a protective fruit coating. The flowering angiosperms, which encase their seeds in fruit, were the last major plant group to evolve. They first appeared around 135 million years ago.

GYMNOSPERMS

Gymnosperm seeds generally rely on the wind for transport. The male gamete is carried on the wind, with the help of specialized structures that promote floating, to an egg. After pollination, a seed forms around the zygote, creating a protective shell for the growing embryo. With a seed capable of transportation via wind, gymnosperms radiated across long distances.

Gymnosperms comprise four divisions:

1. **Division Coniferphyta:** Evergreen trees, including pines, firs, and spruces

2. **Division Cycadophyta:** Palmlike trees

3. **Division Ghenophyta:** Some shrubs and vines

4. **Division Ginkophyta:** Deciduous trees with fanlike leaves

ANGIOSPERMS

Angiosperms produce flowers and seeds encased in fruit. The adaptations of angiosperms have been so successful that this group comprises more living species than all other plant groups combined.

Angiosperms employ highly diverse means of reproduction, including the use of wind, insects, and water as mediums for pollination. The angiosperms are known as the flowering plants because they attract pollinators with their flowers. The ovules of these plants are fully enclosed, and after fertilization the seed or seeds become encased in a fruit. The fruit of the angiosperms is not only protective, but also edible, enabling animals to disperse the seeds via digestion. Despite the abundance of species in this group, the angiosperms are made up of only one division, the Anthophyta.

> *EXAMPLE:* The flowers of many angiosperms have coevolved with a specific animal or insect that pollinates it. Some bee-pollinated flowers have developed lipped designs, where one petal forms a lip that acts as a landing pad for the bee and supports it as it accesses the pollen. Other flowers have colors, shapes, or patterns designed to attract certain insects or birds.

Summary

Kingdom Plantae

- Nearly all plants share several common characteristics, including multicellularity, eukaryotic cells, chloroplasts, the ability to perform photosynthesis, cell walls, and the lack of mobility.

Plant Adaptations

- Plant evolution is marked by the appearance of four major adaptations: **alternation of generations, vascular tissue,** the **seed,** and flowers and **fruits.**

- Plants evolved to prevent water loss, absorb minerals from the soil, and reproduce on land.

Nonvascular Plants

- Primitive plants lacked vascular tissue, were small, and could only grow close to the ground.

- Nonvascular plants are dependent on water for reproduction, as they reproduce via flagellated sperm that swim to and fertilize eggs.

- The life cycle of nonvascular plants is dominated by the haploid stage.

- Mosses, hornworts, and liverworts compose the nonvascular plant group.

CHAPTER 10
Land Plants

Seedless Vascular Plants

- Vascular tissue, including a **root** system, evolved in the descendents of the nonvascular plants.

- Vascular plants grew much taller than their nonvascular ancestors, but their flagellated sperm still required water for reproduction.

- The diploid stage of the seedless vascular plant is dominant, as it is for all subsequent groups that evolved.

- Ferns and club mosses compose the seedless vascular plant group.

Seed Vascular Plants

- The seed vascular plant group is made up of the **gymnosperms** and the **angiosperms.**

- Gymnosperms are seed-bearing plants whose seeds lack a fruit covering.

- Angiosperms are seed-bearing plants that have flowers and seeds encased in fruit. The protective seeds encasing the embryos allow seeds to be dispersed farther than those of other seed-bearing plants.

- Water is not necessary for reproduction, as these plants have developed adaptations that make use of wind and animals to aid in pollination. This break from water dependence allowed seed vascular plants to radiate away from the moist environments.

Sample Test Questions

1. Discuss three challenges facing plants as they moved onto land. What adaptations did they develop to counter these challenges?

2. Compare and contrast major characteristics of mosses and ferns.

3. Describe three methods plants use for pollination and/or fertilization.

4. Which kingdom are plants members of?

    A. Monera
    B. Animalia
    C. Plantae
    D. Fungi
    E. Protista

5. What is the name for the waxy covering of a plant that acts as a water barrier?

    A. Thorn
    B. Flower
    C. Seed
    D. Stomata
    E. Cuticle

6. What is the life cycle of change between haploid and diploid stages referred to in plants?

    A. Diploid cycle
    B. Haploid cycle
    C. Vascularization
    D. Wind pollination
    E. Alternation of generations

7. All of the following are characteristics of plants EXCEPT

    A. Eukaryotic
    B. Multicellular
    C. Nucleoid
    D. Chloroplasts
    E. Cell walls

CHAPTER 10
Land Plants

8. In which of the following plants is the haploid phase the dominant stage in the life cycle?

    A. Bryophytes
    B. Seedless vascular plants
    C. Gymnosperms
    D. Angiosperms
    E. Flowering plants

9. Which feature of an angiosperm is an adaptation designed to attract animals for assistance in pollination?

    A. Seed
    B. Flower
    C. Fruit
    D. Root
    E. Stem

10. What are the diploid individuals called in the alternation of generations?

    A. Stomata
    B. Gametophytes
    C. Vascular tissue
    D. Gametes
    E. Sporophytes

11. What type of plant is a pine tree?

    A. Protist
    B. Bryophyte
    C. Seedless vascular plant
    D. Gymnosperm
    E. Angiosperm

12. What adaptation allows the passage of materials through the plant cuticle?

    A. Stomata
    B. Gametophytes
    C. Seeds
    D. Chloroplasts
    E. Vascular tissue

13. Which of the following types of plants have flowers?

    A. Gymnosperms
    B. Seedless vascular plants
    C. Angiosperms
    D. Bryophytes
    E. Conifers

14. Which of the following types of plants do NOT contain real vascular tissue?

    A. Angiosperms
    B. Gymnosperms
    C. Seeded vascular plants
    D. Seedless vascular plants
    E. Bryophytes

15. What is the name for single reproductive cells that grow into a new generation of plants through cell division?

    A. Spores
    B. Seeds
    C. Fruit
    D. Haploid
    E. Diploid

ANSWERS

1. As they moved onto land, plants had to cope with the loss of water to the environment, reproduction without water, and mineral absorption. To cope with water loss, plants evolved stomata, which open and close the pores on plant leaves to prevent excessive evaporation as photosynthesis is taking place. Plants also developed a thick, waxy coating, called the cuticle, to keep water in. The earliest land plants grew near swampy areas, where enough water was available for reproduction. Later plants evolved the seed and means of pollination that uses wind or animals to carry sperm. Finally, the adaptations of a vascular system, particularly the development of a root system, allowed plants to absorb minerals from the soil.

2. Ferns and mosses both depend on water for reproduction because their sperm can only swim to the egg. Both types of plants are restricted to highly moist, swampy environments. However, they differ on several

levels. Mosses must still absorb nutrients via diffusion or osmosis and are therefore small in size and grow close to the ground. Ferns possess vascular tissue that can pull minerals and nutrients from the soil, enabling them to grow very tall. Finally, mosses have a dominant haploid stage, whereas ferns have evolved to have a dominant diploid stage, which allows ferns to disperse farther over land than moss.

3. Flagellated plant sperm can swim through water to fertilize an egg. Among the seed-bearing plants, some seeds have structures that allow them to be carried easily by the wind, and other seeds are contained in fruits, which can be eaten and excreted by animals.

4. C   Plants are members of the kingdom Plantae.

5. F   Plants evolved a cuticle, a waxy outer covering that acts as a water barrier.

6. E   In alternation of generations, the plant cycles from haploid to diploid life stages in its lifetime.

7. C   Nucleoid are found in prokaryotic cells. They do not exist in plants, which are eukaryotic.

8. A   Only in nonvascular plants, such as bryophytes (mosses), is the haploid the dominant stage in the life cycle.

9. B   The flower developed in angiosperms specifically to attract pollinators to assist in reproduction.

10. E   In alternation of generations, the diploid individuals are called sporophytes.

11. D   A pine tree is a type of gymnosperm.

12. A   To allow vital materials such as water to pass through the cuticle, plants evolved specialized openings called stomata.

13. C   Angiosperms are flowering plants.

14. E   Bryophytes lack true vascular tissues.

15. A   Spores grow into a new generation of plants through mitosis.

PLANT STRUCTURE AND GROWTH

Plant Structure

Plant Tissue

Plant Growth

Monocots and Dicots

11

Plant Structure

The plant body is divided into two main parts:

1. **Roots:** The structure that anchors the plant to the ground and absorbs water and minerals from the surrounding environment

2. **Shoots:** The aboveground plant structure that includes the **stem,** which is a framework for leaves, flowers, and fruits

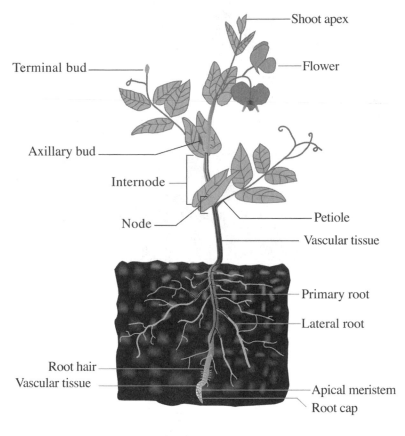

Terminal bud

Axillary bud

Internode

Node

Shoot apex

Flower

Petiole

Vascular tissue

Primary root

Lateral root

Root hair
Vascular tissue

Apical meristem
Root cap

Basic Plant Structure

ROOTS

Roots are the descending portion of the plant body, generally located below ground. Roots anchor the plant and are responsible for the absorption of water, minerals, and other important materials. Root systems come in two forms: taproot systems and fibrous root systems.

- **Taproot systems:** Composed of a large central root, or primary root, off of which smaller roots and root hairs grow, taproot systems have a relatively small surface area and so are not as effective at absorbing water and nutrients from the soil. However, the large taproot can store nutrients and water—an advantage to plants that grow in regions of minimal water or little sunlight for photosynthesis. Carrots and oak trees are examples of plants with taproot systems.

- **Fibrous root systems:** Composed of a diffuse system of roots and root hairs, with no single major root, fibrous roots systems have a greater surface area and can extend away from the plant, allowing the plant to absorb more water and nutrients. Ferns, grasses, and chives are examples of plants with fibrous root systems.

EXAMPLE: Most trees use a taproot system early in their development. This allows the young tree to establish itself in the ground, setting the stage for greater growth to come.

Features of a Root System

The features of a root system are summarized in the table below.

Plant Structure	Description / Function
Primary root	The central root of the plant, extending from the plant's stem.
Lateral roots	Roots growing directly off of the primary root.
Apical root meristem	The zone of cell growth at the tip of the primary root. The apical meristem produces new cells required for growth of the plant body.
Root hairs	Hairs that extend from the root and greatly increase the surface area for the absorption of water and nutrients.
Root cap	Structure at the far end of the apical meristem that provides protection as the apical meristem pushes through the soil. Cells at the root cap are continually lost and replenished as the plant grows.
Vascular tissue	Cells arranged to form tubes through which water, minerals, and the products of photosynthesis (sugars) can flow through the plant. **Xylem** conveys water and minerals up from the roots; **phloem** conveys the products of photosynthesis down from the leaves.

THE STEM

The stem acts as the structural support for the plant and provides the framework for leaves, flowers, and fruits. The features of a plant stem are summarized in the table below.

Plant Structure	Description / Function
Vascular tissue	Cells arranged to form tubes through which water, minerals, and the products of photosynthesis (sugars) can flow through the plant. Xylem conveys water and minerals up from the roots; phloem conveys the products of photosynthesis down from the leaves.
Nodes / Internodes	**Nodes** are the points at which leaves are connected to the stem; **internodes** are stretches of stem between the nodes.
Terminal buds	The undeveloped shoot at the tip of the stem. Terminal buds can remain dormant or grow into a shoot at a later time.
Axillary buds	Undeveloped shoot located where a **petiole,** or leaf stalk, meets the stem. These buds usually remain dormant but can grow into a side stem. **Apical dominance** refers to a process in which growth of the main stem is primary, and growth of side stems is inhibited.
Apical shoot meristem	The zone of cell growth at the tip of the stem. The apical meristem produces new cells required for growth of the plant body.

The Structure of Leaves

Leaves extend from a plant's apical shoot meristem. From species to species, leaves can take a variety of forms, sizes, and arrangements and can vary greatly in internal structure. All leaves serve the primary function as the principal sites of photosynthetic reactions. However, this variation from plant to plant is largely responsible for the enormous diversity and adaptability of the plant kingdom.

In spite of their great variety, most leaves share the basic features summarized in the table on the following page.

Plant Structure	Description / Function
Leaf	The primary sites of photosynthesis. Leaves are made up of a flat **blade** and a petiole, which joins the stem to the leaf.
Cuticle	Waxy coating on aboveground plant structures that help to prevent loss of water to the air. Cuticles also protect the plant from damage and contaminants, such as bacteria, viruses, and dust.
Stomata	Tiny pores on the leaf surface that allow substances such as water, oxygen, and carbon dioxide to pass through as they either enter or leave the plant.

Plant Tissue

Plants are composed of a combination of four main tissue types, each made up of different types of cells:

- **Epidermal tissue**

- **Vascular tissue**

- **Ground tissue**

- **Meristem tissue**

EPIDERMAL TISSUE

Epidermal tissue is made up of flattened epidermal cells and constitutes the outer skin layer of the plant. Epidermal cells are often covered by a waxy coat known as the **cuticle,** which helps prevent water loss from the plant. Epidermal tissue also includes **guard cells,** which are paired cells with an opening between them, called the **stomata.** Stomata allow substances such as water, oxygen, and carbon dioxide to pass through them and either enter or leave the plant.

CHAPTER 11
Plant Structure and Growth

VASCULAR TISSUE

Vascular tissue allows the transport of water and minerals, as well as sugars, the products of photosynthesis, throughout the plant body. Xylem and phloem are the two types of plant vascular tissue, and they work together to circulate the necessary ingredients and products of photosynthesis and respiration to the cells, as well as transport waste to be disposed.

Xylem

Water and other necessary materials are moved up from the plant's root system via the xylem. This movement is achieved through the physical process known as the transpiration-cohesion-tension mechanism. Water molecules in the xylem cells come together by forming hydrogen bonds in a process known as cohesion. These water molecules are drawn to the walls of the xylem cells by a force known as adhesion. Together these two forces create a column of water that is pulled up the xylem, along with minerals and other materials. The water then evaporates as water vapor through stomata in the leaves in a process known as **transpiration.** It is the evaporation of water vapor through transpiration that creates the negative pressure that allows the water column to be continually drawn up through the plant.

Phloem

Phloem is sieve-like tissue located toward the outer portion of a plant's roots and stems. It transports sugars produced during photosynthesis (a plant's main food source), hormones, and other materials from a plant's leaves to the rest of the plant body.

Features of Xylem and Phloem

The features of xylem cells and phloem cells are summarized in the table on the following page.

Cell Type	Description
Xylem cells	There are two types of xylem cells: **tracheids,** long cells with tapered ends, and **vessel elements,** which are shorter and more square. Tracheids and vessel elements line up to create a channel for water to flow through. Tracheids are found in gymnosperms and some primitive angiosperms. Vessel elements, which transport water more effectively than tracheids, are found in the majority of angiosperms.
Phloem cells	Phloem cells include **sieve cells** and **sieve-tube members.** Sieve cells are found in the more primitive seedless vascular plants and gymnosperms, while sieve-tube members are found in most angiosperms. Sieve-tube members are arranged end to end and are separated by a sieve plate, through which sugars and other compounds move as they travel through the plant body. Companion cells grow adjacent to sieve-tube members and carry out certain metabolic functions for the cell.

GROUND TISSUE

Ground tissue exists in the space between the epidermal tissue and the vascular tissue. Ground tissue stores food and water and carries out the functions of photosynthesis.

Ground tissue is made up of three types of cells, described in the table below:

Cell Type	Description
Parenchyma cells	Parenchyma cells are the most common cells found in the majority of plant species. The cortex of stems and roots, the body of fruits, and the areas of leaves that carry out photosynthesis are all made up of parenchyma cells. They are alive in the mature plant, which means that they continue to divide throughout the plant's lifetime, making them central to cell regeneration and wound healing.
Collenchyma cells	Collenchyma cells have single, thick outer cell walls. They elongate to provide support during plant growth.
Sclerenchyma cells	Sclerenchyma cells have two tough outer cell walls. They are incapable of elongation and therefore exist only in tissue where growth has stopped. These cells are usually dead in mature plants. Sclerenchyma cells function as support cells, providing structure and strength to the plant body.

MERISTEM TISSUE

Meristem tissue comprises groups of undifferentiated cells. Its only function is the creation of new cells that will eventually differentiate into the other three tissue types, similar to stem cells in animals.

Plant Growth

Meristem tissue is the ultimate source for new plant cells. New cells are created in meristem tissue via mitosis before differentiating and growing into the three other common tissue types. This growth can occur in one of two methods: primary growth and secondary growth.

PRIMARY GROWTH

Primary growth occurs as cells in the **apical meristem,** the tissue located at the tips of roots and shoots, divide. The tip of each plant root ends in a cone-shaped structure called a **root cap,** which covers and protects the cells that compose the apical meristem. Cells of the root cap are scraped away as the root pushes through the ground, and the apical meristem produces new cells that replace these lost ones. At the tip of each plant shoot, new cells emerge from the apical meristem and are differentiated into ground tissue, dermal tissue, or vascular tissue. As these new cells differentiate, they undergo significant elongation, pushing the apical meristem upward and contributing significantly to the vertical growth typically associated with plants.

SECONDARY GROWTH

A plant can undergo **secondary growth,** during which the plant expands in girth or bulk, while primary growth is still continuing. Secondary growth takes place among the cells that have differentiated into epidermal, vascular, and ground tissue.

Secondary growth occurs in two different types of meristem tissue: vascular cambium and cork cambium.

Vascular Cambium

Vascular cambium is meristem tissue resulting from the initial differentiation of cells produced by the apical meristem. Vascular cambium, in turn, produces new tissues called secondary xylem and secondary phloem. Secondary phloem forms from the vascular cambium cells that divide outwardly. Secondary xylem forms from the vascular cambium cells that divide inwardly. Secondary xylem is the wood portion of a plant.

Cork Cambium

Cork cambium is formed from the parenchyma cells of the plant cortex. The cork cambium is made up of plates of dividing cells that produce two layers of cells: an inward layer of parenchyma cells and an outward layer of dead cork cells. The cork cambium, cork layer, and parenchyma cells collectively form the plant's outer protective coating and are known as the periderm.

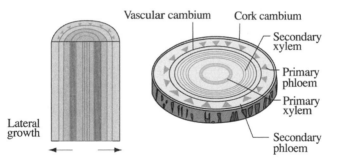

Primary and Secondary Plant Growth

Monocots and Dicots

Angiosperms, or the flowering plants that make up the largest and most diverse phyla in the plant kingdom, are divided by structure into two main classes: **monocots** and **dicots.** While differences exist among the species in the monocot and dicot classes, several major structural differences distinguish the monocots and dicots from each other.

CHAPTER 11
Plant Structure and Growth

Characteristic	Monocot	Dicot
Number of cotyledons (food-bearing leaves in a seed)	Monocots are monocotyledons, meaning they have one cotyledon.	Dicots are dicotyledons, meaning they have two cotyledons.
Leaves	Monocot leaves have parallel veins.	Dicot leaves have reticulate, or netted, veins.
Flowers	Monocot flower parts are typically divided into groups of threes, such as three petals or two groups of three petals.	Dicot flower parts are typically divided into fours or fives.
Stems	Vascular tissue of the stem is scattered.	Vascular tissue of the stem is arranged in rings.
Roots	Monocots usually have fibrous root systems.	Dicots usually have taproot root systems.

MONOCOTS

One cotyledon

Parallel veins

Floral parts divided into groups of three

Vascular tissue scattered

Fibrous root system

| Embryos | Leaf Venation | Flowers | Stems | Roots |

DICOTS

Two cotyledons

Netted veins

Floral parts divided into groups of four or five

Vascular tissue arranged in rings

Taproot system

Monocots vs. Dicots

Summary

Plant Structure

• Plants are composed of **roots, shoots,** and **leaves.**

- Plant roots come in two forms: **fibrous root systems** comprising a diffuse system of roots and root hairs, and taproot systems comprising of a large central root off of which smaller roots and root hairs grow.

- Water moves up the **xylem,** from roots to shoots, while sugars produced in the leaves move down the **phloem.**

- **Leaves** are specialist organs used for photosynthesis.

Plant Tissue

- Plants are composed of four main tissue systems: **epidermal tissue** system, which constitutes the outer skin layer; **vascular tissue** system, which allows the transport of water, minerals, and sugars throughout the plant body; the **ground tissue** system, which exists in the space between the epidermal tissue and the vascular tissue; and the **meristem tissue,** the source of all new plant cell growth.

- Epidermal tissue is made up mainly of epidermal cells but also of **guard cells** that form **stomata.**

- Vascular tissue is made up of xylem cells and phloem cells.

- Ground tissue is made up of **parenchyma cells, collenchyma cells,** and **sclerenchyma cells.**

Plant Growth

- Undifferentiated cells produced through mitosis within the meristem tissue are the source of all new plant tissue.

- Two main processes characterize plant growth: **primary growth,** in which the shoots and roots of the plant lengthen; and **secondary growth,** in which the plant increases in girth.

- In primary growth, cells in the **apical meristem** divide. After division, they differentiate into the three main tissue types. After differentiation, cell elongation causes an increase in plant size.

- Secondary growth occurs via two different types of meristem tissue: **vascular cambium** and **cork cambium.**

- Vascular cambium results from the initial differentiation of new cells produced by the apical meristem, which then produces new tissues called secondary xylem and secondary phloem.

- Cork cambium is formed from parenchyma cells of the cortex. Cork cambium produces cork, which protects the vascular system of the plant.

Monocots and Dicots

- Plants can be divided by structure.

- The two main classes of angiosperms, the largest phyla in the plant kingdom, are **monocots** and **dicots.**

- Monocots have one cotyledon as embryos, leaves with parallel veins, and usually fibrous root systems.

- Dicots, which constitute most of the angiosperms, have two cotyledons as embryos, leaves with a cross-hatched pattern of veins, and taproot root systems.

Sample Test Questions

1. Explain the process of secondary growth and how it differs from primary growth.

2. Describe the advantages of both a fibrous root system and a taproot system.

3. What role does hydrogen bonding play in the movement of water and minerals through the xylem?

4. What category of plant has leaves with parallel vein patterns?

    A. Monocots
    B. Dicots
    C. Angiosperms
    D. Cotyledons
    E. Sieve-tube members

5. What is plant wood also known as?

    A. Primary phloem
    B. Secondary phloem
    C. Primary xylem
    D. Secondary xylem
    E. Vascular cambium

6. Meristem tissue is composed of what kind of cells?

    A. Undifferentiated cells
    B. Collenchyma cells
    C. Sclerenchyma cells
    D. Companion cells
    E. Parenchyma cells

7. Which plant tissue system includes the cuticle?

    A. Vascular
    B. Ground
    C. Epidermal
    D. Cambrial
    E. Root

8. What structure connects the leaf of a plant to the stem?

    A. Terminal bud
    B. Node
    C. Internode
    D. Axial bud
    E. Petiole

CHAPTER 11
Plant Structure and Growth

9. The parenchyma cells of the cortex give rise to which of the following structures?

  A. Vascular cambium
  B. Cork cambium
  C. Secondary xylem
  D. Secondary phloem
  E. Root cap

10. Which of the following types of cells are part of dermal tissue?

  A. Guard cells
  B. Parenchyma cells
  C. Collenchyma cells
  D. Phloem cells
  E. Xylem cells

11. Which group of plant cells is central to cell regeneration and wound healing?

  A. Sclerenchyma
  B. Collenchyma
  C. Parenchyma
  D. Cork cambium
  E. Vascular cambium

12. Which of the following structures lies between sieve-tube members in phloem?

  A. Sieve plate
  B. Sieve tube
  C. Companion cell
  D. Tracheid
  E. Vessel element

13. Sugars are created during what important process in plants?

  A. Primary growth
  B. Secondary growth
  C. Transpiration
  D. Photosynthesis
  E. Meristem division

14. What is the name of the phenomenon in which the terminal bud inhibits the growth of the axillary buds?

A. Primary meristem
B. Terminal bud dominance
C. Apical dominance
D. Bud dominance
E. Terminal dominance

15. Which of the following is a common feature of a dicot?

A. Seeds have one cotyledon
B. Reticulate leaves
C. Scattered vascular tissue in stem
D. Parallel leaf veins
E. Flowers have three petals

ANSWERS

1. Primary growth is the vertical growth of a plant through the development of its roots and stems. Secondary growth is an increase in plant girth. The apical meristem produces new cells, which differentiate into the various tissue types. As the cells differentiate in primary growth, they also elongate, a significant contributor to plant growth both downward in the root and upward in shoots. Secondary growth takes place in the newly differentiated cells originally produced by the apical meristem. Secondary type of growth occurs via two different types of meristem tissue: vascular cambium and cork cambium. On the outermost section of the tree, a layer of cork is produced by the cork cambium and acts as a tough layer of cells protecting the vascular system of the plant.

2. Some plants need to store nutrients more than others: the taproot can be used to store large amounts of starch during events such as flowering and fruit production. This is advantageous for plants that grow in areas that have light periodic rain or little sunshine. At times when resources are scarce or photosynthesis is not occurring at a sufficient rate, the plant can draw from the food stored in its large taproot tuber. The fibrous root system provides a large network of roots that allow a plant to draw minerals and nutrients from a larger surface area than a taproot. This large root system allows these plants to exploit a wider range of the soil in search of water and minerals.

3. In hydrogen bonding, a hydrogen atom covalently bonded to an electro-negative atom is attracted to another electronegative atom—a common bond among water molecules. Hydrogen bonding causes water molecules in the plant's xylem cells to stick to one another (cohesion) and to the walls of the xylem cells (adhesion). These forces combine to create a column of water that is pulled up the xylem by the force of transpiration, the evaporation of water through the leaves of the plant.

4. **A**   Plants with leaves that have parallel veins are called monocots. Monocots have one cotyledon as embryos, leaves with parallel veins, and usually fibrous root systems. Monocots are grasses, grains, palms, and bamboos.

5. **D**   Wood is also known as secondary xylem. Wood is formed when new cells produced by the apical meristem differentiate to produce the vascular cambium, which in turn produces new tissues called secondary xylem and secondary phloem.

6. **A**   Meristem tissue is composed of undifferentiated cells. These cells divide via mitosis to produce new cells that eventually differentiate into the other tissue types.

7. **C**   The epidermal tissue system constitutes the outer skin layer of the plant and includes the cuticle, which is a waxy coat that helps prevent water loss from the plant.

8. **E**   The structure that joins the leaf to the stem of a plant is called a petiole.

9. **B**   Cork cambium is formed from parenchyma cells of the cortex. When the epidermal tissue layer and the cortex begin to be sloughed off, a layer of cork, produced by the cork cambium, replaces them and acts as a tough layer of cells protecting the vascular system of the plant.

10. **A**   Guard cells are specialized cells in dermal tissue. Paired guard cells form stomata.

11. **C**   Parenchyma cells in a plant's ground tissue are important in cell regeneration and wound healing.

12. **A**   Between each sieve-tube member is a sieve plate through which sugars and other compounds move as they travel through the plant body.

13. **D**   Photosynthesis is the process by which plants harness the energy of the sun to create their own food in the form of sugars.

14. **C**   The terminal bud inhibits the growth of the axillary buds in the phenomenon known as *apical dominance*.

15. **B**   Reticulate leaves are common to dicots. Monocot leaves have parallel veins.

PLANT REPRODUCTION, DEVELOPMENT, AND BEHAVIOR

Plant Reproduction

Indeterminate Growth
and Life Cycles

Plant Hormones
and Rhythms

12

Plant Reproduction

Plants are capable of both sexual and asexual reproduction. While asexual reproduction eliminates the need to search for a mate, sexual reproduction provides the advantage of genetic variation through the combination of gametes from two separate individuals. Genetic variation has paved the way for the evolution of new plant structures that, in turn, have resulted in further adaptations of sexual reproductive methods.

SEXUAL REPRODUCTION

Sexual reproduction in plants evolved over time to adapt to the unique barriers and hazards of life on land. Three major evolutionary innovations have resulted in landmark changes in plant sexual reproduction.

1. The alternation of generations

2. The development of seeds

3. The development of the flower

Alternation of Generations

The earliest plant forms, including many still found on Earth today, such as mosses, required water for reproduction. For fertilization to occur in water, a sperm needed to swim to an egg (a jacket of cells and water). However, the evolution of **alternation of generations** led to the extension of the diploid stage of a plant, during which time the plant contains double its genetic material. The longer diploid stage led to the development of reproductive cells called **spores,** which are highly resistant to drying out and are released by plants during this stage. The development of spores meant that plants cleared the first evolutionary hurdle, breaking their dependence on water and moving onto land.

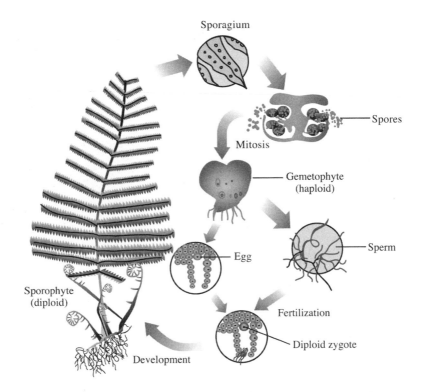

Sporagium

Spores

Mitosis

Gemetophyte
(haploid)

Egg

Sperm

Sporophyte
(diploid)

Fertilization

Diploid zygote

Development

The Alternation of Generations in a Fern: The diagram above outlines
the life cycle of a fern undergoing reproduction through the production of
spores. A gametophyte (the organism's haploid generation) produces both
sperm and eggs and releases them into the surrounding environment.
This phase of the reproductive cycle requires a moist environment
through which the sperm swim to the eggs. Once the egg is fertilized, it
enters the diploid generation, also known as the sporophyte. In modern
plants, the diploid phase is the dominant, adult stage of the plant. The
sporophyte produces a spore-containing sporangium, which releases
spores. These spores develop via mitosis into gametophytes, which begin
the cycle again. This method of reproduction is common among nonvascular and seedless vascular plants such as ferns.

The Seed

The development of the seed led to the rise of the **gymnosperms,** or seed-producing plants. Reproduction in gymnosperms involves the development of two gametophytes, a male
(a **pollen** grain) and a female (an **ovule**), produced in the
sporophyte. The male gametophytes are transferred to other
plants through the process of **pollination,** in which insects,
wind, or animals transport pollen grains from one individual
to the ovule of another. These methods of transportation further diminish the reliance on water for reproduction.

Once fertilization occurs through pollination, a plant embryo
encased in a **seed** develops. Seeds protect the embryo and
provide it with a source of food until adequate external conditions prompt it to **germinate,** or grow. Seeds are advantageous
to gymnosperms because they can be widely dispersed, remain
dormant until conditions are sufficient for growth, and provide
nourishment and protection for the young plant.

The Flower

The most recent evolutionary development in plant reproduction is the flower. Flowers are exclusively produced by plants in
the **angiosperm** phyla. Fertilization among the angiosperms
occurs within the flower. This increases the efficiency of pollination, because insects carrying gametes are more likely to
travel to flowers of the same species. Insects carrying gametes
from the nonflowering gymnosperms randomly disperse the
gametes, resulting in a less efficient rate of pollination.

Angiosperms constitute the largest and most diverse phyla in
the plant kingdom, with roughly 255,000 known species today.
Over the course of their estimated 150- to 200-million-year
history, angiosperms have developed several adaptations of the
basic flower structure to take advantage of a wide variety of
environmental niches. The basic structure of the flower and its
major features are described in the table on the following page.

Flower Structure	Function
Sepals	Green, leaflike structures that enclose the flower and protect it when closed.
Petals	Colorful structures that attract pollinators to the flower.
Stamens	Male organs of a flower; at the ends of the stamen are **anthers,** the sites of meiosis and pollen grain development.
Carpel	Female organ made up of the **stigma,** a sticky or feathery surface that receives pollen grains during pollination; the ovary, the site of the reproductive ovules; and the ovule, which houses the developing egg.

POLLINATION

Pollination is more complex in angiosperms than in gymnosperms. The process of pollination in angiosperms can be summarized in an eight-step process.

	Step	Description of Major Event
1.	**Formation of pollen grains in the anther of the flower**	Diploid cells undergo meiosis to produce four haploid spores. Each spore then undergoes mitosis to produce two haploid cells: a tube cell and a generative cell. A wall forms around the two cells to form the immature pollen grain.
2.	**Formation of embryo sac in the ovary of the flower**	A cell in the ovary undergoes meiosis to produce four haploid spores. Three of the spores degenerate, but one enlarges and undergoes three rounds of mitosis to produce an embryo sac, which is then enclosed in a protective cell layer. The embryo sac is made up of one large central cell containing eight haploid nuclei, one of which will be fertilized.
3.	**Pollination**	**Pollination** takes place as pollen grains are transferred from the anther of one plant to the stigma on the carpel of another. Upon landing on the stigma, the pollen grain germinates, and its tube cell transforms into a pollen tube, which grows down the stigma to the plant's ovary. At the same time, the generative cell of the pollen grain undergoes mitosis to produce two sperm cells.

CONTINUED

	Step	Description of Major Event
4.	**Double fertilization**	When the pollen tube reaches the ovary, one sperm fertilizes the egg cell to produce a zygote. The other sperm combines with two central haploid nuclei to produce a $3n$ cell, in a process called **double fertilization.**
5.	**Seed development**	The ovule develops into a seed. The zygote develops into an embryo and the $3n$ cell develops into an **endosperm,** which remains in the seed to provide the embryo with nourishment. The embryo and endosperm are contained within a seed coat, which protects the embryo and its food supply.
6.	**Dormancy**	The seed lies dormant—it does not grow or continue to develop—until outside conditions are favorable for germination.
7.	**Fruit development**	The plant's flower loses its petals and the ovary walls begin to increase in thickness, creating a fleshy case, or fruit, around the fertilized seeds. The seeds are then dispersed as the fruit is transported by wind or water or is eaten by an animal, which then drops the seeds in its feces. The seeds remain protected throughout the process of dispersal.
8.	**Germination**	Germination takes place when the seed reaches soil with adequate conditions and water. An embryonic root emerges and grows downward into the ground, while an embryonic shoot emerges and grows upward. Growth continues until the plant reaches maturity.

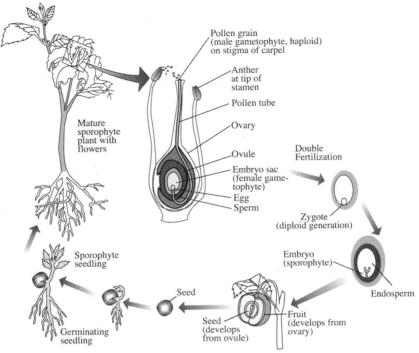

Pollen grain
(male gametophyte, haploid)
on stigma of carpel

Anther
at tip of
stamen

Pollen tube

Ovary

Mature
sporophyte
plant with
flowers

Ovule

Double
Fertilization

Embryo sac
(female game-
tophyte)

Egg
Sperm

Zygote
(diploid generation)

Sporophyte
seedling

Embryo
(sporophyte)

Endosperm

Seed

Seed
(develops
from ovule)

Fruit
(develops from
ovary)

Germinating
seedling

Major Steps in Pollination

ASEXUAL REPRODUCTION

Some plants, such as garlic and certain grasses and trees, can reproduce asexually through the production of **clones:** individuals genetically identical to their parent. Asexual reproduction in plants can take several different forms.

Vegetative Reproduction

Vegetative reproduction is a common form of asexual reproduction in which pieces of a parent plant form cloned individuals. Although there are many variations, five main methods of vegetative reproduction exist among plants.

1. **Fragmentation:** Pieces of a parent plant break off and give rise to a new plant.

2. **Runners:** Horizontal stems radiate out from the parent plant and grow along the surface of the ground. New plants form from nodes in these stems.

3. **Rhizomes:** Horizontal stems, similar to runners, grow underground, generally close to the surface, from a parent plant. Nodes on the horizontal stem give rise to new plants.

4. **Suckers:** Suckers, or sprouts, are produced at the roots of a parent plant. These suckers can develop into new plants.

5. **Adventitious plantlets:** Meristematic tissue located along notches on the parent plant's leaves either drop or are broken off and take root in the soil to form a new plant.

Apomixis

Apomixis is a form of asexual reproduction in which seeds genetically identical to the parent are formed. Plants reproducing by this method, such as Kentucky bluegrass and dandelions, take advantage of the protective and dispersal properties of seeds, an adaptation generally exclusive to sexually reproducing plants. This form of reproduction is most common in harsh environments, where genetic variation can be a disadvantage, and where embryo protection is vital.

EXAMPLE: Asexual reproduction is particularly advantageous for food scientists working with genetically engineered plants. The exploitation of asexual reproductive methods has led to the development of monocultures, farms containing a single, genetically identical crop. Monocultures ensure that the crop will grow at a uniform rate, allowing the entire crop to be harvested at one time. However, the lack of genetic variation leaves monoculture crops more vulnerable to outside threats, such as disease, which can wipe out an entire crop.

CHAPTER 12
Plant Reproduction

Indeterminate Growth and Life Cycles

Most plant species, whether reproducing sexually or asexually, continue to grow throughout their lifetimes in a process called indeterminate growth. (Animals, by contrast, undergo determinate growth, in which growth stops once the individual reaches a certain size.) Through the process of indeterminate growth, an individual plant can constantly increase the amount of air, water, and sunlight it receives over the course of its life.

Indeterminate growth does not translate into immortality. All plants follow one of three specific life cycles from germination to death:

1. **Annuals** complete their entire life cycle within a single year or growing season. Many plants farmed as crops, such as corn, wheat, and soybean, are annuals. Growth can be rapid in favorable conditions, but annuals tend to die after they flower.

2. **Biennials** are uncommon compared with annuals. They complete their life cycle within two years: They store food and nutrients in the first year, then produce flowers using those resources in the second year. Some biennials actually spend three or more years storing food. As with two-year biennials, they die after flowering.

3. **Perennials,** a group that includes most vascular plants such as trees and shrubs, continue to grow over the course of many years. In most perennial plants, flowering and seed production occurs continually for an indefinite number of seasons.

Plant Hormones and Rhythms

Plant hormones, like animal hormones, are chemical messengers released in one part of the organism that cause a change in another part. The following table identifies the major plant hormones and their function.

Plant Hormone or Hormone Class	Site of Production	Resulting Activity
Auxins	Roots, shoots, and young leaves	Promote stem elongation and are involved in overall growth and the dropping, or abscission, of leaves, and in the differentiation of vascular tissue cells (xylem and phloem)
Gibberellins	Apical meristem, seeds, and young leaves	Promote stem elongation and the germination of seeds and are involved in the growth of fruit
Cytokinins	Roots	Stimulate cell division in plant growth and are involved in the differentiation of plant cells into tissues
Ethylene	Most plant tissues	Slows the lateral growth of buds, promotes leaf abscission, and induces fruit ripening
Abscisic acid	Leaves, roots, and fruit	Regulates leaf openings, or stomata, in times of drought; balances growth hormones; involved in seed dormancy

EXAMPLE: Ethylene, aided by other hormones, begins the process of fruit ripening by breaking down the fruit's cell walls. This softens the fruit; initiates the recycling of chlorophyll and the production of other pigments, which give the fruit its ripe coloring; and breaks down complex sugars into simple sugars, which make fruit taste sweet. In addition to these changes, chemical compound byproducts are produced, which make the fruit smell and taste good. All of these changes result in a fruit enticing to animals, which in turn eat the fruit and spread the seeds.

Because the plant hormone ethylene is a gas, it can spread from one individual to influence the ripening of others in close proximity. If a bag of apples contains one very ripe apple, the ethylene produced by the ripe apple will spread to the others, causing them to ripen more quickly. The same effect would hold if a ripe apple were placed in a bag full of bananas, for example.

CHAPTER 12
Plant Reproduction

TROPISM

Plants respond to external signals through the mechanism of **tropism,** the turning or bending movement toward or away from an external stimulus, such as light, heat, gravity, or touch. There are three main plant tropisms, each of which are facilitated by the production of specific hormones:

1. **Phototropism:** Plants grow, or bend, toward light. A type of auxin, a plant hormone called IAA (indole-3-acetic acid), is involved in phototropism.

2. **Gravitropism:** The shoots or roots of a plant bend in response to the pull of gravity. IAA is also involved in gravitropism.

3. **Thigmotropism:** Plants grow in response to touch, as when a vine grows up a fence or wall. Thigmotropism is controlled by the production of IAA and ethylene.

INTERNAL CLOCKS OF PLANTS

Plants operate on both twenty-four-hour cycles called **circadian rhythms,** which continue even in the absence of environmental cues to guide them, and **biological clocks,** internal monitors that depend on the environmental cues of daytime and nighttime. Biological clocks are thought to influence major events in a plant's life cycle, such as flowering, growth of stems, loss of leaves, and seed germination.

Plants use the **photoperiod,** the duration of the day and night, to detect the season. The flowering in angiosperms provides a clear example of a plant's reliance on the photoperiod. Biologists divide angiosperms into three categories based on when they flower:

1. **Short-day plants,** such as strawberries and tobacco, flower in the late summer, fall, or winter, when the duration of daylight is shorter.

2. **Long-day plants,** such as clover, flower in the summer or early spring, when the duration of daylight is longer.

3. **Day-neutral plants,** such as roses, flower regardless of day length, provided there is sufficient light for normal growth.

Some flowering plants do not fit into any of these three categories. These plants, such as ivy, flower at two photoperiods during each year, when the day is not too long or too short.

Photoreceptors

Biologists believe that a class of photoreceptor pigments, known as **phytochromes,** are responsible for detecting daylight and setting the biological clocks of plants. Phytochromes take on two forms: one that stimulates the occurrence of a set biological response and another that inhibits that response. The levels of each pigment form fluctuate as the length of daylight changes over the course of the year. These fluctuations result in different responses, such as flowering, for individual plants throughout the year. Photoreception is a vital mechanism by which a plant can interact with its environment, performing certain time-sensitive functions when environmental conditions are suitable, animals and insects that act as pollinators are available, and competition with other plants for resources is minimal.

Summary

Plant Reproduction

- Plant reproduction can occur through both asexual and sexual methods.

- **Sexual reproduction** in the plant kingdom has evolved over time as key development occurred in plant structure.

- The development of the **alternation of generations** led to the production of spores capable of withstanding the harsh environment as they dispersed over greater distances.

- The development of the **seed** and **pollination** provided means other than water for the dispersal of **gametes.** Seeds also remain dormant until

CHAPTER 12
Plant Reproduction

conditions are sufficient for growth and provide nourishment for young plants.

- The development of the flower provided a site for reproduction in plants, leading to more complex methods of pollination and **fertilization.**

- **Asexual reproduction** results in **clones** genetically identical to the parent plant. Asexual reproductive methods include various forms of vegetative reproduction and **apomixes.**

Interdeterminate Growth and Life Cycles

- Plants grow continuously throughout their lives, in a process called indeterminate growth.

- All plants follow one of three possible life cycles: **annual, biennial,** or **perennial.**

Plant Hormones and Rhythms

- Plant hormones are chemical messengers that are released in one part of the plant body and cause change in another part.

- The major plant hormones include auxins, gibberellins, cytokinins, ethylene, and abscisic acid.

- **Tropism** is the mechanism by which a plant moves toward or away from an external stimulus, such as light, gravity, heat, or touch.

- The three main plant tropisms are **phototropism, gravitropism,** and **thigmotropism.** Specific hormones facilitate plant tropisms.

- Plants operate on **circadian rhythms,** daily cycles that require no environmental cues, and **biological clocks,** seasonal monitors cued by the length of day and night.

- Biologists believe biological clocks influence major events in plants, such as germination and flowering.

- Plants rely on the **photoperiod,** the length of day and night at a particular time of year.

- Angiosperms (flowering plants) are broken into three groups based on their photoperiod: **short-day plants,** which flower when daylight is short; **long-day plants,** which flower when daylight is long; and day-neutral plants, which flower whenever sufficient light is available.

- **Phytochromes** are photoreceptor pigments thought to play a role in daylight detection in plants. Variations in pigment levels result in different biological responses in the plant.

Sample Test Questions

1. Describe three ways in which pollination can occur.

2. Name the three key developments in plant evolution and describe their specific relation to sexual reproduction.

3. Describe the pros and cons of asexual reproduction in plants.

4. What structures are responsible for protecting the flower bud before it opens?

    A. Sepals
    B. Petals
    C. Stamens
    D. Carpels
    E. Shoots

5. What form of reproduction is at work when part of the parent plant breaks off to form new plants?

    A. Diversification
    B. Sexual reproduction
    C. Germination
    D. Fragmentation
    E. Diploidy

CHAPTER 12
Plant Reproduction

6. Which of the following hormones regulates the opening and closing of the stomata during times of drought?

    A. Gibberellins
    B. Ethylene
    C. Auxins
    D. Cytokinins
    E. Abscisic acid

7. A maple tree is an example of which of the following plant types?

    A. Diurnal
    B. Annual
    C. Biennial
    D. Nocturnal
    E. Perennial

8. A long-day plant is most likely to flower during which of the following months?

    A. July
    B. September
    C. January
    D. March
    E. November

9. In which of the following structures does the male gametophyte develop?

    A. Stigma
    B. Ovary
    C. Anther
    D. Ovule
    E. Sepal

10. Which of the following hormones or classes of hormones is involved in seed germination?

    A. Auxins
    B. Cytokinins
    C. Gibberellins
    D. Abscisic acid
    E. Ethylene

CHAPTER 12
Plant Reproduction

11. Which tropism occurs in response to gravity?

    A. Thigmotropism
    B. Phototropism
    C. Gyrotropism
    D. Gravitropism
    E. Planetropism

12. What is the sticky structure that a pollen grain lands on when pollinating another plant?

    A. Ovary
    B. Anther
    C. Stamen
    D. Stigma
    E. Petal

13. What is the name of the twenty-four-hour cycle in which plants operate?

    A. Biological clocks
    B. Timepieces
    C. Phytoperiods
    D. Photoperiods
    E. Circadian rhythms

14. Which of the following plant structures develops into the fruit?

    A. Stigma
    B. Ovary
    C. Anther
    D. Ovule
    E. Sepal

15. Which plant hormone or class of hormones is responsible for phototropism?

    A. Gibberellins
    B. Cytokinins
    C. Abscisic acids
    D. Ethylenes
    E. Auxins

CHAPTER 12
Plant Reproduction

ANSWERS

1. Pollination can occur through dispersal by the wind, where pollen grains are blown from the anther of one plant to the stigma of another. Insects can also be plant pollinators. When an insect such as a bee lands on a flower, pollen grains collect on the insect's body. The insect will carry those pollen grains to another flower, where they can drop and stick to the stigma. Other animal pollinators, such as hummingbirds, can spread pollen grains as they feed from plant to plant.

2. The alternation of generations led to the development of reproductive cells known as spores, which are resistant to drying out and are able to withstand the harsh environment as they disperse from the parent plant. This development allowed plants to begin colonizing land. Next, the development of the seed and the use of pollination for gamete dispersal further broke the dependence on water for the transportation of gametes for fertilization. Seeds can be widely dispersed and provide protection and nutrition for embryos until they are ready to germinate. Finally, the development of the flower resulted in a site for sexual reproduction, increasing the accuracy with which gametes could be transferred from plant to plant. Furthermore, fruit on flowering plants encourages animals to eat and disperse the seeds.

3. Asexual reproduction produces clones genetically identical to the parent, ensuring that the offspring is ideally suited to its environment. Asexual reproduction is also not dependent on outside forces, such as pollinators, to facilitate fertilization. Asexual reproduction does come with the disadvantage of little genetic variation, resulting in slow or nonexistent evolutionary development. As a consequence, plants that are ideally suited to a specific environment would be less able to adapt to change. Also, the emergence of outside threats, such as disease, is more likely to devastate populations of asexually reproducing plants.

4. **A**   Sepals, green, leaflike structures, enclose the flower and protect it before it opens. Sepals are visible at the base of a flower that is open.

5. **D**   Many plants, such as garlic, undergo a type of asexual reproduction known as fragmentation, in which parts of the parent plant break off and form mature plants.

6. E Abscisic acid regulates leaf openings, or stomata, in times of drought. This plant hormone is also involved in seed dormancy and in balancing the effects of growth hormones.

7. E Maple trees are perennials, plants whose life cycle spans the course of many years.

8. A A long-day plant is most likely to flower in the summer or early spring, when the duration of daylight is larger.

9. C Pollen grains (the male gametophyte) are formed in the anther of the flower.

10. C The class of plant hormones known as gibberellins promotes stem elongation and is involved in the growth of fruit and the germination of seeds.

11. D Gravitropism is the mechanism by which plants grow in response to gravity.

12. D Pollen grains are received by the stigma, the sticky or feathery surface on the carpel, the female sexual organ in plants.

13. E Plants operate in twenty-four-hour cycles called circadian rhythms. These cycles occur independent of environmental cues.

14. B The plant structure that develops into the fruit is the ovary. As the fruit develops, the flower loses its petals, and the walls of the ovary begin to increase in thickness.

15. E The plant hormone IAA (indole-3-acetic acid) is a type of auxin that is involved in phototropic movements, in which a plant grows, or bends, toward light.

CHAPTER 12
Plant Reproduction

ANIMAL EVOLUTION AND DIVERSITY

Characteristics of Animals

Animal Origins and Evolution

Animal Diversity

Human Evolution

13

Characteristics of Animals

Although the kingdom **Animalia** is diverse, all animals share four defining characteristics:

1. Comprised of eukaryotic cells

2. Contain cells that lack cell walls

3. Are multicellular

4. Are **heterotrophic,** meaning they ingest energy in the form of food, rather than producing it from a nonorganic source, as plants do through photosynthesis

In addition to these defining characteristics, most, but not all, animals also share the following characteristics:

• Reproduce sexually

• Capable of independent movement

• Body plans are symmetrical

• Bodies contain **tissues,** groups of cells enclosed in a membrane and organized to carry out a specific function

• Undergo unique embryonic developmental stages not shared by all invertebrates, beginning with the formation of a **blastula,** a hollow ball of cells that develops from the **zygote,** or fertilized egg

Animal Origins and Evolution

Most scientists believe that the kingdom Animalia is **monophyletic,** meaning that all animals descended from a single common ancestor, likely a colonial protist. Although the exact date of Animalia origin is unknown, the earliest animal fossils date to the late Precambrian era, approximately 575 million years ago.

Paleontologists believe that approximately 545 million years ago, in the early part of the Cambrian era, the animal kingdom underwent a major radiation, increasing greatly in both number and diversity. As proof of this radiation, known as the Cambrian explosion, paleontologists point to the great number and diversity of fossilized animals dating to this period. However, some argue that the fossil record does not necessarily point to an explosion of organisms in the animal kingdom. Instead, these scientists say that suitable conditions for fossilization that did not exist on Earth prior to the Cambrian era explain the increased number of fossils.

In either case, by the end of the Cambrian era, approximately 488 million years ago, all major innovations in animal body plans had developed.

THE EVOLUTION OF BODY PLANS

Key innovations in the evolution of animal body plans include the evolution of tissues, structural symmetry, body cavities, embryonic developmental plans, and segmented bodies. These innovations and their significance are summarized in the table below.

KEY INNOVATIONS IN ANIMAL EVOLUTION	
Innovation	Significance
Tissues	Tissues are groups of cells contained in a membrane. They carry out vital functions in the body, provide protection and structural support for the animal, and combine with other tissues to form complex organs.
Structural symmetry	Symmetry describes the form an animal's body takes. Most animals are either radial or bilateral. **Radial symmetry** describes a circular structure developed around a point in the center of the body. **Bilateral symmetry** describes a structure developed along a single plane, such that the animal is composed of two mirror-image halves.
Body cavity	The body cavity, or space surrounded by tissue, allows the development of a complex digestive system and provides space and protection for other internal organs.
Embryonic developmental plans	Beginning with the development of the blastula in the egg, and the formation of the blastopore opening and primitive gut cavity known as the archenteron, animal development evolved to be increasingly complex.
Segmented bodies	Animal bodies may be divided into segmented parts, which may look alike but have specialized functions. Some animals are fully segmented, while others are only partly segmented. Segmentation increases locomotion abilities and allows the duplication of identical organ systems in numerous segments in some organisms.

Evolution of Tissues

The first evolutionary division in the animal tree is based on the presence of tissues. Tissues are formed when groups of cells, designed to carry out specific functions, are separated from other groups by membranes. True tissues are absent in the most primitive animals, known as **Parazoa** ("near animals"). Each parazoan cell carries out all of the body's functions. Animals known as **Eumetazoa** ("true animals") possess tissues that carry out particular functions in the body, play a protective or structural role in the body, and can combine to form organs, leading to even more complex organisms.

Evolution of Symmetry

The next division in the animal tree occurs in the Eumetazoans branch and results in the separation of animals into two distinct groups based on their body symmetry: radial and bilateral.

- **Radiata (radially symmetrical animals)** are symmetrical around the center point of the body. Jellyfish are examples of animals displaying radial symmetry.

- **Bilateria (bilaterally symmetrical animals)** are symmetrical along a single, central axis. Most Eumetazoans, such as fish, birds, and mammals, exhibit bilateral symmetry.

The development of bilateral symmetry provided a major boost to animal evolution. It allowed different, specialized tissues and organs to develop in different parts of the body. It led to the development of sensory organs in more strategic locations on the body and provided a means for greater and more efficient mobility in the environment than radial symmetry allowed. For example, some bilateral animal species developed a distinct top (dorsal) and bottom (ventral) and front (anterior) and back (posterior), allowing movement in any chosen direction. By contrast, radial animals tend to be sessile (attached to a base) or passively floating.

Evolution of a Body Cavity

The development of the body cavity marks the next evolutionary divide in the animal kingdom. Body cavities paved the way for the development of complex digestive systems by providing protection and space for interactions among specialized internal organs and allowing the distribution of materials throughout the body.

Those animals possessing no body cavity are known as **acoelomates.** This group contains the radiata, such as jellyfish and comb jellies, as well as some species of bilateria. The bilateria are further separated into two groups possessing body cavities: the **pseudocoelomates** and the **coelomates.** These groups differ primarily based on the tissue from which their body cavity is formed.

The bodies of most animals are divided into three tissue layers: an outer **ectoderm;** an inner **endoderm,** which lines the internal organs; and a third layer between the two called the **mesoderm.** In all pseudocoelomates, a body cavity known as a pseudocoel is formed between the mesoderm and endoderm. In coelomates, a body cavity known as a coelom is formed entirely inside of the fluid-filled mesoderm, with specialized organs such as the heart suspended within. Because the endoderm and mesoderm are separated in the pseudocoel, interaction between these layers is limited. In the coelom, the mesoderm and endoderm remain in contact with each other after the body has developed. This structural advantage allows interactions between these two tissues.

Evolution of the Developmental Plan

The bilateria can be divided into two groups based on their embryonic developmental plans: the **protostomes** and the **deuterostomes.**

All animals develop from a single, fertilized egg, called a **zygote.** In most animals, with the exception of simple, primitive animals such as sponges, the zygote divides multiple times to form a hollow ball of cells called a **blastula.** The blastula folds in on itself in a process called **gastrulation** and forms a structure known as the **gastrula.** The gastrula is made up of a distinct front and back and three layers of cells: the endoderm, mesoderm, and ectoderm. An opening, called the **blastophore,** develops into the mouth, in protostomes, or the anus, in deuterostomes.

It is at this stage in the embryo's development that the plans for protostome development and for deuterostome development diverge. The differences between the plans are outlined in the following table.

KEY DIFFERENCES IN THE DEVELOPMENTAL PLAN OF PROTOSTOMES AND DEUTEROSTOMES	
Protostomes	Deuterostomes
Undergo **spiral** cleavage, in which cells divide parallel to and at right angles from a polar axis, leading to a relatively loose arrangement of cells	Undergo **radial** cleavage, in which cells divide in a spiral pattern outward from a polar axis, leading to a tightly packaged arrangement of cells
The blastopore becomes the animal's **mouth,** and the **anus** develops at the opposite end	The blastopore becomes the animal's **anus,** and the **mouth** develops at the opposite end
The developmental state of each embryonic cell is **determinate,** meaning the specific tissue that each embryonic cell will develop into is determined early in development	The developmental state of each embryonic cell is considered flexible, or **indeterminate,** meaning the signals that promote the differentiation of embryonic cells occur late in the embryo's development

Protostomes include bilateria such as the mollusks (for example, snails), annelids (for example, earthworms), and arthropods (for example, crustaceans and insects). Deuterostomes include all other bilateria, such as the echinoderms (for example, starfish) and chordates (for example, mammals).

CHAPTER 13
Animal Evolution & Diversity

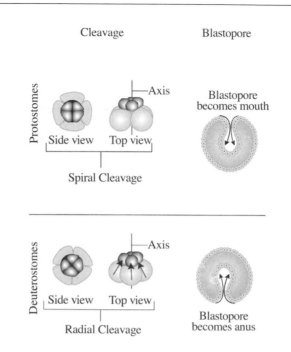

Cleavage   Blastopore

Protostomes

Axis

Side view   Top view

Spiral Cleavage

Blastopore
becomes mouth

Deuterostomes

Axis

Side view   Top view

Radial Cleavage

Blastopore
becomes anus

Protostome Development vs. Deuterostome Development

Evolution of Segmented Bodies

The development of segmentation, or a body plan composed of repeated body units, is the next major division in the animal phyla. In some animals, such as earthworms, the body is completely segmented. In others, such as vertebrates, only some parts of the body are segmented, such as the backbone. Segmentation provides two major benefits:

1. **Locomotion:** The ability to move individual segments of the body makes the animal more flexible and increases overall locomotive efficiency. For example, the segmented body of the earthworm allows its peristaltic, wavelike motion. Peristaltic waves are created as different parts of the body pulse at different times, resulting in a ripple effect along the body. This method of movement enables earthworms to burrow underground.

2. **Duplication:** In animals possessing a high degree of segmentation, such as the annelids, each segment can contain an almost complete set of the animal's organs. Damage to one segment will not be fatal to the animal, as there are more segments that can duplicate the same functions.

Segmentation is characterized by three key features:

1. **Repeated segments:** Numerous individual segments that make up the body. Excretory and locomotion organs are repeated in each segment, providing the means for each segment to remove waste and move independently.

2. **Specialized segments:** Front segments are specialized for sensory purposes. Front segments of some annelids are light-sensitive, others have elaborately formed eyes, and others even contain a well-developed brain.

3. **Connections:** Neural connections link one segment to the adjoining segment, allowing blood and nerves to travel through the body.

Animal Diversity

Biologists consider animals the most diverse of the four eukaryotic kingdoms, ahead of protists, fungi, and plants. The kingdom Animalia contains the greatest number of identified species, estimated at well over a million divided into approximately thirty different phyla. (However, since most of the Earth's animal diversity has yet to be fully documented and described, some scientists estimate the actual number of animal species to be around 30 million.) **Invertebrates,** animals that lack a backbone, make up the most diverse and abundant animal group, comprising around 97 percent of all species in the animal kingdom. The **vertebrates,** animals with a backbone, are found only in the phylum **Chordata.**

What follows are descriptions of the nine phyla that together account for most of the animal species on Earth. These phyla are listed roughly in order of their development based on the evolutionary benchmarks noted above.

PORIFERA

The phylum Porifera, which includes the sponges, is one of the most ancient animal lineages. Porifera lack true tissues or body symmetry. They are **sessile,** or nonmoving, filter feeders who collect their food from water that passes through their bodies. Most Porifera live in a marine, or ocean, environment.

CNIDARIA

The phylum Cnidaria includes jellyfish, coral, and sea anemones. Cnidarians have true tissues and radial symmetry and may be either sessile or mobile. One unique characteristic of Cnidarians is the presence of **cnidocytes,** stinging cells on the surface of tentacles that assist in capturing prey. Most cnidarians are marine animals.

PLATYHELMINTHES

The phylum Platyhelminthes includes the flatworms, such as tapeworms and flukes. Platyhelminthes have true tissues and bilateral symmetry. They lack an internal cavity, placing them in the acoelomate group. Platyhelminthes can live in freshwater or marine environments, and many are parasitic.

NEMATODA (ASCHELMINTHES)

The phylum Nematoda, also known as Aschelminthes, includes the round worms, such as pinworms and hookworms. Nematodes have true tissues, bilateral symmetry, and protostome development. Nematodes possess a pseudocoelom, a primitive circulatory system, and a complete digestive tract from mouth to anus. Nematodes live in both terrestrial and aquatic soils, and some are parasitic.

MOLLUSCA

The phylum Mollusca includes snails, oysters, clams, mussels, octopuses, and squid. Mollusca have true tissues, bilateral symmetry, protostome development, and a coelom. The mollusk body plan is divided into three parts: a muscular head or foot specialized for locomotion and/or the capture of food; a central section that houses the primitive internal organs; and a **mantle,** a heavy fold of tissue that protects the central section and contains the gills or lungs. Most mollusks have an open circulatory system with a three-chambered heart, although the cephalopods (for example, squid and octopus), have a closed circulatory system similar to annelids and vertebrates.

Mollusks are among one of the first animal groups to develop an efficient excretory system for waste disposal. This system consists of **nephridia,** tubular structures that collect waste from the coelum and expel it into the mantel cavity. Mollusks live in either terrestrial or aquatic environments.

ANNELIDA

The phylum Annelida includes the segmented worms, such as earthworms, polychaetes, and leeches. Annelids have true tissues, bilateral symmetry, protostome development, and a coelom. Annelids also mark the earliest development of segmentation.

In most annelids, blood circulates through a closed circulatory system, similar to that found in vertebrates. Annelids have an excretory system similar to the mollusks', in which nephridia collect and remove waste products from the body cavity. About two-thirds of all annelids are marine; earthworms, which constitute the remainder, are terrestrial.

ARTHROPODA

The phylum Arthropoda includes insects, crabs, spiders, and shrimp. Arthropods have true tissues, bilateral symmetry, protostome development, and a coelom. Like annelids and chordates, they also have a segmented body plan. In addition, arthropods developed jointed appendages that are specialized for movement, feeding, defense, and/or reproduction. The bodies of arthropods are covered by an **exoskeleton,** a hard structure made of chitin that protects the body.

CHAPTER 13
Animal Evolution & Diversity

The circulatory system of an arthropod is open: The flow of hemolymph (blood-like fluid) is controlled by a longitudinal vessel known as the heart. The transport of oxygen throughout the body is facilitated by the respiratory system rather than by the circulatory system. Arthropods contain air ducts known as trachea that transport air throughout the body. A variety of excretory systems have developed in the Arthropoda phylum.

The arthropods encompass many well-known animal groups living in both terrestrial and aquatic environments. Arthropod groups include the crustaceans (such as shrimps, crabs, and barnacles), the arachnids (such as spiders, mites, ticks, and scorpions), and the myriapods (centipedes and millipedes). The arthropod phylum also contains the largest group of organisms on Earth: the insects. Insects account for more than half of the known animal species, with many species still unidentified by biologists. Arthropods live in both terrestrial and aquatic environments.

ECHINODERMATA

The phylum Echinodermata includes sea stars, sea urchins, and sea cucumbers. Echinoderms have true tissues, deuterostome development, and a coelom. A unique characteristic of echinoderms is their hard **endoskeleton** formed of calcium that exists just beneath their skin. In addition, they have a **water-vascular system,** which flushes water through the internal cavity. Echinodermata also possess **tube feet,** appendages covered in small suckers that collect food from the water and transport it to the mouth. Most echinoderms exhibit bilateral symmetry in their mobile juvenile stages of development and radial symmetry in their more sessile adult form. Echinoderms live in marine environments, often along the sea floor.

EXAMPLE: Most echinoderms are capable of regenerating lost body parts. A familiar example is the sea star, which can regrow an arm that has been removed.

CHORDATA

The phylum Chordata includes all the vertebrates as well as lancelets and sea squirts. Chordates have true tissues, deuterostome development, and a coelom. All chordates have a **notochord,** a flexible rod that runs along the back of the body and serves as an attachment for muscles. In vertebrates, the notochord develops into the backbone as the organism matures into adulthood.

Other key characteristics of chordates include the presence, at some stage of development, of the following:

- **Dorsal nerve:** A single, hollow nerve cord that runs just under the **dorsal** (or back) surface of the animal.

- **Pharyngeal slits:** Slits that form a passageway between the mouth and the **pharynx,** a muscular tube connecting the mouth to the digestive tract and windpipe.

- **Postanal tail:** A tail that extends beyond the anus. The postanal tail is usually present in embryonic form.

Some of these characteristics are only present in embryonic stages in certain chordates, such as humans. Chordates live in both terrestrial and aquatic environments.

Major Classes of Vertebrates

All vertebrates possess a backbone made of segmented vertebrae, and most also possess a bony skull and hinged jaws. The five major classes of vertebrates are fish, amphibians, reptiles, birds, and mammals.

- **Fish** complete their entire life cycle in water. **Gills** allow them to extract oxygen from water and regulate the **osmolarity** (or the concentration of solutes, such as salt) in their bodies. The two main groups of fish are the cartilaginous fishes—such as sharks and rays, whose skeletons are composed of cartilage—and the bony fishes.

- **Amphibians** are the first vertebrates to inhabit land, and they possess adaptations suitable for both terrestrial and aquatic environments. Amphibians can take in oxygen through gills, lungs, and their skin. All amphibians require aquatic habitats for some life stages, such as for the laying of eggs or larvae.

- **Reptiles** are the first vertebrates to complete their entire life cycle in terrestrial environments. Their dry, usually scaly skin helps prevent dehydration. Additionally, the production of **amniotic eggs,** which are shelled and contain a nutrient-filled amniotic sac, prevents dehydration of the embryo and provides protection and nourishment.

- **Birds** possess feathers, which many biologists believe to be derived from reptilian scales. Birds produce amniotic eggs, have scaled legs, and are **endothermic,** meaning they generate heat internally to maintain a relatively high body temperature.

- **Mammals** are characterized by the presence of hair, which provides insulation; a middle ear, which amplifies vibrations on the eardrum; and mammary glands, which produce milk to feed young. Like birds, mammals are also endothermic. The three main groups of mammals are the **monotremes,** or egg-laying mammals; the **marsupials,** which house their young in a pouch as they complete development; and **eutherians,** whose young develop in the womb and are nourished by a placenta.

Human Evolution

Humans belong to the order known as **Primates,** in the eutherian group of mammals. Primates evolved from small, arboreal (tree-dwelling) mammals known as Archonta about 65 million years ago and are characterized by grasping hands and eyes located at the front of the head to provide depth perception. Primates can be subdivided into two groups:

1. **Prosimians,** which include tarsiers, lorises, and lemurs, are nocturnal (active at night) and resemble a cross between a squirrel and a monkey.

2. **Anthropoids,** which include monkeys, apes, and humans, are diurnal (active during the day) and have advanced color vision, an extended developmental juvenile period, and relatively large brains.

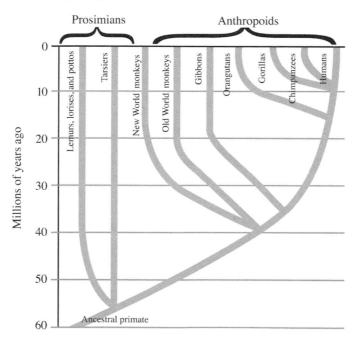

Primate Evolutionary Tree

HOMO SAPIENS

Humans, or *Homo sapiens*, belong to the family Hominidae and are closely related to apes. In fact, humans share about 99 percent of their genes with chimpanzees. All humans share the following four characteristics, some of which may have been present in other hominid forms:

1. Bipedal locomotion (upright movement using two feet)

2. Reduced jaw muscles and digestive tract

3. Large brains capable of abstract thought

4. An exceptionally long juvenile period

Although *Homo sapiens* is the only extant (still-living) member of the family Hominidae, other species existed in the past, some of which gave rise to modern humans. Earlier species of hominids include the following:

- The *Australopithecus* species existed approximately 3 million years ago. This species was the first primate to walk upright, a characteristic termed bipedal. They were smaller than modern humans and had substantially smaller brains.

- *Homo habilis* existed approximately 2 million years ago and was the first species of the genus *Homo. Homo habilis* were similar to *Australopithecus* but had larger brains. They are the first hominid to use tools—a trait believed to result from their larger brain size.

- *Homo rudolfensis* existed 1.9 million years ago. Their skull shows a larger brain cavity than *Homo habilis,* suggesting that they were a separate species.

- *Homo ergaster* existed approximately 1–2 million years ago. Compared to *Homo habilis, Homo ergaster* was tall, with legs, hands, feet, and teeth more similar in shape and proportion to modern humans. Sexual size dimorphism, the difference in size between males and females, was also reduced compared to *Homo habilis* and was more similar to modern humans.

- *Homo erectus* existed approximately 1.6 million to 400,000 years ago. They possessed brains larger than *Homo ergaster,* and many features similar to modern humans, such as small teeth, a less sloping forehead, and a smaller size difference between males and females. *Homo erectus* employed more advanced tools than previous hominids, indicating a higher capacity for abstract thought.

- *Homo neanderthalensis* (Neanderthals) existed 500,000–30,000 years ago, simultaneously with early *Homo sapiens.* Compared to *Homo sapiens,* Neanderthals were stockier and had heavier brows. Neanderthals had large brains, comparable to those of the *Homo sapiens,* and are believed to have been capable of language and complicated tool use. Neanderthals went extinct, and many biologists believe that they were cousins of modern humans, rather than ancestors, and did not contribute to our gene pool. Investigators are divided over the classification of modern humans following *Homo erectus.* While some biologists classify Neanderthals as a species of modern human separate from *Homo sapiens,* others lump these two hominids together, along with an earlier form, *Homo heidelbergensis,* into one species, *Homo sapiens.*

- *Homo sapiens* (modern humans) are believed to have evolved 130,000 years ago in Africa. Modern humans spread across Siberia and reached North America at least 13,000 years ago, although debate exists about the exact timeframe. Humans are capable of high levels of conceptual thought and use highly advanced tools. Humans are the only animals to formulate a symbolic language.

Early Human Phylogeny

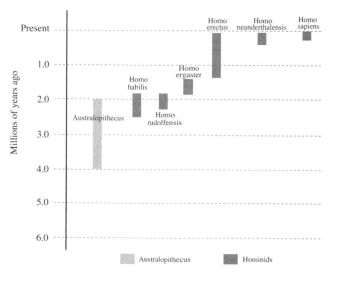

Evolution of the *Homo* Genus

Summary

Characteristics of Animals

- All animals are eukaryotic, multicellular, heterotrophic, and lack cell walls.

Animal Origins and Evolution

- Animals evolved from a colonial protist more than 575 million years ago and diversified during the Cambrian era (540 million years ago).

- The period known as the Cambrian explosion is believed to mark the origin of a significant number of diverse animal species.

- Key innovations in the evolution of animal body plans include the development of specialized tissues, symmetry, the body cavity, segmented bodies, and the embryonic developmental plan.

Animal Diversity

- The nine phyla that comprise most animal species include Porifera, Cnidaria, Platyhelminthes, Nematoda, Mollusca, Annelida, Arthropoda, Echinodermata, and Chordata.

- Porifera lack true tissues and symmetry.

- Cnidarian have radial symmetry.

- Platyhelminthes have bilateral symmetry and lack an internal body cavity.

- Nematodes have **protostome** development and a pseudocoelom formed from the **blastocoel.**

- Molluscs have protostome development and a coelom formed from **mesoderm.**

- Annelids have protostome development and a coelom. They are the earliest segmented animals.

- Arthropods have protostome development, a coelom, and a hard exoskeleton made of chitin.

- Echinoderms have **deuterostome** development, a coelom, and a hard endoskeleton made of calcium.

- Chordata have deuterostome development, a coelom, and a **notochord** (a flexible rod that runs along the back of the body).

Human Evolution

- Humans, known as *Homo sapiens*, belong to the family Hominidae in the order **Primates.**

- Several hominid forms existed prior to the evolution of *Homo sapiens*. Some, though not all, of these species gave rise to modern humans.

Sample Test Questions

1. Which form of symmetry is generally associated with a sessile lifestyle, and which form is generally associated with a motile lifestyle? What are some reasons for the connection between body symmetry and movement?

2. Compare the adaptations possessed by amphibians and reptiles. How can these adaptations explain the suitability of each to its environment?

3. Describe the characteristics that distinguish humans from other mammals. Explain the animal group that first exhibited each characteristic: primates, anthropoids, *Australopithecus,* or *Homo.*

4. The defining characteristics of animals include all of the following EXCEPT:

  A. Eukaryotic cells
  B. No cell walls
  C. Movement
  D. Multicellularity
  E. Heterotrophism

5. Which group of animal develops an anus in the area where the blastula folds inward?

   A. Parazoa
   B. Eumetazoa
   C. Acoelomates
   D. Protostomes
   E. Deuterostomes

6. Humans belong to which of the following families?

   A. Homo
   B. Hominidae
   C. Australopithecus
   D. Anthropoid
   E. Prosimian

7. Egg-laying mammals are known as which of the following?

   A. Monotremes
   B. Marsupials
   C. Eutherians
   D. Parazoa
   E. Eumetazoa

8. Which of the following phyla have members with no true tissues or symmetry?

   A. Arthropoda
   B. Chordata
   C. Cnidaria
   D. Porifera
   E. Annelida

9. Which of the following key innovations in body plan design was the last to develop?

   A. Symmetry
   B. Body cavities
   C. Tissues
   D. Developmental plans
   E. Segmentation

10. Which of the following phyla possess a coelom?

    A. Platyhelminthes
    B. Porifera
    C. Nematoda
    D. Chordata
    E. Cnidaria

11. Insects, spiders, and crabs belong to which phyla?

    A. Nematoda
    B. Arthropoda
    C. Cnidaria
    D. Annelida
    E. Echinodermata

12. What type of symmetry is exhibited by the wheel of a car?

    A. Bilateral symmetry
    B. Unilateral symmetry
    C. Radial symmetry
    D. Bilateral and radial symmetry
    E. Unilateral and radial symmetry

13. Which of the following phyla has a three-part body plan consisting of a head or foot, a central section, and a mantle?

    A. Mollusca
    B. Nematoda
    C. Platyhelminthes
    D. Annelida
    E. Arthropoda

14. Characteristics that separate humans from other apes include all of the following EXCEPT which?

    A. Larger brain
    B. Ability to perform symbolic thought
    C. Enlarged digestive tract
    D. Reduced jaw muscles
    E. Extended juvenile period

CHAPTER 13
Animal Evolution & Diversity

15. Which of the following phyla possesses an endoskeleton made of calcium plates?

   A. Arthropoda
   B. Chordata
   C. Echinodermata
   D. Cnidaria
   E. Annelida

ANSWERS

1. Radial symmetry, found in cnidarians and the adult form of most echinoderms, tends to be associated with a sessile lifestyle, whereas bilateral symmetry tends to be associated with a motile lifestyle. Bilaterally symmetric organisms have a distinct front and back, allowing sensory and motor systems to be oriented to work best in a wide range of directions. Radial symmetry is better suited for sessile organisms because it allows the organism to grow equally in all directions to better fill up its available space.

2. Amphibians are adapted for a life spent between water and land. Reptiles are adapted for an entirely terrestrial life. At different stages of development, amphibians possess both gills and lungs, enabling them to extract oxygen in either terrestrial or aquatic environments. Amphibians must remain in close proximity to water, however, to protect their moist skin and unshelled eggs. Reptiles have lungs, dry scaly skin, and shelled eggs, all of which enable them to live far from bodies of water.

3. The characteristics that distinguish humans from other mammals include enhanced color vision and depth perception, grasping hands, large brains capable of symbolic thought, an extended juvenile period, reduced jaw and digestive systems, and bipedal locomotion. Grasping hands and depth perception can be seen in all primates, and all anthropoids have well-developed color vision. In addition, large brain size and extended juvenile periods can be seen in anthropoids, although these features are not present to the same extent that they are in humans. Bipedal locomotion was first observed in Australopithecus, but this group is similar to other anthropoids in all other areas of development. Only in *Homo* have all characteristics of modern humans been fully developed.

4. C The defining characteristics of animals include eukaryotic cells, a lack of cell walls, multicellularity, and heterotrophism. Although most animals move, movement is not a defining characteristic of animals.

5. E In deuterostomes, an anus forms in the area where the blastula has folded inward.

6. B Humans belong to the family Hominidae.

7. A Egg-laying mammals are known as monotremes.

8. D Porifera have no true tissues or symmetry.

9. E Segmentation was the last of the key innovations to develop in body plan design.

10. D Only relatively recent phyla, such as Chordata, possess coeloms.

11. B Insects, spiders, and crabs are arthropods.

12. D The wheel of a car has both bilateral and radial symmetry.

13. A Molluscs have a three-part body plan consisting of a head or foot, a central section, and a mantle.

14. C Characteristics that separate humans from other apes include a larger brain capable of symbolic thought, reduced jaw muscles, reduced digestive tract, and an extended juvenile period.

15. C Echinoderms possess an endoskeleton made of calcium plates.

CHAPTER 13
Animal Evolution & Diversity

INTRODUCTION TO ANIMAL PHYSIOLOGY

Levels of Organization

Tissue Types

Organs and Organ
    Systems

Homeostasis

14

Levels of Organization

The bodies of all living organisms, including plants, fungi, and animals, adhere to a distinct hierarchical organization, in which small, relatively simple structures combine to form larger, more complex structures. The structures can include the following:

- **Atoms:** The smallest units that possess the characteristics of an element. Hydrogen is an example of an atom.

- **Molecules:** Structures made up of two or more atoms. DNA is an example of a molecule.

- **Organelles:** Groups of molecules organized to perform specific cellular functions. Mitochondria are organelles.

- **Cells:** The smallest units that can carry out all of life's processes. A single cell can be a complete organism, as in bacteria, or a part of a larger organism, as in urinary epithelium cells.

- **Tissues:** Groups of cells organized to carry out a particular function in a multicellular organism, such as the epithelial tissue, which provides covering or lining to internal and external structures.

- **Organs:** Groups of tissues organized to carry out a particular function. The urinary bladder, which is composed of urinary epithelial and muscle tissue, is an organ.

- **Organ systems:** Groups of organs designed to carry out a particular task. The urinary system, which includes the urinary bladder, the kidney, and the ureter, is an organ system.

All animal bodies are organized to perform the same fundamental tasks: respiration and circulation, digestion, waste removal, structural support, movement, defense against disease, response to internal and external conditions, coordination of body activities, and reproduction. **Physiology** is the branch of biology that studies these fundamental biological tasks, and the processes by which animals carry them out.

STRUCTURAL ORGANIZATION AMONG INVERTEBRATES

Among the members of the animal kingdom, the body plans of invertebrates are very simple compared with vertebrates. The more primitive invertebrates, such as sponges and jellyfish, do not have organs at all. The table on the following page briefly outlines the levels of organization in the most common invertebrate groups.

CHAPTER 14
Animal Physiology

Invertebrate Phylum	Highest Levels of Organization
Sponges	Several cell types but possessing no organized tissues or organs
Cnidarians	Cells organized into specialized tissues but no true organs
Ctenophores	Cells organized into specialized tissues but no true organs
Flatworms	The simplest animals, in which tissues are formed into simple organs
Nematodes	Tissues formed into simple organs
Mollusks	Tissues formed into specialized organs, as well as some basic organ systems, such as an excretory system
Annelids	Tissues formed into specialized organs, as well as some basic organ systems
Arthropods	Tissues formed into specialized organs, as well as some basic organ systems
Lophophorates	Tissues formed into specialized organs; appear to be related to the mollusks, annelids, and nematodes
Echinoderms	Tissues formed into specialized organs, as well as some basic organ systems

Tissue Types

Tissues are composed of groups of specialized cells and are found in four basic types across members of the animal kingdom: epithelial, connective, muscle, and nervous.

EPITHELIAL TISSUE

Epithelial tissue takes the form of a tight sheet that lines the internal and external surfaces of the body, including the skin, digestive tract, and body cavities. Epithelial tissue usually acts as a selective barrier, allowing the passage of some, but not all, materials.

Epithelial tissue may be simple, comprising a single layer of cells; **stratified,** comprising multiple cell layers; or pseudostratified, comprising a single layer of cells structurally resembling stratified tissue. Most epithelial tissue falls into one of the five basic types described in the chart below.

	Tissue Type	Function	Location
1.	Pseudostratified columnar	Protection and secretion	Parts of the respiratory tract
2.	Stratified squamous	Protection	Skin and mouth
3.	Simple columnar	Protection, secretion, and absorption	The digestive system and parts of the respiratory tract
4.	Simple cuboidal	Secretion and absorption	Glands and kidney tubules
5.	Simple squamous	Thin barrier that allows easy passage of molecules	Lining of lungs and blood vessels

CONNECTIVE TISSUE

Connective tissue supports, stabilizes, and protects the body's many organs. All connective tissue consists of cells surrounded by fluids or materials called **ground substance,** which are proteins produced and secreted by the cells and fibroblasts around it. There are five basic types of connective tissue, described in the chart below.

	Tissue Type	Function	Location
1.	Dense connective	Connects organs	Tendons, around muscles, under skin
2.	Loose connective	Supports, insulates, and nourishes organs	Between organs and under skin
3.	Blood	Transports oxygen, nutrients, wastes, and other chemicals throughout the body	The circulatory system
4.	Cartilage	Absorbs shock and friction	Joints, ear, nose, trachea
5.	Bone	Provides framework for body and site for muscle attachment	The skeletal system

CHAPTER 14
Animal Physiology

MUSCLE TISSUE

Muscle tissue facilitates movement throughout the body by contracting. There are three basic types of muscle tissue, described below.

	Tissue Type	Function	Location
1.	**Skeletal**	Voluntary movement	Attached to bones in the skeleton
2.	**Smooth**	Involuntary movement; assists digestion and circulation	The digestive system and walls of veins and arteries
3.	**Cardiac**	Involuntary movement; responsible for the heartbeat	The heart

NERVOUS TISSUE

Nervous tissue produces and conducts electrochemical signals between organs of the body and the brain. There are four main types of nervous system tissue, as described in the table on the following page.

	Tissue Type	Function	Location
1.	**Motor neurons**	Send information from the brain to the muscles and glands	Spinal cord, brain, and body
2.	**Sensory neurons**	Send information gathered from internal and external stimuli to the brain	Skin, eyes, ears
3.	**Association neurons (also called interneurons)**	Integrate and relay information between neurons	Spinal cord and brain
4.	**Glial cells (also called neuroglia)**	Provide structural support and facilitate information transfer along and between neurons	Spinal cord and brain

Organs and Organ Systems

Organs are composed of multiple tissue types organized to carry out a specific function. Examples of organs include the heart, the brain, the pancreas, blood vessels, bones, and skin. Groups of multiple organs working together to carry out a major bodily function are called **organ systems.**

Any animal more complex than the cnidarians and ctenophores (jellyfish and comb jellies) uses one or more organ systems to perform the body's necessary functions. Each organ system has evolved within a species to help keep the particular animal functional. In general, more highly evolved animals require more complex organ systems than their primitive ancestors. For example, Platyhelminthes (flatworms), which have no body cavity, use a urinary system for the removal of waste but lack a circulatory system for the transport of oxygen and nutrients. In contrast, vertebrates have several organ systems, including one designed to circulate blood around the various body cavities.

All vertebrates possess the same eleven principal organ systems that facilitate all of life's major functions.

	Organ System	Function	Component Organs
1.	**Skeletal**	Structural support and the site of muscle attachment	Bones, cartilage, ligaments
2.	**Muscular**	Movement	Muscles
3.	**Integumentary**	Protects the body and regulates body temperature	Skin, sweat glands, nails, hair
4.	**Respiratory**	Facilitates the intake of oxygen and removal of carbon dioxide	Lungs or gills, trachea, skin
5.	**Circulatory**	Transports materials such as nutrients, waste products, carbon dioxide, and oxygen	Heart, blood vessels, blood
6.	**Digestive**	Breaks down food for the acquisition of nutrients	Mouth, stomach, intestines, liver, pancreas

CHAPTER 14
Animal Physiology

CONTINUED

	Organ System	Function	Component Organs
7.	Urinary	Removes waste from the blood	Kidneys, bladder, ureters, urethra
8.	Immune	Provides defense against pathogens (disease-causing agents)	White blood cells, lymph nodes, lymph vessels, spleen, thymus
9.	Endocrine	Controls and regulates bodily functions through chemical communication between the brain and organs and aids the nervous system in integrating the activities of all bodily systems	Glands
10.	Nervous	Detects internal and external stimuli and aids in controlling and coordinating responses to stimuli via electrochemical communication between the brain and body; also aids the endocrine system in integrating the activities of all bodily systems	Nerves, brain, spinal cord, sensory organs
11.	Reproductive	Replicates genetic material to be passed on to organisms' offspring	Testes, ovaries, penis, uterus, vagina

Homeostasis

To stay alive, all animals need to maintain steady internal conditions, such as body temperature and the levels of water, salts, nutrients, oxygen, and waste in the body fluids and cells. All organ systems participate in the maintenance of these stable internal conditions through a process known as **homeostasis,** meaning "steady state." Although the acceptable ranges for all of these factors vary from species to species, no animal can survive unless its internal environment is kept within tolerable limits.

The most common regulatory tool used to maintain homeostasis is the **negative feedback** loop. In a negative feedback loop, the end point, or product, of a specific process regulates the beginning of that same process. If levels of the product are low in an animal, the process increases; if levels of the product are high, the process stops. Often the body has two negative feedback loops for each process: one to raise levels and another to lower them.

> *EXAMPLE:* The body must maintain a specific level of blood calcium for the proper functioning of muscles. If blood calcium levels are low, the parathyroid gland releases a hormone that pulls calcium from the bones and releases it into the bloodstream. When the concentration of calcium in the blood reaches the proper level, the parathyroid gland stops releasing the hormone. On the other hand, if blood calcium levels are high, the thyroid gland releases a hormone that causes calcium to be drawn out of the bloodstream and deposited into the bones. Once the blood calcium concentration drops to the proper level, the thyroid gland stops releasing the hormone.

Although less common than negative feedback loops, **positive feedback** loops also work to regulate functions in the body. In a positive feedback loop, a specific response amplifies until a proper level is reached, at which point the response is reversed or stopped.

> *EXAMPLE:* During childbirth, pressure on the mother's uterus wall and cervix causes the secretion of the hormone oxytocin. This hormone stimulates contraction of the uterus walls, which in turn exert pressure on the fetus. Pressure from the fetus increases as a result, further stimulating the release of oxytoxin. A positive feedback loop of increasing pressure contin-ues until the fetus is born, at which point stimulation of both the fetus and uterus wall ceases, completing the cycle.

CHAPTER 14
Animal Physiology

Summary

Levels of Organization

- All living organisms possess a distinct hierarchical organization, in which small, relatively simple structures combine to form larger, more complex structures.

- The structures common to all vertebrates include **atoms, molecules, organelles, cells, tissues, organs,** and **organ systems.**

- Invertebrates have much simpler body plans than vertebrates.

- Primitive invertebrates do not contain any organs.

- Sophisticated invertebrates may contain organs and some organ systems.

Tissue Types

- Tissues are composed of groups of specialized cells. Vertebrates have four basic tissue types: epithelial, connective, muscle, and nervous.

- **Epithelial tissue** lines the surfaces of the body and usually acts as a selective barrier that allows the passage of some, but not all, materials. The five basic types of epithelial tissue are pseudostratified columnar, stratified squamous, simple columnar, simple cuboidal, and simple squamous.

- **Connective tissues** support, stabilize, and protect the organs of the body. All connective tissues consist of cells surrounded by fluids or materials called **ground substance.** The five basic types of connective tissue are dense connective, loose connective, blood, cartilage, and bone.

- **Muscle tissue** has the ability to contract, allowing organisms to move. The three basic types of muscle tissue are skeletal, smooth, and cardiac.

- **Nervous tissue** produces and conducts electrochemical signals between organs of the body and the brain. The four main types of nervous tissue are **motor neurons, sensory neurons,** association neurons (interneurons), and glial cells (neuroglia).

Organs and Organ Systems

- **Organs** are composed of multiple tissue types organized to carry out a specific bodily function. Multiple organs work together as **organ systems** to carry out the major functions of the body.

- Vertebrates possess eleven organ systems: skeletal, muscular, integumentary, respiratory, circulatory, digestive, urinary, immune, endocrine, nervous, and reproductive.

Homeostasis

- **Homeostasis** is the process by which an organism maintains relative stability of internal conditions.

- Homeostasis is maintained through **negative feedback** loops and **positive feedback** loops.

Sample Test Questions

1. Describe the epithelial tissue and explain how its composition relates to its function.

2. Explain how the respiratory, urinary, digestive, and immune systems contribute to homeostasis.

3. Explain how a household thermostat can be described as a negative feedback mechanism.

4. The skeletal system comprises which of the following organs?

    A. Bones
    B. Ligaments
    C. Cartilage
    D. All of the above
    E. A and C only

**CHAPTER 14
Animal Physiology**

5. Which type of muscle tissue is found in blood vessels and the digestive track?

   A. Skeletal
   B. Striated
   C. Smooth
   D. Rough
   E. Cardiac

6. What is the name for a group of cells organized to carry out a specific function?

   A. Tissues
   B. Organelles
   C. Organs
   D. Organ systems
   E. Organ groups

7. If a person feels cold, her body will initiate a number of mechanisms, such as shivering, designed to increase body temperature. Shivering and other temperature-raising mechanisms will stop once the body reaches a normal temperature. This is an example of what process?

   A. Positive feedback
   B. Neutral feedback
   C. Negative feedback
   D. Positive regulating
   E. Negative regulating

8. Which organ system is responsible for chemical communication between various parts of the body?

   A. Immune system
   B. Nervous system
   C. Reproductive system
   D. Respiratory system
   E. Endocrine system

9. Which type of nervous tissue is responsible for providing structural support and facilitating information transfer along and between neurons?

   A. Motor neurons
   B. Sensory neurons
   C. Interneurons
   D. Accessory neurons
   E. Glial cells

10. Which type of tissue is composed of cells suspended in ground substance?

   A. Epithelial tissue
   B. Connective tissue
   C. Ground tissue
   D. Nervous tissue
   E. Muscle tissue

11. What is the name for the thin, single layer of tissue that lines the lungs and blood vessels?

   A. Simple squamous epithelial
   B. Simple cuboidal epithelial
   C. Stratified squamous epithelial
   D. Stratified columnar epithelial
   E. Pseudostratified columnar epithelial

12. Which of the following is NOT considered part of the immune system?

   A. White blood cells
   B. Lymph nodes
   C. Spleen
   D. Pancreas
   E. Thymus

13. What is the primary type of tissue found in the heart?

   A. Smooth muscle tissue
   B. Skeletal muscle tissue
   C. Cardiac muscle tissue
   D. Connective tissue
   E. Epithelial tissue

CHAPTER 14
Animal Physiology

14. Which type of nervous tissue integrates and relays information between neurons?

    A. Glial cells
    B. Motor neurons
    C. Sensory neurons
    D. Association neurons
    E. Contact neurons

15. Which organ system includes the skin, nails, hair, and sweat glands?

    A. Integumentary system
    B. Skeletal system
    C. Muscular system
    D. Endocrine system
    E. Circulatory system

ANSWERS

1. Epithelial tissue often takes the form of a tight sheet, with very little space between the cells. Epithelial tissue lines the surfaces of the body, acting as a selective barrier through which only certain substances pass. The small spaces between cells in the epithelial tissue allow only the smallest molecules to diffuse past the tissue without assistance from the body.

2. The respiratory system contributes to homeostasis by keeping the body's oxygen and carbon dioxide levels within acceptable limits. Homeostasis is also maintained by the urinary system, which keeps the level of wastes in the bloodstream at low levels. The digestive system aids homeostasis by providing the body with the proper nutrients. Finally, the immune system contributes to homeostasis by fighting disease-causing agents called pathogens.

3. Similar to the negative feedback loop used by animal bodies, a household thermostat regulates the process of heating or cooling to maintain a suitable end product, the temperature. A heater will automatically turn on if the temperature drops below a certain level. The system will turn off once the temperature reaches a predetermined acceptable level.

4. D The skeletal system is composed of bones, ligaments, and cartilage.

CHAPTER 14
Animal Physiology

5. C   Smooth muscle tissue is found in blood vessels and the digestive track.

6. A   Groups of cells organized to carry out a specific function are known as tissues.

7. C   Shivering and other temperature-raising mechanisms are examples of a negative feedback mechanism, in which the end point (body temperature) controls whether the process continues.

8. E   The endocrine system is responsible for chemical communication between various parts of the body.

9. E   Glial cells are responsible for providing structural support and facilitating information transfer along and between neurons.

10. B   Connective tissue is composed of cells suspended in ground substance.

11. A   The single thin layer of tissue that lines the lungs and blood vessels is known as simple squamous epithelial.

12. D   The pancreas is not considered part of the immune system.

13. C   The primary type of tissue in the heart is cardiac muscle tissue.

14. D   Association neurons integrate and relay information between neurons.

15. A   The integumentary system includes the skin, nails, hair, and sweat glands.

THE INTEGUMENTARY, SKELETAL, AND MUSCULAR SYSTEMS

15

The Integumentary System

The Skeletal System

The Muscular System

The Integumentary System

Many complex animals are highly mobile and continually exposed to a variety of terrestrial and aquatic environments. For their bodies to function properly, they must protect themselves from a barrage of external forces and process a vast array of stimuli. The **integumentary system** is an animal's first way of defending against and interacting with the outside world. This organ system is composed of the protective layer, **skin,** as well as additional structures such as hair, nails, feathers, scales, and glands. The integumentary system not only provides a barrier between the inner workings of the animal and the outside environment, but it also facilitates temperature control and the movement of important molecules such as water and carbon dioxide into and out of the animal.

SKIN

Skin is an organ composed of three types of tissue—epithelial, connective, and nerve—arranged in two main layers. The epidermis forms the thin outer layer of the skin. The thicker layers of tissue, called the **dermis** and the **hypodermis,** lie underneath. Intersecting these layers is an assortment of other cell types including hair follicles, blood vessels, and nerve cells.

The primary function of skin is to separate an animal's internal organs from the outside environment. It also performs several other key functions:

- Produces vital vitamin D nutrients that result from the reaction of compounds in the skin with ultraviolet light from the sun.

- Senses heat, cold, pressure, and touch and relays information to illicit a response from other parts of the body.

- Protects against UV radiation and prevention of physical injury to internal structures.

- Forms a barrier against bacteria and protects the body from infection.

- Contains fat necessary for bodily functions and insulation.

- Regulates body temperature by monitoring internal and external heat.

- Secretes waste from the body.

CHAPTER 15
The Integumentary, Skeletal, and Muscular Systems

The Epidermis

The first layer of skin is the epidermis. It is the thinner of the two layers of skin tissue and is composed of epithelial tissue. Epidermal cells divide mitotically on the bottom of the epidermis, moving closer to the surface as they age. Epidermal cells divide rapidly, continually creating a fresh set of cells that push up. As they move upward, epidermal cells synthesize **keratin,** the protein that makes up nails and hair. When they reach the surface of the epidermis, the cells flatten out and stop dividing. Because they continually move farther away from vital blood vessels close to the dermis layer, epidermal cells die as they reach the epidermis surface. They are then brushed off and replaced by the next layer of cells.

The Dermis

The thicker second layer of skin, known as the dermis, consists primarily of connective tissue interspersed with a variety of other cells, such as the nerve endings necessary for the sense of touch. Unlike cells in the epidermis, dermis cells are replaced very slowly.

Interlaced with the dermis is a network of blood vessels that play an important role in thermoregulation, the regulation of body temperature. When an animal is overheated, the blood vessels in the dermis dilate, allowing rapid dissipation of heat from the blood. When cold, these same blood vessels will constrict and keep the heat of the blood in the body.

The Hypodermis

Beneath the skin is a third layer of tissue known as the hypodermis. The **hypodermis,** sometimes called the subcutaneous (sub-skin) layer, is the connective tissue attaching the skin to internal organs. Much of the body's fat is stored in the hypodermis.

ADDITIONAL STRUCTURES

In addition to skin, the integumentary system may contain additional body structures such as hair, scales, feathers, nails, and glands.

Hair

Hair is entirely composed of keratin protein. It has no nerve tissue and is not considered living. Hair grows outward from a hair follicle embedded in the dermis. A single hair shaft will grow for a period of time before falling out and being replaced by a new hair. This process will continue unless the follicle shrinks, preventing any new hair growth.

Although the hair shaft itself cannot relay senses from the outside environment, the hair follicle can. Each hair follicle is **vascularized,** supplied with blood vessels, and contains nerve endings. The follicle senses the hair's movement, making the organism aware of touch or sensation. In addition to relaying sensations from the outside environment, hair provides an added degree of warmth to the animal.

Scales

Scales are rigid plates composed of a variety of substances, including keratin protein, and can vary in shape, size, and structure across different species of animals. Scales grow out of the skin and provide additional protection against physical force, as well as dehydration.

Feathers

Feathers, the defining feature of all birds, are believed by biologists to have derived from reptilian scales. They are produced by cells in the epidermis and are composed of keratin proteins. Feathers provide insulation and are necessary for flight.

Nails

Nails, also made from keratin, grow out from the fingers and toes and are connected to nerve endings at the base of the nail. The free edge of the nail extends past the finger and has no nerve endings. Nails are useful for picking and scratching and protection of the fingers and toes.

Glands

The term **gland** characterizes any organ in an animal body that produces and secretes a substance, such as a hormone. **Exocrine** glands, such as sweat glands in the skin, secrete these substances through tubelike ducts. There are two types of skin glands:

1. **Sebaceous glands** produce the oily **sebum** that lubricates hair and skin and can prevent bacterial growth.

2. **Sweat glands** aid in thermoregulation by producing sweat from two subdivisions of glands. The merocrine glands produce the salty sweat that facilitates evaporation and the subsequent cooling of an organism. The apocrine glands secrete fluid into hair follicles rather than through ducts. This secretion contains a milky fluid of unknown purpose as well as clear, salty sweat. This milky fluid is only secreted in the armpits, groin, and anal area of humans and causes body odor.

The Skeletal System

Muscles need a framework around which they can direct movement. In most animals, with the notable exception of those in the Porifera, which have no muscle tissue, muscles are attached to a skeleton that provides the support, structure, and tension necessary for muscles to generate motion. Animal skeletons come in three basic types:

1. **Hydraulic skeletons** are found in soft-bodied invertebrates and use hydrostatic pressure (pressure generated by
   fluids) for movement. Hydraulic skeletons alter the pressure
   of fluid-filled cavities in the body to maintain the organism's shape or to enable movement.

   EXAMPLE: In annelids, such as earthworms, muscle fibers
   surround each fluid-filled cavity along the segmented body.
   Contraction of the fibers causes the cavity to constrict and
   the fluid pressure to increase. Contractions occur in a series
   along the worm's entire body, resulting in the wavelike
   action that promotes movement. In addition to annelids,
   cnidarians (jellyfish), flatworms, and nematodes also have
   hydraulic skeletons.

2. **Exoskeletons** are hard external shells with internal muscle
   attachments. An animal will continually shed and replace
   its exoskeleton as it grows. Arthropods, such as spiders,
   insects, and crustaceans, all develop exoskeletons.

3. **Endoskeletons** are rigid skeletons attached internally to
   muscle tissue. Some invertebrates have endoskeletons made
   of a form of calcium carbonate. This type of skeleton is
   found in echinoderms such as starfish, sand dollars, and
   sea urchins. Vertebrate endoskeletons may consist of strong,
   flexible cartilage, as is found in sharks, or a combination of
   cartilage and bone, which is stronger and less flexible.

VERTEBRATE ENDOSKELETONS

Four forms of connective tissue are found in vertebrate endoskeletons:

1. **Bone:** The main structural tissue found in most vertebrates. In addition to forming the skeletal framework,
   bones are the site of blood cell production.

2. **Cartilage:** Padding for joints in most vertebrates, and the
   main structural tissue for some aquatic organisms that do
   not possess bones, such as sharks.

CHAPTER 15 The Integumentary, Skeletal, and Muscular Systems

3. **Tendons:** Tissue that joins bone to the muscle.

4. **Ligaments:** Tissue that connects bones to one another.

Bone

Bone is a living organ: It grows, is vascularized (consisting of vessels through which fluid can flow), can store materials such as minerals, and includes nerve tissue. Bone is made up of mineralized osseous tissue that forms a honeycomb-like internal structure, making bones rigid. In addition, tissues such as **marrow, nerves, blood vessels,** and **cartilage** are present in bones.

Three different cell types make up the osseous tissue in bones:

1. **Osteoblasts** secrete the calcium and collagen used to create new bone. Osteoblast secretions collect in greater density than the cells within the bone.

2. The new cells that mature from osteoblast secretions are known as **osteocytes.** Osteocyte cells aid in bone formation, maintenance of bone structure, and control of calcium levels in the bone.

3. Bone changes continually throughout an organism's lifetime; **osteoclast cells** are largely responsible for regulating these changes. Osteoclasts facilitate bone resorption, the remodeling process of new bone to create the porous cavity through which vessels can flow. Resorption reduces the volume of bone, resulting in the release of calcium into the body's circulation. Osteoclasts also help control calcium levels in the bone itself.

Bone forms in concentric layers surrounding Haversian canals, which house the blood vessels and nerve fibers that keep the bone alive. Bone develops in either a web of connective tissue, as in the flat bones of the skull, or from cartilage molds, as in the bones of the limbs, which are replaced by more rigid bone as the organism grows.

Bone Structure

Bone consists of both a hard and a soft component. Solid bone is made up of **osteons,** tubes composed of cylindrical layers of bone. In the center of these tubes are blood vessels and nerve fibers. The soft interior portion of the bone is composed of the highly porous **spongy bone.** Within the pores of the spongy bone is a network of two types of marrow tissue: **red marrow,** the site of blood cell production, and **yellow marrow,** comprising fat cells that store energy.

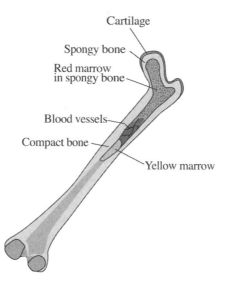

Cartilage

Spongy bone

Red marrow
in spongy bone

Blood vessels

Compact bone

Yellow marrow

Bone Structure

Skeleton Structure

Vertebrate endoskeletons, such as those of humans, are divided into two distinct sections:

1. The **axial skeleton** comprises the head and torso region and includes the skull, spinal column, and rib cage.

2. The **appendicular skeleton** contains the remaining bones, including the appendages, pelvic girdle, and shoulder blades.

Points of flexibility where bones meet are called **joints.** Joints maximize the necessary movement of the connected bones and minimize the shock and impact that movement creates. Cartilage, tendons, and ligaments all play a vital role in reducing the impact and tension produced by the movement of bones.

Joint type determines the potential range of movement connected bones can achieve. Some joints, such as the pelvic joint connecting the hipbone to the femur (upper thigh), have a wide range of motion necessary for much of an animal's daily movement and endure incredible stress as a result. These highly moveable joints are encased in capsules filled with lubricating fluid that ease friction. On the other hand, the joints that connect the various bone plates in the skull are bridged by cartilage, allowing a minimal amount of movement.

The Muscular System

The muscular system is attached to the skeletal system and provides motion through a series of contractions that pull the bones. In addition to these locomotive functions, the muscular system also enables important movement of organs within the organism, such as the pumping of the heart that sends blood through the circulatory system and the movement of food through the digestive system.

Three muscle types make up the tissues in the muscular system. Each muscle type performs a specific method of movement:

1. **Skeletal muscle,** which accounts for the majority of the muscle tissue and much of the weight of an organism, is attached to the bones by tendons. Contractions pull on the skeleton to move the organism through its environment.

2. **Cardiac muscle** pumps the heart and circulates blood through the animal.

3. **Smooth muscle** performs many functions, including the rhythmic pumping that moves food through the digestive system and causes the contractions of the eye's iris and blood vessels throughout the body.

STRUCTURE OF MUSCLE TISSUE

Microfilaments are found in vertebrate cells throughout the body. They are composed of two protein filaments, **actin** and **myosin,** which cause the fibers to shorten as they slide together. The action of these two proteins produces movement in non-muscle cells. In muscle cells, microfilaments appear in proportionally high numbers and take up most of the cell's space. The large number of microfilament fibers gathered together in the many cells of muscle tissue can produce the large contractible force, which allows movement in the body.

In the muscle cells of skeletal and cardiac muscle tissue, microfilaments are densely packed into fibers called **myofibrils,** giving them their characteristic banded (striated) appearance. In contrast, smooth muscle cells are loosely packed with microfilaments and do not have a striated appearance.

Muscle Contraction

All muscle contraction, whether voluntary (consciously initiated by the organism) or involuntary (the muscles receive stimulation from a nerve or hormonal stimulus) is initiated by nerve cells. The nerve cell secretes a neurotransmitter chemical across the gap between nerve and muscle cell, called the neuromuscular junction. The neurotransmitter chemical binds to the muscle tissue, causing it to release calcium ions. Calcium ions in turn allow the head of the myosin filament to bind to the actin filament, producing the contraction.

Each muscle type is capable of a different method of contraction to facilitate its function in the body:

- **Smooth muscles** perform involuntary movements, such as the constriction and expansion of blood vessels and the contractions of the stomach and intestines. Nerve impulses and hormonal stimulus stimulate the activity of smooth muscle cells. In some cases, as in the wall of the gut, individual cells spontaneously contract, causing the entire muscle to contract.

- **Cardiac muscles** carry out the involuntary pumping action of the heart. Cardiac muscle cells are arranged next to one another and linked by gap junctions that permit the passage of substances between the cells. Specialized cardiac muscle cells initiate electric pulses that travel from cell to cell through the gap junctions, causing all cardiac muscle cells to contract as a unit.

- **Skeletal muscles** perform the voluntary contractions necessary for the movement of the organism through its environment. Muscles are joined to bones by connective tissue known as tendons. Motion, such as the straightening of an arm, is achieved as flexor muscles on one side of the bone relax and extensor muscles on the other contract. Contractions shorten the muscles, causing the attached bone to move around the origin, the stable point where one bone attaches to another (for example, the elbow). The attached bone that actually moves as a result of this contraction is called the insertion (for example, the radius and ulna).

The contractions of skeletal muscles are only capable of pulling a bone; they cannot push. Smooth and cardiac muscles do not require a point of pulling because their pulse action enables the contraction that moves objects, such as blood and food, through the body.

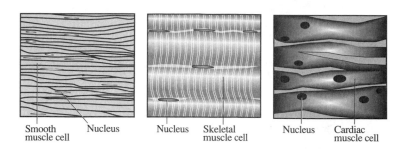

Smooth muscle cell   Nucleus   Nucleus   Skeletal muscle cell   Nucleus   Cardiac muscle cell

Types of Muscle Tissue

Summary

The Integumentary System

- The **integumentary system** separates the internal environment of the organism from the outside. It is composed of **skin, hair, scales, feathers, nails,** and **glands.**

- The integumentary system protects the internal organs, aids in thermoregulation, produces vitamin D, and communicates sensations from external stimuli through nerve endings.

- Skin is composed of a thin **epidermis** layer over a thick, vascularized **dermis** layer. A third layer, the **hypodermis,** connects the skin to the internal tissue and stores fats.

- Hair, scales, feathers, and nails are all made of protein. Hair is common in mammals, scales in reptiles and fish, feathers in birds, and nails in a wide variety of animals.

- **Exocrine glands** in the skin secrete substances through tubelike ducts. These glands come in two types: sebaceous and sweat glands.

- **Sebaceous glands** secrete necessary oil into hair and onto skin.

- **Sweat glands** secrete sweat for thermoregulation. Some specialized sweat glands secrete a milky substance responsible for body odor.

CHAPTER 15
The Integumentary, Skeletal, and Muscular Systems

The Skeletal System

- Skeletons form the framework necessary for muscles to act on and move an animal.

- There are three types of skeleton: **hydraulic skeleton, exoskeleton,** and **endoskeleton.**

- All vertebrates have endoskeletons, which are composed of the connective tissues of **bone, cartilage, tendons,** and **ligaments.**

- Bone is a vascularized, innervated organ that is made up of relatively few cells and many tissues.

- There are three types of bone cells: **osteoblasts,** which secrete calcium and collagen to create new bone; **osteocytes,** which form as the new bone cells mature; and **osteoclasts,** which release calcium from bones to create the space necessary for blood vessels and nerve fiber.

- Bones form either from a web of connective tissue or from a cartilage model.

- Bones are made up of both a hard component of **osteons** and a soft component of spongy bone.

- **Spongy bone** is porous and contains networks of red and yellow marrow. **Red marrow** produces blood cells; **yellow marrow** stores fat cells.

- The vertebrate skeleton is divided into two parts: the **axial skeleton** and the **appendicular skeleton.**

- Two bones meet and move around a **joint.** Cartilage, ligaments, and tendons stabilize bones and ease movement.

The Muscular System

- The muscular system is made up of three different types of muscle tissue: **skeletal, smooth,** and **cardiac muscles.**

- Muscle cells consist of a large number of **microfilaments.** Microfilaments are composed of **actin** and **myosin** protein filaments. Contractions of muscle cells result from the sliding together of actin and myosin.

- Skeletal and cardiac muscle cells contain densely packed microfilaments, giving them a striated appearance. Smooth muscle cells are loosely packed and are not striated.

- Skeletal muscle is attached to bone. Skeletal muscles contract voluntarily, pulling the attached bone to create movement.

- Smooth muscles perform involuntary contractions primarily in response to nerve or hormonal stimuli.

- Involuntary cardiac muscle contractions occur when specialized muscle cells send electric pulses through gap junctions to the rest of the cardiac muscles.

Sample Test Questions

1. Compare and contrast the basic form and function of the three types of muscle tissue.

2. Compare and contrast the basic functions of the three types of bone cells.

3. What important functions does skin perform in the overall maintenance of the body?

4. The contraction of a microfilament is caused by the sliding together of actin and which other protein?

   A. Hemoglobin
   B. Keratin
   C. Myosin
   D. Collagen
   E. Calcium

CHAPTER 15
The Integumentary, Skeletal, and Muscular Systems

5. Which muscle tissue type is under voluntary control?

A. Skeletal
B. Tubular
C. Smooth
D. Cardiac
E. Striated

6. Bone is primarily made up of which types of cells?

A. Osteocytes
B. Osteoclasts
C. Osteoblasts
D. A calcium/collagen material
E. Epidermal

7. The ribcage is part of which section of the human skeleton?

A. The sternum
B. The appendicular skeleton
C. The axial skeleton
D. The breast skeleton
E. The hydraulic skeleton

8. Blood cells are produced in which part of the skeletal system?

A. Cartilage
B. Marrow
C. Tendons
D. Solid bone
E. Osteons

9. Which kinds of glands are responsible for body odor?

A. Sebaceous
B. Sweat
C. Hair
D. Merocrine
E. Endocrine

10. Sebaceous glands are part of which of the following organ systems?

    A. Muscular
    B. Integumentary
    C. Skeletal
    D. Digestive
    E. Reproductive

11. Skin's structure does NOT include which of the following?

    A. Hypodermis
    B. Dermis
    C. Epidermis
    D. Blood vessels
    E. Smooth muscle

12. Which of the following characteristics distinguishes epidermis tissue from dermis tissue?

    A. The presence of cells.
    B. Epidermis tissue is part of skin; dermis tissue is not.
    C. Epidermis tissue is constantly regenerating itself.
    D. Dermis tissue dies very easily.
    E. Dermis tissue is part of the muscle system.

13. What tissue connects skeletal muscles to the bone?

    A. Tendons
    B. Myosin
    C. Actin
    D. Joints
    E. Ligaments

14. Which of the following muscles are striated?

    A. Smooth, skeletal, and cardiac
    B. Skeletal and cardiac
    C. Smooth and cardiac
    D. Only skeletal
    E. None of the above

CHAPTER 15
The Integumentary, Skeletal, and Muscular Systems

15. Which cells turn into osteocytes once they are coated in bone secretion?

    A. Osteoclasts
    B. Muscle cells
    C. Myosin filaments
    D. Osteoblasts
    E. Stem cells

ANSWERS

1. The three different muscle tissue types are skeletal, cardiac, and smooth. Both skeletal and cardiac muscles are composed of long, fibrous cells, giving them a striated appearance. Smooth muscle is composed of sheets of muscle cells and has no striations. Skeletal muscle is the only muscle type under voluntary control; cardiac and smooth muscle react in response to involuntary nerve and hormone stimuli. Skeletal muscle is used to pull bone and move the animal through its environment. Cardiac muscle generates the pumping action of the heart to move blood throughout the body. Smooth muscle is responsible for all other contractions in the body, such as the movement of the eyes and digestive system.

2. The osteoblasts secrete a collagen/calcium material that makes up the majority of the bone structure. Osteocytes help to maintain a bone's structure, assist with bone formation, and aid in controlling the calcium levels of the bone. Osteoclasts are responsible for resorption of calcium, the process by which calcium is released from the bone into the body system.

3. Skin is first and foremost a barrier between the internal organs of the organism and the outside environment. Skin also produces vitamin D when compounds in the skin react with sunlight, secretes wastes from the body, provides the first point of contact with tactile stimuli, stores a large proportion of the body's fat, and assists with thermoregulation through the constriction and dilation of blood vessels within the dermis.

4. **C**   Myosin fibers attach to actin fibers, causing them to slide together and shorten the fiber length. This movement leads to contraction of the muscle cell.

5. **A**   Skeletal muscle is under voluntary, conscious control.

CHAPTER 15
The Integumentary, Skeletal, and Muscular Systems

6. D Bone is primarily made up of a calcium/collagen material secreted by the far less abundant bone cells.

7. C The ribcage is part of the axial portion of the skeleton.

8. B Blood cells are produced in bone's red marrow.

9. B A special type of sweat gland, the apocrine gland, secretes a substance responsible for body odor.

10. B Sebaceous glands, as well as sweat glands, are part of the integumentary system.

11. A The hypodermis is not part of the skin; it is the connective tissue that attaches skin to internal tissue.

12. C Epidermis tissue is constantly regenerating itself via mitosis as dead epidermis cells are brushed off of the surface.

13. A Tendons attach skeletal muscles to bones.

14. B Both skeletal muscle and cardiac muscle have cells densely packed with microfilaments, giving them a banded, or striated, appearance.

15. D Osteoblasts develop into osteocytes once they are coated in bone secretion.

GAS EXCHANGE AND CIRCULATION

16

Respiration: Gas Exchange

All animal cells undergo cellular respiration, which is the breakdown of energy into a usable form to provide power to perform cellular functions. Oxygen is an essential component of cellular respiration, resulting in carbon dioxide as a waste product. Animals, therefore, use an organ system in which the exchange of oxygen and carbon dioxide can take place with the external environment, as well as a system to transport these gases throughout the body. The respiratory and circulatory systems perform these two intertwined tasks. The respiratory system exchanges gases with the environment, while the circulatory system transfers those gases throughout the body.

In the simplest animals, gases diffuse directly across the cell membrane of primitive single-celled organisms. While this method is sufficient for simple organisms, as the size of animals increases, a more efficient method is necessary to get gases to flow throughout the body. Specialized structures consisting of a **respiratory surface** through which gases can diffuse are required. More complex invertebrates such as the mollusks, arthropods, and echinoderms, as well as the vertebrates, employ four types of specialized structures for respiration: tracheae, skin, gills, and lungs.

FROM WATER TO LAND

All gas exchange requires a moist medium for the gases to move from surface to surface. This requirement poses no problem for animals living only in aquatic environments, as gills and skin absorb gases directly from the surrounding water. However, as animals evolved to live on land, they developed tracheae and lungs that contained moist surfaces for this purpose. The following table describes each respiratory structure in more detail.

Respiratory Surface	Description	Example
Skin	In a process known as cutaneous respiration, the skin, which is moist and well-vascularized, exchanges gas directly with the environment through its blood vessels.	Amphibians, such as frogs and salamanders, as well as some marine snakes, respire through their skin. This usually takes place in conjunction with respiration through lungs.
Gills	Gills are extensions of internal tissues that project into the water, where gases can diffuse. Gills are heavily branched, providing a large amount of surface area for gas to exchange, making them more efficient than skin. Gills can be external (not enclosed in the body), which requires constant movement to ensure contact with fresh water, or internal, enclosed within branchial chambers where water is pulled in and out, creating currents over the gill tissue.	External gills are found on the larvae of many fish and amphibians, as well as adult amphibians such as the axolotl. Internal gills are found in arthropods such as crustaceans and spiders, mollusks, echinoderms many adult amphibians, and fish.
Tracheae	Tracheae are air-filled tubular passages that form an extensive network in the body. These passages connect the body surface with all internal structures. Oxygen is passed straight from the tracheae into cells. Insects rely solely on this respiratory system, rather than a circulatory system, to transfer oxygen throughout their bodies. Insects have external structures on their exoskeleton called spiracles that open and close to prevent water loss during respiration.	Tracheae are the respiratory system of most terrestrial arthropods, including all insects, some spiders, mites and ticks, millipedes, and centipedes.
Lungs	Air moves into and out of the body through branched tubular passageways, which moisten the air in the process. Eventually moist air reaches a thin, wet membrane that permits exchange of gases within the lungs, where it can be circulated throughout the body.	Lungs are found in some arthropods, as well as all amphibians, reptiles, birds, and mammals.

The Human Respiratory System

Lungs are the respiratory structure found in many terrestrial animals. Although lungs serve the same basic purpose as other respiratory systems, they are structurally and functionally different.

Human bodies contain two lungs, both located in the chest cavity, one on each side of the body. They are surrounded by a ribcage and are situated above a large muscle, the **diaphragm.** Throughout the human lung are numerous tiny grape-shaped sacs called alveoli, which give the lung its spongy texture. Alveoli provide a large surface area over which gases can diffuse and are connected to a series of passages through which air passes during breathing.

BREATHING

Air moves into and out of the lungs via the process of **breathing.** The chest cavity expands as the diaphragm contracts (moves downward), air is drawn in, or **inhaled,** into the lungs. When the diaphragm relaxes, the chest cavity decreases in volume, causing air to be **exhaled,** or released from the body.

During the process of breathing, air moves through the structures of the respiratory system in the following order:

1. Air is taken in through the mouth and nasal passages.

2. Air passes through the **pharynx.**

3. Air moves into the **larynx,** or voice box, where it passes through the glottis, an opening in the vocal cords.

4. Air travels through the windpipe, or **trachea.** This structure is supported by rings of cartilage, which prevent collapse.

5. The trachea forks into two smaller tubes called **bronchi,** each of which leads to one of the two lungs.

6. The bronchi branch into increasingly smaller tubes called bronchioles within the lungs.

7. The bronchioles deliver the air into the alveoli.

Alveoli

Gas exchange takes place over the surface of the alveoli. The alveoli are covered in dense capillary beds in which blood is constantly circulated. Oxygen diffuses from the air in the alveoli into these capillaries, where it then enters the blood system. As oxygen enters, carbon dioxide diffuses out of the blood system and into the air in the alveoli. Oxygen is then transported throughout the body by the **circulatory system,** while the carbon dioxide waste is exhaled back to the external environment through the same route that the oxygen entered.

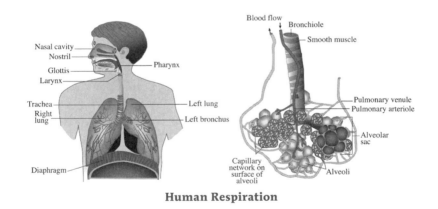

Human Respiration

Circulation

Respiration replenishes oxygen stores and removes carbon dioxide from the body. The circulatory system connects with the respiratory system and transports oxygen and carbon dioxide between the respiratory system and cells of the body.

Circulation is the movement of blood throughout the body. Blood carries many substances vital for cellular functioning, including nutrients, hormones, and gases. The circulation of blood also allows the removal of cellular waste products, which are exchanged with the blood during diffusion.

The **circulatory system** consists of three components:

1. A pumping organ, the heart

2. Blood, which carries substances as it travels through the circulatory system

3. Vessels through which blood can travel

Just as respiratory systems vary in form and function from animal to animal, so does the circulatory system. In fact, the most simple, primitive animals, such as flatworms and some cnidarians, do not have a circulatory system at all. Instead, they exchange gases with the external environment via a **gastrovascular cavity.** All other animals, however, have some form of circulatory system for the exchange of gases. All circulatory systems can be divided into two major categories: open or closed.

OPEN AND CLOSED CIRCULATORY SYSTEMS

Blood is pumped through the body in two ways.

- **Open circulatory system:** No distinction exists between blood and extracellular fluid (the fluid surrounding body tissues). Instead, all fluids are mixed in a common fluid called **hemolymph.** Hemolymph moves through the body through open-ended channels connected to a **heart,** which pumps it through the body. These channels carry blood from the heart, where it empties into the body cavities. Hemolymph is then drained from these cavities back to the heart, where the cycle repeats itself.

- **Closed circulatory system** (also called the cardiovascular system): Blood is kept separate from extracellular fluid in closed tubes called **vessels.** Vessels form a network for blood, pumped from the heart, to travel throughout the body. While hemolymph in the open circulatory system drains back to the heart from cavities around the body, in the closed system, these vessels also transport the blood back to the heart. Those vessels that take blood away from the heart and to the cells are called **arteries**; those that return blood to the heart are **veins**. **Capillaries** are small vessels that exchange substances with tissues and move blood between veins and arteries.

EXAMPLE: Small animals, such as mollusks and arthropods, tend to have open circulatory systems, while larger animals, including all vertebrates, tend to have closed circulatory systems. In a smaller organism, where tissues tend to be gathered in the same body area, the body's general movement is adequate to circulate body fluids. In larger, more complex organisms, tissues may be many layers thick, with cells that are far from the surface or located far from the digestive system where nutrients are processed for use by cells. In these animals, a closed system of vessels is necessary to transport nutrients to all cells of the body.

COMPONENTS OF THE CARDIOVASCULAR SYSTEM

The cardiovascular system is the circulatory system used by annelids and all vertebrates. Significant differences distinguish the systems found among the different groups of animals, but they all contain similar components: the heart, the blood, and the vessels.

- **Heart:** A muscular organ that pumps blood throughout vessels in the body. The heart consists of a series of chambers through which blood is passed.

- **Blood:** A fluid composed of plasma, platelets, and cells. **Plasma** is made up of solutes, such as metabolites; wastes; hormones; ions; and proteins, which are exchanged via diffusion with tissues. **Platelets** allow the blood to clot, which reduces blood leakage when vessels break or tear. Two types

of cells are present in blood: **white** and **red blood cells.**
White blood cells, also known as **leukocytes,** defend the
body from foreign agents, while red blood cells transport
oxygen and carbon dioxide. Red blood cells, also known as
erythrocytes, contain a pigment called hemoglobin that
enables them to bind with oxygen and transport it to other
cells.

• **Vessels:** The network of tubes, such as arteries, arterioles,
  veins, venules, and capillaries, through which blood travels.

Blood

Blood is composed of cells suspended in an aqueous solution
called plasma. Blood cells make up about 45 percent of the
total blood volume, with plasma accounting for the remaining
55 percent.

Plasma is 90 percent water. The other 10 percent is composed
of the following three components:

1. A mixture of salts that help maintain osmotic balance,
   which allows the movement of fluids into the capillaries,
   and proper pH levels

2. Plasma proteins, which also maintain osmotic balance and
   pH, but play an additional role in blood clotting and the
   immune system

3. Substances transported by the blood, such as CO_2, O_2,
   nutrients, and waste products

Blood Cells

The three types of blood cells include red blood cells, white
blood cells, and platelets.

Gas Exchange and Circulation   CHAPTER 16

Type of Blood Cell	Characteristics	Function
Red blood cells (erythrocytes)	Small concave disks that lack a nucleus and mitochondria	Transport oxygen and carbon dioxide
White blood cells (leukocytes), including lymphocytes, monocytes, basophils, eosinophils, and neutrophils	Have nuclei and are different shapes depending on type; usually found in the interstitial fluid (fluid surrounding cells and tissues) and in the lymph system	Aid in immunity and fight infections
Platelets	When blood vessels are damaged, platelets trigger the soluble form of a protein called fibrinogen to be converted to the insoluble form of fibrinogen, which plugs the break	Aid in blood clotting

Blood Vessels

With the exception of capillaries, all blood vessels consist of an endothelium layer covered by a thin layer of elastic fibers, which is itself covered in smooth muscle and connective tissue. This multilayer structure prevents blood from diffusing directly to the surrounding tissues. Capillaries, however, are thin and consist only of an endothelium that allows diffusion to occur through the capillary walls.

Blood is transported from the heart through large vessels called arteries. Arteries connect to smaller arterioles that carry blood to smaller parts of the body, where they eventually enter the even smaller capillaries. Beds of capillary cover body tissues and allow an efficient exchange of substances such as nutrients, oxygen, and waste products. Once oxygen leaves the blood and the waste product carbon dioxide enters, the blood begins its journey back to the heart, traveling from capillaries into larger vessels called venules. Venules lead to veins, which return the deoxygenated blood to the heart.

CARDIOVASCULAR SYSTEMS IN VERTEBRATES

Cardiovascular systems of the vertebrates differ primarily in the structure and functioning of the heart. Main differences are outlined in the table below.

Cardiovascular System	Description
Fish heart	Fish hearts are composed of a four-chambered tube, with each chamber next to the other. Two chambers, called the sinus venosus and the atrium, collect the blood; two chambers, called ventricle and conus arteriosus, pump the blood. Blood leaves the heart and passes through the gills, where it collects oxygen. Blood then travels to the body in a continuous, one-way circle.
Amphibian and reptile heart	Amphibians were the first to develop two separate blood flow systems with a three-chambered heart. The pulmonary system transports blood between the lungs and heart, and the **systemic circulation** transports blood between the heart and the body. To suit this purpose, the structure of the amphibian heart is different compared to the fish. In amphibians, the atrium is split in two—one side receives blood from the lungs, and the other receives blood from the heart. Deoxygenated blood travels from one atrium to the lungs, while oxygenated blood is transported from the other atrium through the body. Channels running through the single-chambered ventricle help minimize the mixing of blood from each atrium. The slightly more advanced reptilian hearts include subdivisions within the ventricle. The crocodile is the only reptile to exhibit complete division of the ventricle, creating a four-chambered heart.
Mammalian and bird hearts	Mammals and birds have a four-chambered heart consisting of two atria and two ventricles. The right atrium receives deoxygenated blood from the body and sends it to the right ventricle, where it is pumped into the lungs. Oxygenated blood from the lungs is received by the left atrium, pumped into the left ventricle, and then pumped out to the body. The heart pumps twice, once to pump blood from the atria into the ventricles, and a second time to pump blood out of the ventricles.

The Human Cardiovascular System

Like all mammals, humans have a four-chambered heart through which two separate blood systems circulate. The **pulmonary circuit** transports blood from the heart to the lungs, while the **systemic circuit** transports blood from the heart to the rest of the body. The two-circuit system prevents deoxygenated blood from diluting the oxygenated bloodstream, enabling a maximum rate of oxygen absorption by the tissues. Deoxygenated blood going to the lungs is not diluted with oxygenated blood, ensuring that the uptake of oxygen from alveoli in the respiratory system will be more efficient.

THE CARDIAC CYCLE

The cardiovascular system operates according to the cardiac cycle. Each beat of the heart represents a complete **cardiac cycle,** during which the heart fills with blood and then pumps it out. Two pairs of valves in the heart prevent the backflow of blood through this system.

- **Atrioventricular (AV) valves** guard the passageway between the atria and ventricles.

- **Semilunar valves** guard the passageway between the ventricles and the arterial system.

These valves open and close during each cardiac cycle, ensuring that blood flows in only one direction through each passageway. The cardiac cycle proceeds as follows:

1. Blood returning to the heart enters each atrium. The AV valves open, allowing blood to flow into the corresponding ventricle. The heart at this stage is at rest **(diastolic).**

2. The ventricles contract, forcing the AV valves to shut, preventing the backflow of blood to the atria. At the same time, the semilunar valves are forced open and blood is released into the arterial system. The heart at this stage is contracting **(systolic).**

3. Relaxation of the ventricles shuts the semilunar valves, preventing backflow of blood to the ventricles.

4. Blood from the right ventricle travels along the **pulmonary arteries** to the lungs. Blood returns through the pulmonary veins into the left atrium, and then into the left ventricle.

5. Blood from the left ventricle leaves the heart via a systemic artery called the **aorta.** Blood from the aorta then flows to the rest of the body.

6. Blood returning from the body through systemic veins drains into one of two main veins. The **superior vena cava** collects blood from the upper body, while the **inferior vena cava** collects blood from the lower body. Blood from these veins enters into the right atrium, completing the cardiac cycle.

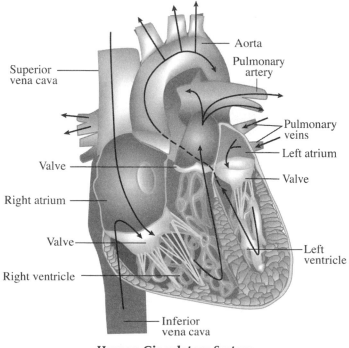

Human Circulatory System

Control of the Heartbeat

The pace of the heartbeat is set by several specialized regions of the heart muscle where groups of cells stimulate the muscle contractions. Contractions in one such region, the **sinoatrial (SA) node,** located in the right atrium wall, set the rate at which all other heart cells contract. The SA node acts like a **pacemaker,** generating electrical signals that travel through the heart muscle wall to the **atrioventricular (AV) node.** A delay in the signal at the AV node allows the atria to empty completely before the ventricles contract. After the delay, the signal travels through the walls of the ventricles, causing them to contract and force blood from the heart.

Summary

Respiration: Gas Exchange

- All animals have a **respiratory surface,** an interface where oxygen (O_2) diffuses in and carbon dioxide (CO_2) diffuses out. The four types of respiratory surfaces seen in animals are **tracheae, skin, gills,** and **lungs.**

- A moist medium is required for the exchange of the respiratory gases oxygen and carbon dioxide.

The Human Respiratory System

- Humans respire using lungs, paired organs surrounded by a ribcage and a **diaphragm.** Lungs are composed of numerous alveoli that connect to tubes through which air is passed.

- **Breathing** is the process of inhaling and exhaling air. During inhalation, the ribcage and the chest cavity expand and the diaphragm contracts. During exhalation, the ribs and diaphragm relax and the chest cavity decreases in volume.

- Air moves through the structures of the respiratory system in the following order: Air is taken in by the mouth or nasal cavity and passes into the **pharynx;** it then travels into the **larynx,** where it crosses the vocal cords and travels through the glottis. Next, air travels through the windpipe, or trachea, which forks into two smaller tubes called **bronchi,** each of which leads to a lung. The bronchi branch into bronchioles, which terminate in alveoli.

- Gas exchange occurs across the membrane surrounding the alveoli. Oxygen in the inhaled air diffuses into capillaries that surround each alveolus. Carbon dioxide travels in the opposite direction out of the body.

Circulation

- Circulation is the movement of blood throughout the body via the **circulatory system,** which is made up of the **heart, blood,** and **vessels.**

- Simple, primitive animals, such as flatworms and some cnidarians, have no circulatory system, exchanging gases instead via a **gastrovascular cavity.**

- Cardiovascular systems can be open or closed.

- In an **open circulatory system,** the heart pumps **hemolymph** out into open-ended vessels that terminate in the body tissue.

- A **closed circulatory system,** or cardiovascular system, consists of a heart and a network of vessels (veins, arteries, and capillaries) through which blood travels.

- The components of the cardiovascular system are the heart, the blood, and the blood vessels.

- Blood is 45 percent blood cells, including **red blood cells (erythrocytes,** responsible for carrying O_2, attached to hemoglobin); white blood cells **(lymphocytes,** involved in immunity); and **platelets,** which are involved in clotting. **Plasma,** which accounts for the other 55 percent of blood volume, is 90 percent water and 10 percent a mixture of salts and proteins. Plasma also contains O_2, CO_2, nutrients, and waste products.

- Cardiovascular systems differ among the vertebrates: Fish have four chambers through which blood flows to the body in a circular cycle; amphibians and reptiles have two separate blood flow systems that have developed in conjunction with a semi-divided heart; and mammals and birds have two blood flow systems and hearts fully divided into four chambers.

CHAPTER 16
Gas Exchange and Circulation

The Human Cardiovascular System

- The heart pumps the blood through the two blood circuits in humans, the **pulmonary circuit** and the **systemic circuit.** The pulmonary circuit moves blood from the heart to the lungs, and the systemic circuit moves blood from the heart to the rest of the body.

- The cardiovascular system operates through the process of the **cardiac cycle.** During a cardiac cycle, the heart beats, filling with blood and pumping it out to the body.

- Two sets of valves in the heart, the **atrioventricular (AV) valves** and the **semilunar valves,** prevent backflow of blood through the heart.

- When the heart is at rest, it is referred to as **diastole.** When it is contracting, it is referred to as **systole.**

- The **sinoatrial (SA) node** sets the rate at which the heart contracts. It generates electrical signals that travel through the heart muscle to the wall of the **atrioventricular (AV) node.** After a delay, the signal causes the ventricles to contact, forcing blood from the heart.

Sample Test Questions

1. Describe the pathway of blood throughout the body during the cardiac cycle. Name the major arteries, veins, and organs used during this process.

2. Describe the passage of a CO_2 molecule as it travels from the cell to the outside of the body. Name all structures and methods for its transport.

3. Compare an open and closed circulatory system, citing the main differences between them.

4. The diaphragm contracts during which process?

    A. Exhalation
    B. Inhalation
    C. Systole
    D. Diastole
    E. Respiration

5. Which of the following structures is known as the pacemaker of the heart?

    A. Right ventricle
    B. Right atrium
    C. SA node
    D. AV node
    E. Pulmonary artery

6. What action moves air into and out of the lungs of humans?

    A. Breathing
    B. Running
    C. Circulation
    D. Diastole
    E. Systole

7. Which of the following structures can most accurately be described as the respiratory surface of humans?

    A. Capillaries
    B. Bronchus
    C. Bronchioles
    D. Nasal cavity
    E. Alveoli

8. Platelets play a role in which of the following processes?

    A. Osmotic balance
    B. pH
    C. Blood clotting
    D. Blood lipid levels
    E. The immune system

CHAPTER 16
Gas Exchange and Circulation

296 • SparkNotes 101: Biology

9. Blood from the legs returns through the _____ and enters the
_____ of the heart.

   A. superior vena cava; right atrium
   B. inferior vena cava; right atrium
   C. superior vena cava; left atrium
   D. inferior vena cava; left atrium
   E. superior vena cava; right ventricle

10. What is the name of the fluid mixture of blood and extracellular fluid in
the bodies of animals with open circulatory systems?

   A. Hemoglobin
   B. Melanin
   C. Lymphocyte
   D. Hemolymph
   E. Interstitial fluid

11. Which type of blood cell transports oxygen around the body?

   A. Esosinophil
   B. Basophil
   C. Erythrocyte
   D. Leukocyte
   E. Monocyte

12. The electrical signal of the heart is momentarily delayed at which of the
following sites in the heart?

   A. The SA node
   B. The left ventricle
   C. The right ventricle
   D. The left atrium
   E. The AV node

13. What is the name of the circuit that moves blood from the heart to the
rest of the body?

   A. Cardiovascular circuit
   B. Pulmonary circuit
   C. Atrioventricular circuit
   D. Systemic circuit
   E. Aortic circuit

CHAPTER 16
Gas Exchange and Circulation

14. A blood clot in the inferior vena cava would prevent blood from reaching which section of the heart?

    A. The left atrium
    B. The right atrium
    C. The aorta
    D. The left ventricle
    E. The right ventricle

15. Which valve opens as blood is pumped into the aorta?

    A. Semilunar valves
    B. Sinoatrial valves
    C. Atrioventricular valves
    D. Cardiac valves
    E. Diastolic valves

ANSWERS

1. Blood from the lungs enters the left atrium; blood from the rest of the body enters the right atrium. The blood is then passed to the corresponding ventricle. When the ventricles contract, blood from the right ventricle is passed into the pulmonary artery and is taken to the lungs. The blood passes into capillary beds within the lungs, where it exchanges gases with alveoli. The blood then travels through veins back to the heart and drains into the left atrium from the pulmonary veins. Blood from the left ventricle is passed into the aorta and travels via arteries and arterioles to capillary beds, where it exchanges substances with tissue cells. After this exchange, blood from the capillaries enters venules and then the veins. Blood traveling through veins in the upper body enters the superior vena cava, while blood traveling through veins in the lower body enters the inferior vena cava. Blood drains from both of these large veins into the right atrium.

2. Carbon dioxide produced by the cell diffuses into blood in the capillaries. Once in the capillary, the CO_2 moves toward the heart through veins, passing through the inferior vena cava and entering the heart first at the right atrium and then through the right ventricle. The CO_2 is pumped out of the heart through the pulmonary artery and into a capillary bed in the lungs. At this point, the CO_2 diffuses from the capillary bed into an alveolus. It moves through the alveolus and into the bronchiole

CHAPTER 16
Gas Exchange and Circulation

system. It then travels through a bronchus, up the trachea, through the pharynx, and out the nasal passages or mouth.

3. An open circulatory system contains a fluid called hemolymph, which is made up of blood and extracellular fluid. Fluid in an open circulatory system travels via open-ended vessels from the heart to cavities through-out the body. Blood from these cavities then drains back into the heart. The blood and extracellular fluid remain separate in a closed circulatory system. Blood in a closed circulatory system travels within a network of closed vessels, which transport blood between the heart and tissues of the body.

4. **B**   During inhalation, the ribcage and the chest cavity expand, and the diaphragm contracts (moves downward) to create more room for the chest cavity. During exhalation, the diaphragm relaxes, causing it to move up and decrease the volume of the chest cavity.

5. **C**   The sinoatrial (SA) node is also known as the pacemaker of the heart. It sets the rate at which the heart contracts, generating electrical signals that travel through the heart muscle wall to the atrioventricular (AV) node.

6. **A**   During breathing, air is inhaled into the lungs via expansion of the chest cavity, and is exhaled from the lungs via compression of the chest cavity.

7. **E**   The alveoli are the sites of gas exchange and the primary respiratory surface of humans.

8. **C**   Platelets help to form blood clots and are important during the repair of broken vessels, as they help stop blood loss.

9. **B**   Blood from the lower body is collected in the inferior vena cava and then emptied into the right atrium of the heart.

10. **D**   Hemolymph is a mixture of blood and extracellular fluid found in an open circulatory system.

11. **C**   Red blood cells, or erythrocytes, transport oxygen and carbon diox-ide in the blood. Red blood cells are small, disc-shaped cells formed in the bone marrow.

12. E   The electrical signal is delayed slightly after reaching the atrioventricular (AV) node, allowing the atria to empty completely before the ventricles contract. After the delay, the signal travels through the walls of the ventricles, causing them to contract and forcing blood out of the heart to the body.

13. D   Mammals have two blood circuits: a pulmonary circuit, which moves blood from the heart to the lungs, and a systemic circuit, which moves blood from the heart to the rest of the body.

14. B   Blood from the upper body is collected in the superior vena cava; blood from the lower body is collected in the inferior vena cava. Both of these large veins empty blood into the right atrium. A clot in the inferior vena cava would prevent blood from returning to the right atrium.

15. A   Blood is pumped into the aorta when the semilunar valves open. In systole, the atria contract so that the ventricles are totally filled with blood. The ventricles then contract, forcing the AV valves closed and the semilunar valves open. Blood is then pumped into the arteries.

NUTRITION AND DIGESTION

Nutrition

Digestion

Vertebrate Digestive Systems

Components of the Vertebrate
Digestive System

Nutrition

All organisms obtain energy from their environment to support
their life functions, such as growth, body maintenance, and
reproduction. Autotrophic organisms, such as plants, algae, and
some archaea, create their own energy via photosynthesis
and other pathways. Heterotrophic organisms, including all
animals, can only obtain the energy for these processes by

consuming other organisms and gaining their energy. **Nutrition** is the study of how the body takes in and uses food for the energy and nutrients necessary for proper functioning.

Basal metabolic rate refers to the minimum rate at which an animal uses energy. It is simply the minimum amount of energy required for an animal to remain alive when in a resting position. Whenever an animal performs a task such as reproduction, foraging, or walking, more energy is required and the metabolic rate rises. A steady intake of food is necessary to provide energy for these tasks.

In addition to energy, food also provides animals with **nutrients,** such as vitamins, amino acids, and minerals, that are necessary for survival but which the animal cannot produce on its own. Not all food contains all the necessary nutrients, and not all animals can break down and absorb nutrients from every type of food. Animals also differ in the types of nutrients they require from food. For example, no vertebrates can produce all required fatty acids, and therefore all vertebrate diets must provide essential fatty acids.

FOOD PROCESSING

Food processing, or the process by which animals convert food into a usable form, takes place in four main stages: ingestion, digestion, absorption, and elimination.

1. **Ingestion** is the consumption of food via the mouth or a mouthlike structure.

2. **Digestion** is the breaking down of food into particles small enough to be absorbed by the lining of the digestive tract. This breakdown is accomplished by methods of **mechanical digestion,** such as chewing or grinding, which divide food into smaller pieces without disrupting chemical bonds, or by methods of **chemical digestion,** such as the breakdown of food's chemical bonds by enzymes. Chemical digestion converts **food polymers** (molecules composed of similar molecular subunits) into smaller constituent parts, called **monomers.**

3. **Absorption** is the movement of small molecules across the mucous membranes that line the digestive tract. Nutrient molecules taken up by cells in the mucous membranes then diffuse into blood vessels in the digestive tissue and enter the bloodstream. Once in the bloodstream, these nutrients travel to cells throughout the body to be used as energy or stored for later use.

4. **Elimination** is the removal of waste products, which are any component of food not needed or usable by the body.

Digestion

Digestion is the process by which food is converted into substances that can be absorbed and assimilated by the body. Most food can be categorized as one of the four main macromolecules vital to bodily functions: nucleic acids, proteins, fats, and carbohydrates. Before they are capable of being absorbed, all of these macromolecules must be broken down into their monomer constituents, the smaller molecules of nucleotides, amino acids, fatty acids and glycerol, and simple sugars that make up the macromolecule. Important minerals, vitamins, and water are also extracted from food during the process of digestion.

The body uses these monomer constituents to rebuild the necessary polymers used in metabolic processes, which are the chemical processes, such as cellular respiration, that occur in a cell.

FOOD SOURCES

Biologists categorize animals into three groups based on their food source.

1. **Herbivores** obtain all food from plants. Examples include cows, horses, and nearly all rodents.

2. **Carnivores** obtain all food from meat. Examples include cats, eagles, wolves, and frogs.

3. **Omnivores** obtain food from both plants and meat. Examples include humans and bears.

An animal's digestive system is specifically suited to processing food obtained from its food source. For example, herbivores' digestive systems are equipped to break down plant material, and carnivores' are not.

DIVERSITY OF ANIMAL DIGESTIVE TRACTS

The complexity of the animal digestive system has evolved over time. Primitive sponges digest food using a method that is simpler than the complex digestive systems found in the vertebrates. The various forms of animal digestion can be divided into three categories:

- **Intracellular digestion,** used by sponges, involves the ingestion of food directly into the cells of the body, without the use of digestive organs or a digestive system.

- **Gastrovascular cavity,** used by the cnidarians, hydra, and some flatworms, is a single hole through which food enters and waste leaves. Enzymes secreted by the cavity lining break down food during digestion. The outside of this cavity is lined with cells that have direct access to the nutrients released.

- **Digestive tract,** used by all other animals, is a tube extending from the animal's mouth to its anus. Food passing through the digestive tract is broken down through mechanical and chemical processes, using enzymes, into simpler chemical compounds that can be readily used by the body.

The digestive tract is the most advanced form of digestion. However, this tract varies widely in complexity from species to species. The most primitive form found in nematodes is simply a tubular gut with no specialized features. The slightly more advanced digestive tract found in earthworms includes specialized areas for ingestion, storage, fragmentation, digestion, and absorption. The most advanced digestive tracts, found among the vertebrates, exhibit similar specialization to the earthworm tract but on a much greater scale.

Vertebrate Digestive Systems

The vertebrate digestive tract, known as the alimentary canal or gastrointestinal tract, is part of a digestive system that includes several accessory organs to aid in digestion. The alimentary canal is a tube, or gut, through which food travels in one direction from the mouth to the anus. The alimentary canal begins at the **mouth** and the **pharynx.** From here, food travels down a muscular tube called the **esophagus** and into the **stomach.** The stomach connects to the **small intestine,** which in turn connects to the **large intestine.** Some form of mechanical and chemical digestion breaks down food at every step along this tube, and nutrients are absorbed across the walls of both intestines. All products remaining after food travels through the large intestine are waste, which exit the body through the **cloaca** in most reptiles and birds and via the **anus** in mammals. The cloaca collects and expels waste from the digestive, urinary, and reproductive tracts. In contrast, the anus expels wastes exclusively from the digestive system.

In addition to the alimentary canal, vertebrate digestive systems include the following accessory organs:

- **The liver** secretes a substance called **bile** into the gallbladder, where it is stored for eventual use in digestion. Bile is a fluid mixture composed of bile pigment and bile salt. Bile pigment is a waste product resulting from destroyed red blood cells. Bile salt plays an important role in preparing fats for digestion.

- **The gallbladder** stores and concentrates bile secreted by the liver. Fatty food in the small intestine triggers contractions in the gallbladder, releasing bile into the bile duct, which transports it into the small intestine.

- **The pancreas** produces a fluid mixture, called pancreatic juice, composed of digestive enzymes and a bicarbonate buffer, which balances the pH levels in the digestive tract. The pancreas secretes pancreatic juice containing digestive enzymes into the small intestine, where it is used to break down proteins, starches, and fats.

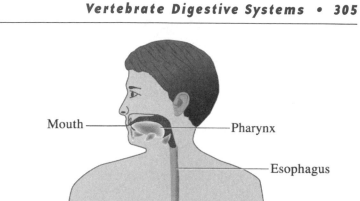

Vertebrate Digestive System

TISSUES OF THE DIGESTIVE TRACT

Four tissue layers make up the digestive tract and play a crucial role in the breakdown, absorption, and transport of food:

1. **Mucosa:** The innermost layer of the digestive tract, composed of epithelial tissue, which comes into direct contact with food as it is digested. Some cells in this layer produce and secrete digestive enzymes that help to break down the food. Other cells secrete mucus that protects the walls of the tract and eases the movement of food through the tract.

2. **Submucosa:** The layer of connective tissue surrounding the mucosa. The submucosa contains blood and lymph

vessels that absorb nutrients during digestion. These vessels transport the nutrients throughout the body via the lymphatic and circulatory systems. In addition, the submucosa houses nerves that regulate gastrointestinal activities.

3. **Muscularis:** A layer of muscle tissue surrounding the submucosa. The muscularis produces wavelike contractions called **peristalsis.** The motion of these contractions mechanically breaks down food and moves it along the alimentary canal.

4. **Serosa:** A thin layer of epithelial tissue covering the entire surface of the gut. The serosa attaches the gut to the body cavity.

Components of the Vertebrate Digestive System

Elements of the digestive system differ slightly between the different vertebrates. However, the general structure and function of their systems is very similar. The major components of the digestive system and their interactions with one another are described here.

THE ORAL CAVITY

The oral cavity consists of the mouth and, in some vertebrates, the teeth. Inside the mouth, the tongue is a muscle used to manipulate food and form it into a **bolus,** a round, easy-to-swallow ball. **Salivary glands** secrete mucous **saliva,** which contains an enzyme called **salivary amylase.** This enzyme breaks down starch into smaller sugars. Saliva also contains buffers, substances that neutralize acidic foods, as well as antibacterial substances that kill bacteria in the food. Saliva also eases the passage of food through the pharynx and esophagus.

The teeth break food into smaller pieces to expose more of the food's surface area to digestive enzymes. The teeth of carnivores are pointed and lack flat surfaces, making them ideal for cutting, shearing, and tearing into prey. Carnivores don't usually need to chew their food, as their digestive enzymes act directly on the animal cells. Herbivore teeth are large and flat with ridges ideal for grinding. The cellulose walls of plant matter need to be ground down before digestion. Omnivores, such as humans, have teeth suited for both carnivorous and herbivorous behavior. Some animals, particularly birds, lack teeth and instead have a muscular chamber called a gizzard, located along the alimentary tract just after the stomach. Stones swallowed by these animals collect in the gizzard and grind down food as it passes through.

THE PHARYNX

The **pharynx,** or throat, connects the oral cavity to the esophagus and the **trachea,** or windpipe. As the bolus moves into the pharynx, a muscular ring called a **sphincter,** which surrounds the esophagus, relaxes to open the esophagus for the bolus to pass. As food passes into the esophagus, a flap of skin, the **epiglottis,** moves to cover the trachea, or windpipe, preventing food from falling into the lungs.

THE ESOPHAGUS

The esophagus shuttles food from the pharynx to the stomach. The muscles surrounding the esophagus perform peristalsis, the rhythmic and stepwise contraction of muscle that forces food to move along the esophageal passage. Many animals, with the notable exception of humans, have a sphincter, called a cardiac orifice, between the esophagus and stomach, which prevents food from regurgitation, or moving back toward the mouth.

THE STOMACH

Food passes from the esophagus into the stomach, a saclike organ. The stomach's inner surface is highly convoluted, allowing it to fold up when empty and expand when full. Digestion that occurs before food reaches the stomach is primarily mechanical, with some minor chemical digestion by saliva. Chemical digestion continues in the stomach as **gastric juice,** an acidic mixture of enzymes and mucus, is secreted by gastric glands

CHAPTER 17
Nutrition and Digestion

located in the stomach lining. Gastric glands are composed of two types of cells:

1. **Parietal cells,** which secrete hydrochloric acid (HCl)

2. **Chief cells,** which secrete **pepsinogen,** the precursor to the enzyme **pepsin**

The acidic environment of the stomach denatures proteins in food—a process in which the protein is made inactive and begins to break down. The HCl, also known as **gastric acid,** secreted by parietal cells lowers the acidity of the gastric juice to a pH level of 2. Low pH levels are necessary for pepsinogen to be converted into the active digestive enzyme pepsin, which further breaks down food proteins. If chief cells were to secrete pepsin directly into the stomach, the pepsin would digest the cells themselves. Additionally, mucus lines the stomach walls and prevents it from being digested.

THE SMALL INTESTINE

A mixture of gastric acid and partially digested food called **chyme** passes out of the stomach and into the small intestine, where most of the chemical breakdown and absorption of food takes place. Bile salts break down fats in the chyme while pancreatic fluid, composed of bicarbonate, neutralizes the acid. The small intestine also contains a host of enzymes that help digest various food molecules. The enzymes present in the small intestine and their functions are outlined in the table below.

Digestive Enzymes	Food Molecule
Pancreatic amylase, maltase, sucrase, lactase	Starch
Trypsin, chymotrypsin, aminopeptidase, carboxypeptidase	Protein
Nucleases	Nucleic acids
Bile salts, lipase	Fats

The small intestine is split into three sections: the **duodenum, jejunum,** and the **ileum.** Chemical digestion takes place in the duodenum and the jejunum. The ileum, the final section of the small intestine, is where most absorption of the nutrients takes place. Small projections called **villi** cover the ileum walls. Cells lining the villi are covered in folds of plasma membrane that form even smaller projections called **microvilli.** Villi and microvilli increase the ileum's surface area, providing more surface across which nutrients can be absorbed. Nutrients pass through capillaries in the lining of the villi and into the bloodstream, where they circulate first to the liver, then throughout the rest of the body. Villi and microvilli projections also contain digestive enzymes to further digest food.

THE LARGE INTESTINE

Food from the small intestine empties into the large intestine, or colon. No digestion and only a small percentage of absorption take place in the large intestine. The large intestine primarily functions to concentrate waste material into a form called feces. Movements of the large intestine compact feces and move it into the rectum, where it is exited from the body through either the anus or the cloaca.

Feces is expelled from the body through the voluntary movements of muscles surrounding the rectum.

Summary

Nutrition

- Animals need energy taken up from food to power their bodily functions.

- Because animals cannot produce **nutrients,** such as vitamins, amino acids, and minerals, on their own, they must get them from food.

- Animals differ in their nutritional requirements.

- Energy and nutrients released from food via ingestion and digestion are absorbed into an animal's body. Waste left over from digestion is removed from the body.

Digestion

- **Digestion** converts food into substances that can be absorbed by the body.

- Food is made up of four main macromolecules: nucleic acids, proteins, fats, and carbohydrates.

- Digestion breaks these macromolecules into their usable monomers: nucleotides, amino acids, fatty acids and glycerol, and simple sugars. The body uses these monomers for metabolic processes.

- Animals are categorized into three groups based on their food source: the plant eaters, **herbivores;** the meat eaters, **carnivores;** and the plant and meat eaters, **omnivores.**

- **Digestive systems** have evolved over time from the simple intracellular digestion to a gastrovascular cavity to the complex specialized **digestive tract.**

Vertebrate Digestive Systems

- All vertebrate digestive systems consist of an alimentary canal and accessory organs. The alimentary canal is a tube divided into several sections: the oral cavity and pharynx, the esophagus, the stomach, the small intestine, the large intestine, and the anus or cloaca.

- Accessory organs of the digestive system include the **liver, gallbladder,** and **pancreas.**

- The digestive tract is made up of four tissue layers: the **mucosa, submucosa, muscularis,** and **serosa.**

Components of the Vertebrate Digestive System

- The digestive system is divided into several sections:

  » Oral cavity, which includes **mouth** and teeth

  » Pharynx, which connects mouth and **esophagus**

  » Esophagus, which moves food from pharynx to **stomach**

» Stomach, where some food is digested by digestive enzymes and gastric acid

» The **small intestine,** where most digestion and absorption occurs

» The **large intestine,** where waste is compacted for removal

Sample Test Questions

1. Forms of digestion differ between simple animals, such as the sponge, and more complex animals, such as vertebrates. Name the three basic forms of digestion, and explain how they differ.

2. What are some major differences between the way herbivores and carnivores process food?

3. Explain how each of the four tissue layers of the stomach plays a role in protecting the stomach and aiding its functioning.

4. Identify the correct order in which food passes through the human alimentary canal.

   A. Pharynx, esophagus, stomach, small intestine, colon
   B. Esophagus, pharynx, stomach, small intestine, colon
   C. Pharynx, esophagus, small intestine, stomach, colon
   D. Pharynx, esophagus, stomach, colon, small intestine
   E. Esophagus, pharynx, small intestine, stomach, colon

5. Which section of the alimentary canal is highly acidic?

   A. Small intestine
   B. Stomach
   C. Mouth
   D. Rectum
   E. Colon

CHAPTER 17
Nutrition and Digestion

6. Which of the following substances does pancreatic amylase break down?

    A. Protein
    B. Fat
    C. Nucleic acid
    D. Polypeptides
    E. Starch

7. What is the innermost lining of the human alimentary canal called?

    A. Serosa
    B. Muscularis
    C. Submucosa
    D. Mucosa
    E. Villa

8. Which of the following helps to break down fat in foods?

    A. Salivary amylase
    B. Pancreatic amylase
    C. Bile
    D. Trypsin
    E. Chymotrypsin

9. What is the primary purpose of the large intestine?

    A. Compact waste and remove water
    B. Break down starches
    C. Absorb nutrients
    D. Mechanically break down foods
    E. Store waste until it is expelled

10. Which accessory gland produces most of the digestive enzymes secreted into the small intestine?

    A. The gallbladder
    B. The pancreas
    C. The liver
    D. The kidney
    E. The salivary glands

11. Where does the breakdown of starch begin?

    A. Mouth
    B. Pharynx
    C. Stomach
    D. Small intestine
    E. Colon

12. All of the following actions represent a mechanism for mechanical digestion EXCEPT

    A. Chewing
    B. Tongue manipulation
    C. Bile secretion
    D. Stomach churning
    E. Peristalsis

13. The duodenum is the upper portion of which of the following organs?

    A. Pharynx
    B. Esophagus
    C. Stomach
    D. Large intestine
    E. Small intestine

14. Which of the following pairs of enzymes causes the breakdown of protein in the small intestine?

    A. Pancreatic and salivary amylase
    B. Trypsin and chymotrypsin
    C. Nucleases and ribonucleases
    D. Bile salts and bile acids
    E. Lipases and maltases

15. Which substance does the chief cells in the stomach secrete?

    A. Pepsinogen
    B. Pepsin
    C. Hydrochloric acid
    D. Mucous
    E. Sodium hydroxide

CHAPTER 17
Nutrition and Digestion

ANSWERS

1. The three forms of digestion exhibited by animals are intracellular digestion, gastrovascular cavity, and digestive tract. Animals using intracellular digestion, such as sponges, digest food directly in their cells. Invertebrates such as some cnidarians and flatworms have a gastro-vascular cavity into which food and waste enter and exit via a single hole. Digestion occurs in this cavity, and each cell of the body lines the cavity and absorbs the products of digestion. Many invertebrates and all vertebrates use a digestive tract for digestion. A digestive tract is a long tube extending through the body. It has two openings: the mouth, in which food enters the body, and the anus or cloaca, through which waste products are removed. The digestive tract differs in complexity from species to species, from the simple forms found in the nematodes to more complex structures found in the vertebrates.

2. Herbivores must break down plant material to extract energy and nu-trients, while carnivores extract their energy and nutrients from meat. Cellulose, a major component of plant material, is tougher to digest than meat. Herbivores use their teeth to grind plant material and break down the cellulose. Birds use gizzard stones to help grind plant material, and other animals, such as cows, have bacteria and multichambered organs to help them digest cellulose. Herbivores also have longer digestive tracts, compared to carnivores, to allow more room for breakdown and absorption. In contrast, carnivores have sharp teeth for tearing meat. They have less complex and shorter digestive tracts, suitable for extract-ing nutrients from meat, which is much softer and easier to break down than cellulose.

3. The four tissue layers from the innermost to the outermost layer are the mucosa, submucosa, muscularis, and serosa. The mucosa produces and secretes digestive enzymes that chemically break down food. The mucosa also secretes mucus, which protects the gut from digestion and eases the passage of food through the digestive system. The submucosa contains blood and lymph vessels to transport nutrients from the food. The submucosa also contains nerves within this layer that regulate the stomach's activities. The muscularis is composed of muscle tissues that contract to push food through, a movement called peristalsis. The serosa encloses the entire gut and attaches it to the body cavity.

4. A Food passes from the oral cavity through the pharynx, to the esophagus, stomach, small intestine, and finally the colon (large intestine), before being expelled through the anus.

5. B The stomach is highly acidic, with a pH of 2.

6. E Pancreatic amylase breaks down starch.

7. D The innermost lining of the alimentary canal is known as the mucosa.

8. C Bile produced by the liver and stored in the gallbladder is used to break down fats.

9. A The primary purpose of the large intestine is to reabsorb water from waste material and compact the waste material in preparation for its expulsion.

10. B Most of the digestive enzymes secreted into the small intestine are produced in the pancreas.

11. A The breakdown of starch begins in the mouth.

12. C Bile aids in the breakdown of fats in the small intestine; this is a form of chemical digestion.

13. E The first section of the small intestine is known as the duodenum.

14. B The breakdown of protein in the small intestine takes place via the actions of trypsin and chymotrypsin.

15. A Chief cells secrete pepsinogen, the precursor to the enzyme pepsin.

CHAPTER 17
Nutrition and Digestion

OSMOREGULATION, THE URINARY SYSTEM, AND THE IMMUNE SYSTEM

18

Osmoregulation

Animals control the entrance and exit of water in their bodies through the process of **osmoregulation.** Almost all chemical reactions in an animal's body involve water. Cells also require water as a medium in which those reactions can occur. Water also constitutes a major component of all bodily fluids, such as blood and cellular fluid. To ensure proper functioning and to prevent dehydration, the body needs to maintain a suitable internal water level.

Water moves through cell membranes via **osmosis,** in which it moves from dilute solutions with a low concentration of solutes to concentrated solutions with high concentrations of solutes. These solutes include ions such as sodium, chloride, magnesium, and potassium, which play a role in various cellular functions. For example, sodium and potassium aid in the transfer of nerve impulses across nerve cells.

OSMOLALITY

Most animals need to maintain a relatively constant **osmolality,** which is the total moles of solute per kilogram of water, to ensure proper functioning of biological systems. The method of exchanging water and solutes with the environment depends on how an organism's osmolality compares with the environment. There are three possibilities:

1. A **hyperosmotic** environment has a higher osmolality than the organism. The organism will lose water to the environment.

2. A **hypoosmotic** environment has a lower osmolality than the organism. The organism absorbs water from the environment.

3. An **isosmotic** environment has the same osmolality as the organism. The organism will neither gain nor lose water.

Specialized osmoregulatory organs prevent dehydration or overhydration by controlling the exchange of water and solutes between the animal's body and the environment.

CHAPTER 18
Urinary and Immune Systems

EXAMPLE: The following table illustrates how the different states of osmolality can influence a frog's interactions with the environment.

Environment	Frog's Reaction
Seawater is hyperosmotic and has a higher osmolality than a frog's internal environment.	A frog will lose water to the environment through its skin via osmosis. The frog will become dehydrated and will need to take in water.
Freshwater is hypoosmotic and has a lower osmolality than the frog's internal environment.	A frog will gain water from the environment via osmosis through its skin. The frog will become overhydrated and will need to remove excess water through urination.
Another fluid is isosmotic and has the same osmolality as the frog's internal environment.	The frog is in a state of equilibrium with its environment and has no net gain or loss of water.

Osmoregulatory Organs

Osmoregulatory organs usually couple with an animal's urinary system in the regulation of water. Osmoregulation in invertebrates can be controlled by a variety of possible organs, depending on the animal: **vacuoles, protonephridia, nephridia,** and **Malpighian tubules.** Osmoregulation in all vertebrates is controlled by the **kidneys.** The following chart describes the various osmoregulatory organs across the animal kingdom.

Animal	Osmoregulatory Organ	Description
Sponges	Vacuoles	Sacs within cell bodies that collect water.
Flatworms	Protonephridia	Excretory tubules that branch throughout the body. Cilia lining the tubules draw water from the body. Water and solutes in the tubules are reabsorbed into the body, and unwanted substances are excreted from the body through the tubules' openings.
Earthworms, mollusks, crustaceans	Nephridia	Tubules that open both internally and externally. Water in the body cavity is drawn into the tubules, and larger molecules are filtered out, leaving filtrate (water that has been filtered). Sodium chloride is actively removed from the filtrate and sent back to the body. The remaining filtrate leaves the body as urine.
Insects	Malpighian tubules	Tubules extending off the digestive tract. Waste molecules and potassium ions are secreted into the tubules, which increases the concentration of fluid in the tubules by drawing in water from the body. Prior to excretion, water and potassium are almost fully reabsorbed into the body, leaving only waste products to be excreted along with digestive feces.
Vertebrates—reptiles, birds, and fish	Gills, salt glands, rectal gland	Fish living in hyperosmotic environments, such as seawater, drink large amounts of water, using their gills to transport excess ions out of the body. Reptiles and birds living in seawater have special salt glands capable of forming a highly concentrated fluid for the removal of sodium and chloride. Sharks and stingrays rely on rectal glands to remove excess ions while in seawater.

CONTINUED

Animal	Osmoregula-tory Organ	Description
Vertebrates	Kidneys	Tubules, called nephrons, filter the blood and create a filtrate containing waste products, as well as small molecules such as glucose, and vitamins, that are useful to the body. Water and useful molecules are reabsorbed back into the body prior to excretion. The remaining filtrate, known as **urine,** is excreted from the body. The kidneys in mammals, and birds to a lesser extent, living in hyperosmotic environments, such as seawater, concentrate urine so that water is conserved. Other vertebrate kidneys are incapable of water concentration and must rely on accessory organs.

The Vertebrate Kidney

Vertebrate kidneys differ in the extent to which they modify urine. All vertebrates can create either an isotonic urine, a urine with the same concentration as their internal fluid, or a hypotonic urine, which is less concentrated than their internal fluid. However, only mammals and birds are able to concentrate their urine and create a hypertonic urine that is denser than their internal fluid, which helps to conserve water in their bodies. Hypertonic urine enables certain species of mammals and birds to occupy habitats that would otherwise be too arid for them. Reptiles and fish kidneys, in contrast, are incapable of concentrating urine. These animals use additional accessory organs that prevent dehydration in arid environments.

The Mammalian Urinary System

Nutrients, waste, and other materials circulate throughout the body through the bloodstream. The urinary system filters the blood so that useful products remain in the body and excess ions and water, along with nitrogenous wastes from cellular processes, are excreted.

THE KIDNEY

The kidney, the principle mammalian urinary organ, is composed of specialized **renal tissue** and divided into two sections: the **renal medulla** and **renal cortex. Nephrons,** numerous identical tubule structures embedded in the renal medulla and cortex, filter blood as it passes through the kidney.

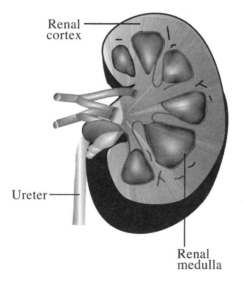

Renal cortex

Ureter

Renal medulla

The Kidney

Blood entering the kidneys passes through several structures during the process of filtration:

1. Blood enters the kidneys via **renal arteries** and passes into the nephrons.

2. Filtrate forms from blood as it passes through the first structure in the nephron, the **glomerulus.** The glomerulus is surrounded by a balloonlike sac called the **Bowman's capsule.**

3. Filtrate exits the glomerulus and enters Bowman's capsule, where it then passes on to the proximal convoluted tubule.

4. Almost all nutrient reabsorption takes place in the proximal convoluted tubule through the energy-consuming process of active transport.

5. Filtrate next enters the section of the nephron known as the **loop of Henle,** a loop of tubule extending from the renal cortex, which houses the remainder of the nephron, down into the renal medulla. Water is removed from filtrate as it descends down the loop into the renal medulla. As a result, the filtrate becomes highly concentrated with NaCl (a common compound in the blood), which is pumped out of the filtrate as it ascends the loop of Henle. The filtrate that reaches the end of the loop of Henle has become dilute throughout this process.

6. Filtrate entering the renal medulla passes through the **distal convoluted tubule.** If the body needs to retain more NaCl, the hormone **aldosterone** stimulates retention of NaCl in this tubule, further diluting the filtrate.

7. From the distal convoluted tubule, the filtrate passes through the collecting ducts, which descend through the renal medulla. The renal medulla is concentrated with NaCl, causing water to diffuse out of the filtrate and into the surrounding tissue, further concentrating the filtrate. If the body requires more water to be retained, as when an animal is dehydrated, the hormone **antidiuretic hormone (ADH)** is released to stimulate increased water reabsorption.

8. Finally, the filtrate, in the form called urine, passes through the **ureters,** which lead from the kidneys into the **bladder.** The bladder stores urine until the animal voluntarily releases it in the **urethra,** an externally opening tube.

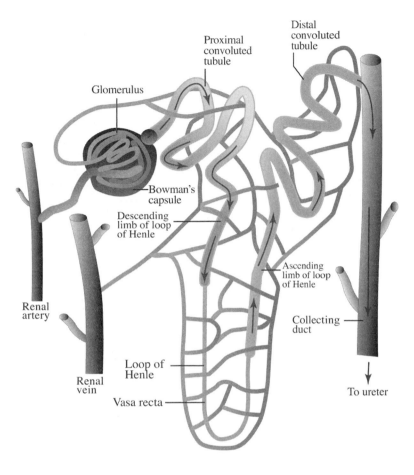

Glomerulus

Proximal convoluted tubule

Distal convoluted tubule

Bowman's capsule

Descending limb of loop of Henle

Ascending limb of loop of Henle

Renal artery

Renal vein

Loop of Henle

Vasa recta

Collecting duct

To ureter

Filtration in the Kidney

Filtration

As described above, several vital tasks are performed as blood passes through the kidney during filtration.

- **Filtrate formation:** Blood is filtered and large molecules, red blood cells, and proteins reenter the circulatory system, while smaller molecules and waste products enter the kidney as a filtrate.

- **Reabsorption:** The kidney reabsorbs useful ions and nutrients from the filtrate, returning them to the circulatory system via either active transport or diffusion.

- **Secretion:** Toxic material, foreign substances, and waste are secreted from the circulatory system into the nephrons. Secretion occurs through both active transport and diffusion.

- **Absorption and reabsorption of water:** Excess water absorbs into the filtrate to be removed from the body as urine. When the body is dehydrated, water may be reabsorbed from the filtrate and reentered into the circulatory system.

The Immune System

The **immune system,** through methods of nonspecific immunity and specific immunity, protects the body from invaders that cause infection, such as bacteria and viruses. Included in the immune system is the **lymphatic system,** which is used in reclaiming water that has left the bloodstream and plays a major role in both the nonspecific and specific immune response. The lymphatic system comprises a network of vessels that transport **lymph,** a fluid that resembles dilute interstitial fluid, connected to the lymph nodes, the **adenoids,** the **tonsils,** the spleen, and the appendix. All of these organs contain large amounts of lymphocytes and macrophages, white blood cells that destroy the viruses and bacteria carried to these sites through the lymph vessels.

NONSPECIFIC IMMUNITY

Nonspecific immunity represents the body's first line of defense against infection. Nonspecific immunity is not selective, acting instead against all foreign invaders. Three methods of nonspecific immunity protect the body: the skin and other protective barriers, cellular counterattack, and inflammation and fever.

Skin and Other Protective Barriers

Skin forms the primary external surfaces that protect the body from the entry of microbes. The skin performs three functions in defense of the body:

1. Provides a physical barrier to prevent bacteria and viruses from entering the body.

2. Secretes chemicals, such as oil and sweat, which produce an acidic environment inhospitable to microbes.

3. Secretes sweat that contains **lysozyme,** an enzyme that destroys bacterial membranes.

In addition to the skin, internal mucous membranes of the respiratory, digestive, and urogenital (genital organs and urinary system) tracts also provide an initial barrier against invading microbes. These internal tracts perform three additional functions in defense of the body:

1. Trap microbes in mucus lining. Mucus then exits the body as waste via the digestive tract.

2. Acidic environment in some internal tracts is hostile to microbes. For example, the stomach is a highly acidic environment in which bacteria in food is destroyed.

3. Enzymes in some tracts kill microbes. For example, enzymes in saliva throughout the digestive tract and in the stomach kill microbes.

Cellular Counterattack

Cells and proteins make up the second line of nonspecific defense by responding to any invading microbes that get past the skin and mucous membranes. White blood cells, also known as **leukocytes,** which constitute the main weapon in cellular defense, come in several basic forms:

- **Macrophages:** Large white blood cells present in extracellular fluid. Macrophages are phagocytic, meaning they can engulf invading viruses and bacteria. Infections too robust for a single macrophage are assisted by other white blood cells, **monocytes,** which are sent to the infection site and transformed into macrophages.

- **Neutrophils:** Phagocytic cells similar to macrophages that engulf invading viruses and bacteria. Neutrophils also produce and release chemicals that can kill bacteria present in their immediate surroundings, destroying the neutrophils themselves in the process.

- **Natural killer cells:** Cells that attack and kill cells that have become infected by viruses. Natural killer cells will also attack and kill cancer cells.

Complement proteins and **interferons** perform the functions of protein defense:

- **Complement proteins** kill microbes by creating an opening in the microbe's cell membrane. Water enters through the opening, causing the cell to expand and eventually burst. Complement proteins also assist in other defense mechanisms. For example, they attract white blood cells to a microbe and alter a microbe's membrane, making it easier for those white blood cells to attack.

- **Interferons** are proteins produced by cells already infected with a virus. Interferons stop the spread of viruses to uninfected cells by preventing infected cells from replicating and repackaging the virus's genetic material.

Inflammation and Fever

Inflammation is the body's response to infected or damaged cells. Damaged cells produce histamine and other chemical signals, such as **prostaglandins,** that cause blood vessels to expand, increasing blood flow to the affected area and resulting in swelling. An increase in blood brings more white blood cells to the area, where they destroy the invading bacteria and viruses and engulf dead cells.

EXAMPLE: The skin surrounding a cut will swell as a result of inflammation. Pus that forms inside the infected area is a combination of dead cells, white blood cells, and dead bacteria and viruses.

Fever results from the detection of infection by the hypothalamus. The hypothalamus sends signals via the nervous system to stimulate an increase in body temperature, which then stimulates the defense activities of white blood cells. Increased body temperature also slows the growth of bacteria, which are typically less effective at growing and duplicating at higher temperatures.

SPECIFIC IMMUNITY

Invading viruses and bacteria that make it past the initial nonspecific line of defense will next encounter the body's specific immune defenses, which recognize and react to specific microbes. In addition to ridding the body of current invaders, the immune system develops specific strategies for attacking bacteria and viruses that it has encountered before. For example, when the chicken pox virus first infects the body, the immune system develops cells to attack it. Those cells remain in the body after the virus has been destroyed. If the chicken pox virus reinfects the body, the immune system releases those specific attack cells, preventing the recurrence of severe illness.

SPECIFIC IMMUNE RESPONSES

Antigens are proteins and sugars located on the surface of foreign cells. In relation to the immune system, antigens are those substances that trigger an immune response from the body. Lymphocyte cells, which are types of leukocytes (white blood cells) produced in bone marrow, recognize invading cells by their antigens, prompting them to attack. Two types of lymphocytes are used in specific immune responses: **T cells** and **B cells.**

T and B cells both originate from immature lymphocyte cells, but each takes a separate path to reach maturity:

- Immature lymphocytes that continue to develop in the bone marrow mature into **B lymphocytes** (B cells).

- Immature lymphocytes transported through the bloodstream to the thymus mature into **T lymphocytes** (T cells).

Mature B and T cells are eventually transported to the lymphatic organs, where they defend the body against infection.

T Cells

T cells perform **cell-mediated immunity.** They reside in the blood and lymph and attack cells already infected with a virus. T cells also attack cells that have become cancerous. The two types of T cells used in cell-mediated immunity are:

- **Cytotoxic T cells** that attack body cells infected with viruses or bacteria

- **Helper T cells** that activate cytotoxic T cells, stimulate B cells, and activate macrophages

Both cytotoxic and helper T cells are involved in the basic steps of cell-mediated immunity:

1. A macrophage ingests virus-infected cells.

2. The macrophage acts as an **antigen-presenting cell,** or **APC,** by using the ingested cell along with its own proteins, called **self proteins,** to outwardly display pieces of the virus on its cell surface.

3. APCs reveal antigens to helper T cells, which can distinguish between the APC's self proteins and the foreign antigens (called nonself proteins). The helper T cell then binds to the APC.

4. Binding to the APC triggers the helper T cell to form a pathway by secreting the protein interleukin-2. The presence of interleukin-2 signals T cells in the lymphatic organs to grow and divide, stimulates cytotoxic T cell activity,

and activates B cells, which trigger the destruction of the infected cell.

5. Cytotoxic T cells then bind to the infected body cell. The T cells secrete a protein called **perforin,** which creates holes in the infected cell's membrane and kills the cell.

B cells

B cells perform **humoral immunity.** Once activated by the release of interleukin 2 by the helper T cells, B cells divide into two B cell types. Long-lived **memory B cells** remain in the system after the infection is neutralized to initiate rapid immune responses when the same antigen reenters the body. Short-lived plasma B cells produce antibodies, which are proteins that attach to infected cells, and signal for their destruction and removal. Plasma B cells release antibodies into the blood, lymph, and other extracellular fluids, where they travel to infected areas and bind to viral proteins. The presence of antibodies triggers proteins and phagocytic cells to destroy the virus.

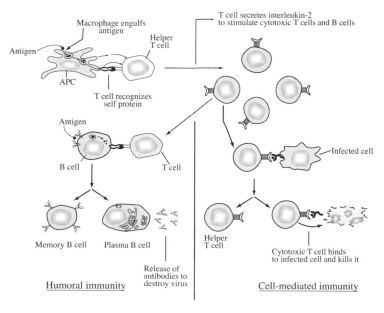

Specific Immunity

CHAPTER 18
Urinary and Immune Systems

EXAMPLE: Vaccinations promote the development of antibodies in the body by using a weakened version of a pathogen. An injection of cowpox, a mild form of pox, prompts the body to develop antibodies that can defend against other forms of pox. Through these types of vaccinations, the effects of epidemic diseases such as smallpox, which in the past was common and deadly, can be controlled and minimized in a population.

ACTIVE AND PASSIVE IMMUNITY

The body's specific immune responses can be further divided into either passive or active immunity. **Active immunity** describes the development of antibodies and memory cells after an infection, which the body can use to defend against future attacks by the same invading microbes. Antibodies can also be obtained from a source outside the body, such as another individual, through the process of **passive immunity.** A pregnant mother, for example, will pass antibodies to her baby via the placenta. Passive immunity does not result in the development of memory cells and therefore will last as long as the antibodies remain in the body.

Clonal Selection in Active Immunity

When initially infected, the body does not contain the millions of specific B and T cells needed to recognize and fight the invading virus. The body instead contains small numbers of various B and T cells to defend against specific antigens found on viruses. When a virus enters the body, these cells will recognize it and begin their cell-mediated and humoral activities. Through the process of **clonal selection,** these cells replicate based on the initial infection. Eventually, the body builds a defense that can be used continuously against the same virus, resulting in immunity against that virus.

Clonal selection follows the following basic outline:

1. An antigen enters the body and a small number of lympho- cytes (B and T cells) recognize it and are stimulated to per- form humoral and cell-mediated immunity. This **primary immune response** is usually weak, due to the low number of cells available for defense.

2. B and T cells activated during primary immune response grow and divide repeatedly, forming an arsenal of special- ized clones that can be stored in the lymphatic organs for use in future defense against that specific antigen.

3. B cells performing humoral immunity during the primary immune response produce plasma cells and memory cells. Once an infection is eliminated, memory cells remain in the body. If the body is reinfected by the same virus, these memory cells recognize the virus and mount a much quicker response using the specialized B and T cells. This **secondary immune response** acts more quickly and ef- fectively than primary immune response, preventing an infection from spreading.

Immune System Defeat

The immune system does not form a perfect barrier against serious illness and infection. Immune system defeat can be brought about by three main causes:

- **Attacks on the immune system:** Some pathogens directly target components of the immune system, rendering it un- able to respond. An example is the AIDS retrovirus, human immunodeficiency virus (HIV). HIV invades T cells and inactivates them, making them incapable of stimulating a response to other infections. Without T cells to react to invading microbes and stimulate the body's other immune defenses, even mild infections can become fatal.

- **Antigen shifting:** Some viruses and bacteria mutate fre- quently, altering their antigens in the process. A virus with mutated antigens may not be recognized by the immune system and will instead by treated as a new virus. The body's primary immune response will therefore be weaker and will take longer, allowing the virus to spread more effectively.

EXAMPLE: The influenza virus undergoes frequent mutations that alter its antigens. Because the antigens are constantly changing along with the mutating virus, any immunization against influenza will be only a short-term solution. Periodic immunizations are administered to protect people from new strains of the virus.

- **Disease brought about by the immune system:** At times, the immune system itself can be the source of disease. **Autoimmune diseases** occur when the immune system cannot recognize the body's own cells and begins to attack them. The immune systems of people suffering from diabetes attack those cells that control insulin secretion. Insulin regulates the sugar levels in the blood, which has various detrimental affects on the body if it is too high or too low. Allergic reactions are the result of abnormal immune system responses to antigens (or allergens, as they are called in these cases). The immune system treats harmless foreign bodies, such as pollen or dust, as potential threats. Hypersensitive responses by the immune system can cause tissue damage along with other common symptoms of allergies, such as itchy reddened skin, watery eyes, respiratory tract constriction, and cramping in the stomach. In the most extreme cases, allergic reactions can cause a drop in blood pressure, a swelling of the epiglottis that may block the trachea, and bronchial construction that prevents air from entering or exiting the lungs.

Summary

Osmoregulation

- **Osmoregulation** controls the entry and exit of water in an animal's body.

- **Osmosis** is the movement of water from dilute to concentrated solutions, across a semipermeable membrane.

- Animals can live in environments in which the outside **osmolality** differs from their blood osmolality. **Hyperosmotic** environments are more concentrated, **hypoosmotic** environments are more dilute, and **isosmotic** environments are the same.

- Osmoregulatory organs assist with osmoregulation. A variety of osmoregulatory organs exist among various animal species: **vacuoles, protonephridia, nephridia, malpighian tubules,** and **kidneys.**

- The vertebrate kidney is the most complex osmoregulatory organ. The kidney filters blood to produce filtrate. Filtrate is subjected to reabsorption of water and molecules as it travels along a tubule through the kidney. The filtrate is excreted as **urine,** a mixture of water and waste products.

- Only mammals and birds have kidneys that concentrate urine. Other vertebrates rely on accessory structures such as salt glands to remove excess ions.

The Mammalian Urinary System

- The kidney is the major component of the urinary system. The kidney consists of **nephrons** embedded in **renal tissue.**

- Blood enters the urinary system via **renal arteries.** After entering the kidneys, blood is filtered by the **glomerulus** and turned into filtrate. Filtrate undergoes further filtration, absorption, and secretion before exiting the kidney via the **ureter.** From the ureter, filtrate is sent to the **bladder,** where it is stored before exiting the body via the **urethra.**

- Specialized areas in the kidney assist in filtration, absorption, and secretion. These specialized areas are the **Bowman's capsule,** proximal convoluted tubule, **loop of Henle, distal convoluted tubule,** and collecting ducts.

The Immune System

- The nonspecific immune system is the body's first line of defense against infection. It consists of three defense mechanisms that attack all foreign microbes: skin, cellular counterattack, and **inflammation** and fever.

CHAPTER 18 Urinary and Immune Systems

- Specific immunity provides defense against specific infections. Cells involved in specific immunity recognize infections and mount responses specific to those microbes.

- The immune response occurs when any **antigen,** which are proteins and sugars coating the outside of a foreign microbe, appear in the body.

- Specific immune responses include both **humoral immunity,** during which **antibodies** attach to antigens, and **cell-mediated immunity,** during which cells directly attack the infectious cells.

- **B cells** are used during humoral immunity. **T cells** activate B cells and are used during cell-mediated immunity.

- During the **primary immune response,** B and T cells are cloned in a process called **clonal selection.** Some cloned cells fight infections currently in the system. Cloned **memory cells** are stored for use when reinfection occurs.

- The **secondary immune response** occurs when memory cells are used to fight reinfection. Secondary immune responses are much stronger and more effective than primary immune responses. Several attacks may result in the defeat of the immune system, when it is unable to protect the body from infection: An infection can attack and disable the immune system; antigens can shift, becoming unrecognizable to memory cells; and diseases can be brought about by the immune system itself.

Sample Test Questions

1. Describe the events that would take place in your body following a cat scratch.

2. The body produces the antidiuretic hormone ADH when dehydrated. Explain how this hormone affects the kidneys to promote water retention.

3. What affect would a disease that affects the monocyteforming cells in the bone marrow have on the immune system?

4. An environment that has a higher osmolality than an organism is said to be

   A. Isoosmotic
   B. Hyperosmotic
   C. Semiosmotic
   D. Hypoosmotic
   E. Superosmotic

5. Which type of cell attacks and kills viruses and cancer cells?

   A. Macrophage
   B. Monocyte
   C. Natural killer cell
   D. Neutrophil
   E. Effector cell

6. You are traveling to a foreign country and receive several immunizations before your trip. What type of immunity does this represent?

   A. Permanent immunity
   B. Passive immunity
   C. Secondary immunity
   D. Dependent immunity
   E. Active immunity

7. Which osmoregulatory organs are made up of tubules extending off the digestive tract?

   A. Malpighian tubules
   B. Nephridia
   C. Vacuoles
   D. Protonephridia
   E. Kidneys

8. The enzyme lysozyme is contained in which of the following?

   A. Digestive fluids
   B. Mucus
   C. Salivary glands
   D. Sweat
   E. Lymph

9. Where do T cells develop?

    A. In the liver
    B. In the thymus
    C. In the bone marrow
    D. In the spleen
    E. In the bloodstream

10. In which part of the kidney does the hormone aldosterone stimulate retention of sodium chloride?

    A. Distal convoluted tubule
    B. Loop of Henle
    C. Bowman's capsule
    D. Glomerulus
    E. Proximal convoluted tubule

11. Which type of animal excretes water from the body through the protonephridia?

    A. Sponge
    B. Flatworm
    C. Crustacean
    D. Insect
    E. Vertebrate

12. Which two types of animals are able to concentrate their urine to conserve water within the body?

    A. Reptiles and amphibians
    B. Fish and amphibians
    C. Birds and fish
    D. Reptiles and mammals
    E. Mammals and birds

13. In which part of the kidneys does the filtrate become highly concentrated with NaCl?

    A. Distal convoluted tubule
    B. Glomerulus
    C. Bowman's capsule
    D. Loop of Henle
    E. Proximal convoluted tubule

14. Secretion occurs when toxic materials are released from the circulatory system into the nephrons. By which two processes can secretion occur?

    A. Absorption and reabsorption
    B. Active transport and diffusion
    C. Osmosis and filtration
    D. Filtrate formation and active transport
    E. Diffusion and reabsorption

15. B cells are involved in which type of immunity?

    A. Humoral immunity
    B. Lymphocyte immunity
    C. Active immunity
    D. Primary immunity
    E. Passive immunity

CHAPTER 18
Urinary and Immune Systems

ANSWERS

1. Immediately following a scratch that punctures the skin, the body initiates an inflammatory response. Damaged skin cells produce histamine and other chemical signals. The presence of the histamine directs the blood vessels to increase flow to the affected area, causing white blood cells to rush in and begin work battling incoming bacteria and viruses. The white blood cells engulf and destroy incoming infecting agents and dispose of any body cells that have been infected or killed. If the wound is severe, dead white blood cells and fluids that result from the body's reaction to the wound build up to form pus. Finally, clotting agents rush to the site to seal the wound and prevent blood loss.

2. The antidiuretic hormone ADH acts on the collecting ducts of the kidneys. When filtrate enters the collecting ducts, a strong osmotic gradient pulls water out of the collecting ducts and into the blood vessels. ADH opens more water channels, making the collecting ducts more permeable so that more water flows from them and into the blood vessels. The body is able to conserve the water, which flows back to the bloodstream, rather than filtering out of the urinary system.

3. Leukocytes, white blood cells, are formed in the bone marrow and can differentiate into several types of cells, including monocytes. Monocytes mature into macrophages, which ingest and kill microbes that invade the body. Monocytes also engulf dead cells at the site of an infection. A disease affecting the production of monocyte in the bone marrow would remove this line of nonspecific immunity.

4. B   An environment that is hyperosmotic has a higher osmolality than an organism's internal environment.

5. C   Natural killer cells attack and kill virus-infected cells and cancer cells.

6. E   Immunization is a form of active immunity. The body receives a less virulent form of an antigen and produces memory cells to combat this antigen if encountered in the future.

7. A   Malpighian tubules are tubules extending off the digestive tract. They are found in insects.

8. D   Human skin secretes sweat, which contains an enzyme called lysozyme, which can destroy bacterial membranes.

9. B   Immature lymphocytes transported via the bloodstream to the thymus mature to become T lymphocytes, or T cells.

10. A   The hormone aldosterone stimulates sodium chloride retention in the distal convoluted tubule.

11. B   Protonephridia are externally opening excretory tubules branched throughout the body. They are found in flatworms.

12. E   Mammals and birds are able to concentrate their urine, allowing them to conserve water.

13. **D**   Filtrate becomes highly concentrated with NaCl in the loop of Henle.

14. **B**   Secretion occurs when toxic materials are excreted from the circulatory system into the nephrons. Secretion can occur either by diffusion or by active transport.

15. **A**   In humoral immunity, B cells secrete antibodies into the blood that defend against bacteria and viruses.

THE ENDOCRINE SYSTEM

Chemical Signals

The Vertebrate Endocrine System

Human Endocrine Glands

19

Chemical Signals

The body releases a variety of chemical signals to aid in the regulation of its functions. These chemical signals, such as hormones, neurotransmitters, and paracrine regulators, are classified according to where they originate, where they go, and how they act. **Hormones** are regulatory chemicals secreted into the bloodstream from one of many endocrine glands. A specific hormone can travel to any cell in the body, but only that hormone's **target cells** respond to it. **Neurotransmitters** are released by neurons in the nervous system to generate

340

a specific response across a synapse between two nerve cells. Other chemicals, called **paracrine regulators,** originate from a particular organ and only act in that organ.

HORMONES

Hormones are secreted by **glands,** which can include organs that function solely as endocrine glands, as well as organs that secrete hormones in addition to their other bodily functions. Glands secrete hormones directly into capillaries surrounding these organs, where they then enter the bloodstream. Four groups of hormones are secreted by organs in the endocrine system:

1. **Polypeptides:** Chains of amino acids, each less than 100 amino acids long. Insulin, which helps control glucose levels in the blood, is a polypeptide hormone.

2. **Glycoproteins:** A polypeptide chain, longer than 100 amino acids, attached to a carbohydrate. Glycoproteins include the follicle-stimulating hormone (FSH) and luteinizing hormone (LH), both secreted in the female reproductive site during sexual reproduction.

3. **Amines:** Hormones derived from the amino acids tyrosine and tryptophan. Catecholamines, which include adrenaline and noradrenaline, are amines often secreted in response to stress.

4. **Steroids:** Lipids derived from cholesterol. The steroid group differs from other endocrine hormones because they are not derived from amino acids. Corticosteroids secreted from the adrenal gland balance solutes, such as glucose and salt, in the body.

CHAPTER 19
The Endocrine System

Interactions with Target Cells

Hormones released into the body by an endocrine gland travel to the tissue housing the target cells. Upon arrival, hormones can induce an effect in two ways, depending on whether the hormone is lipophilic (fat soluble) or lipophobic (water soluble), also called hydrophilic. All steroids, as well as the amine hormone thyroxine, are lipophilic; all other hormones are lipophobic. Lipophilic hormones enter target cells, whereas lipophobic hormones interact with a cell's membrane receptors.

The methods through which lipophilic and lipophobic hormones interact with target cells are described below:

1. **Lipophilic hormones** attach to proteins in the blood and are transported to their target cells. At the target cell, the hormone detaches from the blood protein and passes through the cell's membrane. Once inside the cell, some hormones form hormone-receptor complexes by attaching to **receptor proteins.** These complexes make their way to the cell's nucleus, where they bind to specific regions of DNA, depending on the hormone. Other hormones travel directly to the nucleus without binding to a receptor protein and bind with the DNA. Transcription at that site of DNA, called the hormone receptor element, is either activated or deactivated by the hormone. Proteins resulting from activation influence activity in the target cell, constituting the cell's response to the hormone stimulation.

2. **Lipophobic hormones** are either too large to pass through or are repelled by the target cell's membrane. Instead of entering the cell, lipophobic hormones bind to receptor proteins on the membrane. A **second messenger,** a molecule existing in the cell, interacts with the receptor proteins on the outside to produce an effect in the cell.

PARACRINE REGULATION

Other chemical messengers, called **paracrine regulators,** secreted by organs, regulate functions in those organs. Paracrine regulators are divided into four groups:

*   **Cytokines,** which include chemicals that regulate immune cells

*   **Growth factors** that promote cell division and growth

*   **Neurotrophins** operating in organs of the nervous system

- **Prostaglandins,** a diverse group of paracrine regulators produced in almost every organ to perform a variety of functions

Prostaglandins

The major functions that prostaglandins perform in the various systems of the body are described in the table below.

System	Prostaglandins Function
Immune system	Produce inflammation in response to painful sensations and fever
Reproductive system	Regulate reproductive functions, such as the stimulation of uterine contractions ultimately leading to the birth of the fetus
Digestive system	Influence the movement of the intestines and absorption of fluid during digestion
Respiratory system	Regulate constriction and dilation of blood vessels in the lungs
Circulatory system	Assist platelets during blood clotting
Urinary system	Stimulate vasodilation of blood vessels, which increases blood flow and urine excretion

EXAMPLE: The medical drug aspirin inhibits prostaglandin production, which explains its use to prevent the occurrence of inflammation and pain. However, because prostaglandins are used by other systems in the body, side effects of aspirin use include gastric bleeding and prolonged clotting time.

THE INVERTEBRATE ENDOCRINE SYSTEM

Endocrine systems regulate the bodies of both invertebrates and vertebrates. In simple invertebrates, such as cnidarians, the endocrine system functions separately from other body systems. In more complex invertebrates, such as mollusks and arthropods, the nervous system and endocrine systems share control of reproduction and development.

The Vertebrate Endocrine System

Although the endocrine system is diverse among the invertebrates, the vertebrates' endocrine system, which secretes over fifty different hormones, is much more complex. Some hormones of the vertebrate endocrine system produce an effect in most of the tissues in the body, whereas others only work on a selective number of tissues. Hormones may be secreted from endocrine glands or from organs that secrete hormones in conjunction with their primary functions. They may produce an effect in most tissues of the body, or they may work on only a select number of tissues.

The following table summarizes the major hormone-producing glands of vertebrates:

Organ	Hormone Produced	Effects
Posterior pituitary (the hormones released by this gland are produced by the hypothalamus, an organ connected to the posterior pituitary gland)	Oxytocin	Stimulates uterine contraction during childbirth and initiates the production of milk by the mammary glands
	Antidiuretic hormone (ADH)	Stimulates the kidneys to reabsorb water, which maintains the body's water balance
Adrenal medulla	Epinephrine, norepinephrine	Stimulate the constriction of blood vessels, increase heart rate, and stimulate the secretion of insulin

CONTINUED

Organ	Hormone Produced	Effects
Anterior pituitary	Growth hormone (GH)	Stimulates muscle, bone, and tissue growth and is involved in regulating metabolism
	Follicle-stimulating hormone (FSH)	Involved in the production of sperm and ova
	Luteinizing hormone (LH)	Stimulates the testes and promotes testosterone production in males; triggers ovulation and promotes estrogen and progesterone production in females
	Prolactin (PRL)	Promotes lactation in females
	Thyroid-stimulating hormone (TSH)	Stimulates the thyroid gland to release the thyroid hormone
	Adrenocorticotropic hormone (ACTH)	Stimulates the secretion of glucocorticoids and androgens by the adrenal cortex
Adrenal cortex	Glucocorticoids	Stimulate fat breakdown in the adipose tissue (large groups of fat cells), stimulate the breakdown of muscle protein into amino acids, stimulate the conversion of glucose to amino acids in the liver (gluconeogenesis), and inhibit glucose uptake
	Mineralocorticoids	Stimulate the reabsorption of sodium by the kidneys to maintain the body's salt levels

CONTINUED

Organ	Hormone Produced	Effects
Pancreas	Insulin	Lowers blood glucose levels
	Glucagons	Raise blood glucose levels
Pineal gland	Melatonin	Regulates the body's daily and seasonal rhythmic cycles
Thyroid gland	Thyroxine (T_4), triiodothyronine (T_3)	Stimulate the metabolic processes of all cells in the body
	Calcitonin	Lowers blood calcium levels
Parathyroid glands	Parathyroid hormone (PTH)	Raises blood calcium levels
Thymus gland	Thymosin	Stimulates the development of T cells used by the immune system
Testes	Androgens	Involved in sperm formation and the development and persistence of male secondary sex characteristics, such as deepening of the voice and growth of facial hair
Ovaries	Estrogens	Involved in growth of uterine lining and the development and persistence of female secondary sex characteristics, such as the growth of pubic hair

Human Endocrine Glands

Humans possess the same ten major endocrine glands found in other vertebrates. These major glands produce and secrete the hormones that regulate the human body; however, their function differs slightly from those of other animals.

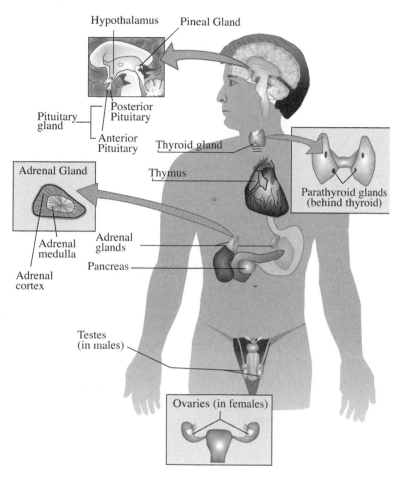

The Human Endocrine System

THE HYPOTHALAMUS

The hypothalamus integrates the endocrine and nervous systems. Nerves throughout the body transmit information about the external and internal environment to the hypothalamus, which is considered the control center of the endocrine system. This gland responds to the information by sending various endocrine signals, generally as part of negative feedback loops, throughout the body in an effort to maintain homeostasis. Though the hypothalamus has widespread functions in regulating the body, many other glands and organs act independently of it.

EXAMPLE: Negative feedback loops are interactions in which the response from one stimulus results in a new stimulus that diminishes the original response. For example, the hypothalamus sends out a hormonal stimulus via the anterior pituitary gland to the thyroid gland. This stimulus causes the thyroid to secrete thyroxine, which is conveyed back to the hypothalamus, resulting in the diminishment of the original stimulus.

Located beneath the hypothalamus are the **posterior** and **anterior pituitary glands,** both operating under control of the hypothalamus. The posterior pituitary gland communicates with the hypothalamus via nerve pathways and with target cells in the body via an endocrine pathway. In contrast, the anterior pituitary gland communicates with both the hypothalamus and its target cells in the body via an endocrine pathway.

Anterior Pituitary

Releasing hormones secreted by the hypothalamus stimulate the anterior pituitary to secrete the hormones it produces. **Inhibiting hormones,** also secreted by the hypothalamus, have the opposite effect on the anterior pituitary. The hormones produced and secreted by the anterior pituitary promote growth in target organs, which includes stimulating those organs to produce and secrete their own hormones. The range of hormones produced by the anterior pituitary gland includes the following:

- **Adrenocorticotropic hormone (ACTH),** which stimulates production of corticosteroids in the adrenal cortex. Cortisol, a corticosteroid, regulates glucose levels in humans.

- **Follicle-stimulating hormone (FSH)** and **luteinizing hormone (LF),** which stimulate spermatogenesis and testosterone production in males and ovulation and ovarian follicle development in females.

- **Growth hormone (GH),** which stimulates growth of muscle, bone, and other tissues. GH also regulates the body's metabolism.

Posterior Pituitary

The posterior pituitary gland is connected to the hypothalamus via nerve fibers. Hormones produced in the hypothalamus travel along these fibers to the posterior pituitary, which then sends these hormones along the bloodstream to target cells in the body. Hormones sent from the pituitary gland include **antidiuretic hormones (ADH),** which stimulate water retention in kidneys in response to increased levels of solutes in the blood, and **oxytocin,** which stimulates the contraction of the uterine walls during childbirth and the secretion of milk from the female nipples.

THE THYROID AND PARATHYROIDS

The thyroid gland produces and secretes two iodine-containing hormones, **thyroxine (T_4)** and triiodothyronine (T_3). Both T_4 and T_3 promote the development of the skeleton and nervous system. In adults, these two hormones help control blood pressure, heart rate, digestion, and certain reproductive functions. In addition, these two hormones stimulate oxidative respiration in most cells, which influences the body's **basal metabolic rate,** the minimum amount of energy required to maintain body functioning.

Four **parathyroid glands** are attached to the thyroid gland. These glands produce parathyroid hormone (PTH), which stimulates the release of calcium from bones and the reabsorption of calcium from urine. Normal calcium levels are required for proper functioning of muscles, which can spasm due to low blood calcium levels, as well as the nerve system and the endocrine system.

THE PANCREAS

The pancreas controls the body's blood glucose levels through the secretion of the antagonistic, or opposing, hormones insulin and glucagon. Increases in glucose levels in the blood stimulate the production and secretion of **insulin,** which causes glucose to be taken up by liver, muscle, and adipose cells. Glucose is then stored either as **glycogen** in the muscle and liver cells or as **fat** in the adipose cells until blood glucose levels drop. Low glucose levels stimulate the production and secretion of **glucagon,** which promotes the breakdown of glycogen and fat, to release glucose into the bloodstream.

EXAMPLE: A high level of glucose in the blood is the result of a low rate of insulin production. Without insulin to stimulate glucose uptake by the liver and adipose cells, glucose remains in the blood. To maintain proper glucose levels, people suffering from the disease diabetes are required to inject insulin obtained from other sources, such as other animals or genetically engineered bacteria, to decrease their glucose levels.

THE ADRENAL GLANDS

The two adrenal glands, the **adrenal medulla** and the **adrenal cortex,** help the body manage stress. The adrenal medulla produces a group of hormones called catecholamines, which include the hormones **epinephrine** (also called adrenaline) and **norepinephrine** (noradrenaline). Nerve cells release the neurotransmitter **acetylcholine** in response to stress, which can be either positive, such as extreme pleasure, or negative, such as exposure to extreme cold. Acetylcholine stimulates the adrenal medulla to secrete epinephrine and norepinephrine. Effects produced by these two hormones include increased basal metabolic rate, increased glycogen breakdown, increased heart rate, and increased blood supply to central organs, depending on the specific stimulus.

Rather than being stimulated by neurotransmitters, the adrenal cortex reacts to stress through stimulation from the endocrine system. The hypothalamus responding to a stress stimulus will stimulate the anterior pituitary gland to secrete **adrenocorti-cotropic hormone (ACTH).** ACTH, in turn, stimulates the adrenal cortex to produce and release corticosteroids.

Among the corticosteroids produced in the mammalian adrenal cortex, **glucocorticoids** increase the breakdown of muscle proteins into amino acids and the conversion of glycogen in the liver to glucose. Mineralocorticoids, other steroids produced by the adrenal cortex, affect salt and water balance in the body. The mineralocorticoid **aldosterone** stimulates the kidney to reabsorb water and sodium ions from kidney filtrate to maintain homeostasis of ion and water levels in the blood.

THE GONADS

The gonads, the testes in males and the ovaries in females, are the glands that produce the sex hormones. The three major categories of sex hormones, **androgens, estrogens,** and **progestins,** are produced in both males and females, although the levels of each hormone differ significantly between the sexes, resulting in varying effects.

The testes in males produce and secrete high levels of androgens, which stimulate the development of the male sexual organs and secondary sex characteristics. **Testosterone,** one of the major androgens, promotes male characteristics such as a low voice and increased body hair growth.

The ovaries in females produce high levels of estrogens and progestins. Estrogens maintain the female reproductive system and are responsible for the development of female secondary sex characteristics. Estradiol, the most important of the estrogens, stimulates female characteristics such as breast development and changes in the body's contours. Progestins, such as **progesterone,** contribute to the preparation of the uterus to house the developing embryo and maintenance of the uterus during pregnancy. Progesterone stimulates the **endometrium** (tissue surrounding the uterus wall) to thicken prior to implantation of a fertilized egg. Low levels of progesterone, resulting from a lack of fertilization, stimulate the breakdown of the endometrium.

THE PINEAL GLAND

The pineal gland contains light-sensitive receptors to regulate the body in response to changes in light exposure. The pineal gland produces the hormone melatonin, which regulates daily cycles and rhythms, such as sleeping and waking patterns. Melatonin is secreted in the dark hours of the night. As a result, during the longer nights of the winter months, the pineal produces and secretes melatonin in higher levels than in summer months.

THE THYMUS

The thymus secretes the hormone **thymosin,** which stimulates maturation of **T cells** used by the immune system to defend the body against disease. The thymus is particularly large in children but shrinks with the onset of puberty.

Summary

Chemical Signals

- **Hormones** are regulatory chemicals secreted into the bloodstream by **endocrine glands.**

- **Neurotransmitters** are regulatory chemicals released into the bloodstream by neurons in the nervous system.

- **Paracrine regulators** are regulatory chemicals released by organs for use in that organ.

- Endocrine glands release hormones. An endocrine gland may be an organ that functions solely to release hormones or an organ that releases hormones in conjunction with performing another bodily function.

- Four groups of peptides are secreted by glands in the endocrine system: **polypeptides, glycoproteins, amines,** and **steroids.**

- Hormones are either lipophilic (fat soluble) or lipophobic (water soluble). Lipophilic hormones enter **target cells** and interact with the cell's contents; lipophobic hormones interact with the cell's membrane receptors.

- The four groups of **paracrine regulators** are **cytokines,** growth factors, neurotrophins, and **prostaglandins.**

- Prostaglandins are the most diverse group of paracrine regulators and are produced in almost every organ for a variety of purposes.

- Invertebrates produce and use many hormones.

- The endocrine system in simple invertebrates works separately from the nervous system. Endocrine and nervous systems often work in conjunction in more complex invertebrates.

The Vertebrate Endocrine System

- Vertebrates use over fifty different hormones.

- Hormones are secreted from endocrine glands, as well as from other organs in the body.

Human Endocrine Glands

- The ten major endocrine glands that regulate bodily functions in humans are the hypothalamus, the **anterior** and posterior **pituitary glands,** the thyroid and **parathyroids,** the **pancreas,** the adrenal glands, the gonads, the pineal gland, and the thymus.

Sample Test Questions

1. Describe a negative feedback loop using an example from the endocrine system.

2. Describe the difference between the three chemical regulators: hormones, neurotransmitters, and paracrine regulators.

3. Explain how the two groups of hormones, lipophilic and lipophobic, interact with their target cells.

4. Diabetes is most likely the result of a disorder in which of the following endocrine glands?

   A. The adrenal glands
   B. The pancreas
   C. The pituitary glands
   D. The pineal gland
   E. The ovaries

5. Which of the following hormone pairs is antagonistic?

   A. Oxytocin and antidiuretic hormone
   B. Melatonin and thymosin
   C. Glucocorticoids and glucagons
   D. Insulin and glucagon
   E. Estrogens and progesterone

6. The hypothalamus produces hormones secreted by which gland?

   A. Anterior pituitary
   B. Adrenal medulla
   C. Posterior pituitary
   D. Adrenal cortex
   E. Pineal

7. Which of the following hormones is most likely to induce the feeling of jet lag?

   A. Thymosin
   B. Insulin
   C. Prolactin
   D. Glucagons
   E. Melatonin

8. Which hormone is responsible for stimulating milk production in females?

   A. Oxytocin
   B. Thymosin
   C. Prolactin
   D. Testosterone
   E. Progesterone

9. The secretion of which of the following hormones results in an increase in blood calcium levels?

    A. Glucose
    B. Insulin
    C. Parathyroid hormone
    D. Calcitonin
    E. Androgens

10. Androgens are involved in which of the following processes?

    A. Circadian rhythms
    B. Development of female secondary sex characteristics
    C. Increased metabolic rate
    D. Development of male secondary sex characteristics
    E. Decreased metabolic rate

11. The lack of sufficient levels of which hormone during development results in certain forms of dwarfism?

    A. Thyroid-stimulating hormone
    B. Follicle-stimulating hormone
    C. Luteinizing hormone
    D. Growth hormone
    E. Elongation hormone

12. Which of the following anterior pituitary hormones stimulates the adrenal cortex to secrete glucocorticoids?

    A. LH
    B. ACTH
    C. TSH
    D. FSH
    E. TRH

13. The loss of which of the following endocrine glands would be particularly detrimental to the immune system?

    A. The pineal gland
    B. The thyroid gland
    C. The parathyroid gland
    D. The adrenal cortex
    E. The thymus

CHAPTER 19
The Endocrine System

14. Which of the following endocrine glands is responsible for responses to stress?

    A. Adrenal glands
    B. Pineal gland
    C. Posterior pituitary gland
    D. Anterior pituitary gland
    E. Thyroid gland

15. Which hormone stimulates the breakdown of glycogen by the liver and the subsequent release of glucose into the blood?

    A. Insulin
    B. Oxytocin
    C. Melatonin
    D. Glucagon
    E. Thyroxine

ANSWERS

1. A negative feedback loop is created when the response produced by one stimulus acts as a stimulus to diminish the initial stimulus. The hypothalamus stimulates the thyroid gland to release thyroxine. High levels of thyroxine in the blood in turn stimulate the hypothalamus to decrease or stop stimulating the thyroid gland.

2. Hormones are secreted from endocrine glands into the bloodstream, where they are transported to cells throughout the body. Neurotransmitters are released by neurons into the bloodstream, where they are transported to cells throughout the body. Paracrine regulators are released by organs in the body and remain to perform functions in those organs.

3. Lipophilic hormones pass through the membranes of their target cells to perform functions in that cell. Lipophobic hormones, in contrast, cannot pass through target cell membranes. Instead, these hormones interact with receptors on the outside of the cell membrane. These receptors communicate with a second messenger in the cell, which then performs the function directed by the hormone.

4. B   Diabetes is most likely the result of a disorder in the pancreas, which regulates glucose levels in the blood through the production of insulin and glucagon.

5. D Insulin and glucagon are antagonistic. When glucose levels in the blood are high, the pancreas secretes insulin to stimulate the liver, and adipose cells store glucose in the form of glycogen and fat cells. When glucose levels are low, the pancreas secretes glucagon to stimulate the release of glucose from glycogen and fat stores.

6. C The hypothalamus produces hormones secreted by the posterior pituitary.

7. E Mclatonin, secreted by the pineal gland, regulates daily or seasonal rhythms based on light levels in the external environment. Jet lag is the result of a disruption of sleep and waking patterns brought on by the effects of melatonin.

8. A Oxytocin, secreted by the posterior pituitary gland, stimulates milk production and the contraction of the uterine walls prior to birth.

9. C The secretion of parathyroid hormone by the parathyroid glands raises blood calcium levels. Parathyroid hormone has an antagonistic relationship with calcitonin, a hormone secreted by the thyroid gland to lower blood calcium levels.

10. D The male testes produce androgens, such as testosterone, to stimulate the development of the male sexual structures and secondary sex characteristics.

11. D Growth hormone is secreted by the anterior pituitary, upon stimulation by the hypothalamus. Certain cases of dwarfism result from a lack of sufficient levels of growth hormone during development.

12. B The hypothalamus stimulates the production and secretion of adrenocorticotropic hormone (ACTH) in the anterior pituitary. ACTH travels to the adrenal cortex, where it helps to stimulate the production and secretion of glucocorticoids.

13. E The loss of the thymus gland would be especially detrimental to the immune system. The thymus produces and secretes the hormone thymosin, which promotes the development of T cells, critical factors in any immune response.

14. A The adrenal medulla, which is part of the adrenal glands, secretes the hormones epinephrine and norepinephrine, two hormones involved in the body's various responses to stress. These hormones stimulate increases in blood pressure and breathing rate, the redirection of blood flow through vessels, and the overall increase in metabolic rate.

15. **D** The hormone glucagon stimulates the breakdown of glycogen by the liver and the subsequent release of glucose into the blood. Glucagon is produced in the pancreas and is in an antagonistic relationship with insulin, which acts to lower blood glucose levels.

THE NERVOUS SYSTEM

Development of the Nervous System

Neurons, Nerve Signals, and Synapses

The Central Nervous System

The Peripheral Nervous System

20

Development of the Nervous System

All animals engage in some degree of interaction with their environment, in which they identify and respond to external stimuli. Identification is generally managed by sensory receptors, and responses are conducted by motor effectors. For both processes to work, information is relayed from one organ to another through a system of nerves. Each nerve is made up of a **neuron,** a cell that transmits **nerve impulses,** and supporting cells that provide the neuron with structural support, nourishment, and insulation.

All members of the animal phylum, with the exception of the Porifera (sponges), contain nerves throughout their bodies. The development of the nervous system from the most primitive cnidarians to the most complex primate, humans, occurred through several major evolutionary advances, including the formation of the nerve net and radial nerves, the division of the central and peripheral components in bilaterally symmetrical organisms, and the grouping of nerves at one end of the body leading to cephalization, or the development of the head.

MAJOR ADVANCES IN THE DEVELOPMENT OF THE NERVOUS SYSTEM

The nervous system found in the cnidarians, such as corals, box jellies, jellyfish, and anemones, is composed of a **nerve net,** a web connecting neurons to one another. Nerve nets, the simplest of nervous systems, convey information to the animal about its environment but do not control activity within the animal's body. The radial nerves found in echinoderms, such as starfish, sea urchins, and sea cucumbers, represent an evolutionary modification over nerve nets. **Radial nerves** spread out from a central cluster of nerves known as a nerve plexus and allow independent movement of echinoderm appendages.

The emergence of the central and peripheral components of the nervous system, first exhibited by the flatworms, accompanied the development of bilateral symmetry. Flatworms have two **nerve cords,** thick bundles of nerves running down the length of the body. Peripheral nerves extend outward from each nerve cord to the muscles of the body. The two nerve cords meet at the head end of the flatworm, forming a primitive central nervous system and marking the first signs of cephalization in animals.

Cephalization

Cephalization describes the formation of a head end where nervous tissue is concentrated. With the development of cephalization, nervous systems became gradually more complex and diversified. The concentration of nerve tissue in the head led to the formation of the brain. The rest of the nervous system continued to differentiate into a central nervous system (CNS), composed of the brain and nerve cord, and peripheral nervous system (PNS), a network of nerves that transmit signals to and from the CNS to the rest of the body.

The most advanced form of cephalization is displayed in the vertebrates. Vertebrate brains are separated into the hindbrain, midbrain, and forebrain, each controlling different functions within the body. In the more advanced vertebrates, these sections are further subdivided, allowing for an even more specialized control over the body's functions.

The nervous system of a flatworm

The nervous system of a vertebrate

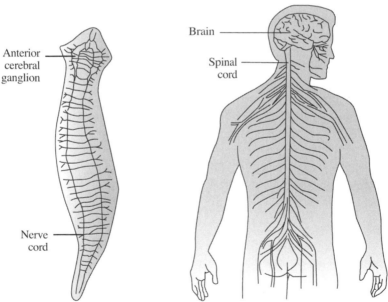

Anterior cerebral ganglion

Brain

Spinal cord

Nerve cord

Cephalization in the Flatworm and Human

Neurons, Nerve Signals, and Synapses

The fundamental component of all nervous systems is the neuron, or nerve cell. Neurons transport impulses throughout the body and relay information between the central and peripheral nervous systems in more advanced animals. Three types of neurons are present in all vertebrates and most invertebrates:

- **Sensory neurons** transmit impulses from **sensory receptors,** specialized structures that pick up external and internal stimuli, to the central nervous system.

- **Interneurons** form a bridge between sensory and motor neurons and transfer nerve impulses between the two. The central nervous system is made up almost entirely of interneurons, which process complex reflexes and allow higher associative functions required for learning and memory.

- **Motor neurons** transmit impulses from the central nervous system to parts of the body, such as muscles and glands, to initiate responses to a particular stimulus.

NEURONS

The majority of neurons share the same basic functional design:

- A cell body containing a **nucleus**

- **Dendrites,** extensions from the cell's body that pick up information from other cells

- A long extension from the cell body, known as the axon, along which impulses travel

- Supporting cells, called **neuroglia,** that remove waste products from the neuron, supply nutrients, and guide impulses along the axon

Many neurons, especially long neurons such as those in the spine or neurons with a large diameter, are supported by special neuroglia called **Schwann cells** (present in the PNS) and oligodendrocytes (present in the CNS). These special support cells wrap around the neuron axon to form numerous layers, called **myelin sheaths.** Myelin sheaths increase the speed of the nerve impulse as it travels along the neuron.

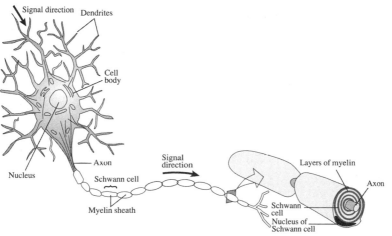

Neuron (Nerve Cell)

NERVE SIGNALS

Neurons communicate with each other through electrical impulses transmitted along the nervous system. The transfer of an impulse throughout the nervous system occurs as an **action potential,** or nerve impulse, travels from cell to cell. All neurons maintain a **resting membrane potential** in which the inside of the cell is negatively charged compared with the outside of the cell. The resting membrane potential of most neurons is between –40 millivolts (mV) and –90 mV, with –70 mV the average. When a neuron responds to a stimulus, its membrane potential is suddenly and temporarily altered. This change results in a nerve impulse that is transferred along the neuron's axon to the area of the neuron body that receives the stimulus.

Stimulating a Nerve Impulse

The following steps describe the process of stimulating a nerve impulse:

1. Before an impulse is sent, the cell maintains its resting membrane potential.

2. **Gated ion channels** in the cell's membrane open in response to a sensory stimulus, such as light or heat, or chemical. This opening allows ions to be transferred across the cell membrane, causing the membrane potential to depolarize, or become less negative.

3. The membrane continues to depolarize until it reaches a **threshold potential.** An action potential (nerve impulse) forms once this threshold is reached.

4. The action potential travels from the neuron cell body, down the length of the axon, and on to another cell. The action potential will continue to pass through the nervous system, from neuron to neuron, until it reaches a target cell, such as a muscle cell, where it will stimulate that cell to perform an action.

SYNAPSES

An impulse travels from neuron to neuron along the nervous system to deliver information throughout the body. A narrow gap, called a **synaptic cleft,** separates a presynaptic cell, the neuron whose axon transmits the impulse, from a postsynaptic cell, the neuron receiving the impulse. A **synapse** at the end of an axon forms the junction, or bridge, over this gap, across which an impulse is transferred. Synapses can be either **electrical,** in which the action potential transmits directly across the synapse, or **chemical,** in which chemicals transfer the impulse across the synapse. Electrical synapses are often found in invertebrates but rare in vertebrates. Chemical synapses, which are a more complex method of transferring impulses, make up the vast majority of vertebrate synapses. Almost all synapses in the central nervous system are chemical.

Chemical Transfers

Any nerve impulse being transferred across a chemical synapse follows the same basic process. Within the presynaptic axon, **synaptic vesicles** merge with the axon membrane and release **neurotransmitters,** chemicals that transfer impulses, into the synaptic cleft. The neurotransmitters bind to **receptor proteins** on the membrane of the postsynaptic cell on the opposite side of the synaptic cleft, causing ion channels on the postsynaptic cell membrane to open.

If the neurotransmitter is a **stimulant,** the open ion channels will allow an influx of ions that move the membrane toward its threshold potential, eventually activating an action potential and passing the impulse on. If the neurotransmitter is an **inhibitor,** the open ion channels will allow an influx of ions that move the membrane away from its threshold potential, preventing the impulse from traveling farther.

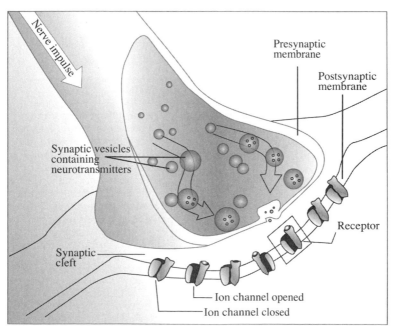

Release of Neurotransmitters

The Central Nervous System

The vertebrate central nervous system consists of a spinal cord and brain. As vertebrates have evolved, the brain has become both larger and more specialized. Primates, including humans, display a high level of specialization, particularly in the cerebrum, which makes up the majority of the brain.

THE HUMAN BRAIN

The diverse range of functions and activities performed by the human body are each controlled by different regions of the brain. Most of these regions are divided between the two halves of the human brain, with each side controlling functions on the opposing side of the body. For example, the right side of the brain instructs movement of the left limbs, and vice versa.

The brain is composed of three major regions: the forebrain, midbrain, and hindbrain. These regions are further divided into separate components, all of which play varying roles in body functioning. Major components of the human brain are described in the following table.

Structure	Position	Functions
Brainstem (the medulla oblongata, pons, and midbrain)	Hindbrain	The medulla oblongata and the pons both control visceral functions, such as breathing, swallowing, and digestion. In addition, these regions integrate sensory and motor nerve signals passing between the brain and the body. The midbrain collects information from sensory neurons and sends it to specialized regions within the forebrain.
Cerebellum	Hindbrain	The cerebellum's primary function is coordination of movement based on information received from joints, muscles, and the auditory and visual senses. In addition, the cerebellum plays a role in both learning and memorizing motor responses.

CONTINUED

Structure	Position	Functions
Thalamus	Diencephalon (part of the forebrain)	The thalamus collects sensory information and relays it to specific regions of the brain, such as auditory information relayed to the temporal region of the cerebral cortex and visual information relayed to the occipital region.
Hypothalamus	Diencephalon (part of the forebrain)	The hypothalamus regulates autonomic functions, such as body temperature, thirst, hunger, and emotional states, which help maintain homeostasis. In addition, the hypothalamus regulates the endocrine system by controlling functions of the pituitary gland.
Cerebrum	Forebrain	The cerebrum processes impulses that influence language and communication, conscious thought, memory, personality, and emotions.

The Cerebrum

Starting with the amphibians and continuing to the reptiles, birds, and mammals, the forebrain evolved to play an increasingly dominant role in processing sensory information. The large mammalian brain, relative to body mass, is attributed to an enlarged cerebrum, which is the region of the forebrain devoted to associative activities such as learning and memorization. In humans specifically, the forebrain surrounds the other areas of the brain and makes up a large majority of the brain's size.

The cerebrum is divided into several regions that control many vital bodily functions. The major regions of the cerebrum and their key functions in the human body include the following:

- The **cerebral cortex** takes up 80 percent of the brain and is its most complex part. It is divided into left and right hemispheres, and each hemisphere is divided into four regions: the parietal, frontal, occipital, and temporal lobes. Each lobe monitors specific body functions: The occipital lobe controls vision, the temporal lobe controls hearing and smell, the parietal lobe integrates sensory information, and the frontal lobe controls motor functions.

- The **corpus callosum** is a thick band of nerve fibers that connect the left and right hemispheres of the cerebral cortex. Information is passed between the two hemispheres via this band.

- The **basal ganglia** receive sensory information from the body and motor commands from other regions of the brain. They output commands that are then sent down the spinal cord to stimulate body movement.

- The **hippocampus** is responsible for emotional responses and memory formation and recall.

- Sensory data from sources, such as the thalamus and brainstem, converges in the **amygdala.** This region also organizes emotional information and sends emotional impulses to the body.

- The **reticular formation** is a collection of neurons that controls sleep and arousal via the **reticular activating system (RAS).** The RAS wakes the body when stimulated by outside factors, such as the presence of light. The release of the neurotransmitter serotonin from cells in the brainstem decreases the activities of the RAS, resulting in sleep.

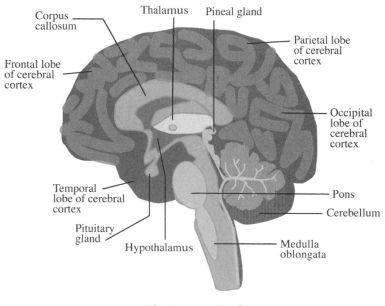

Corpus callosum

Thalamus

Pineal gland

Parietal lobe of cerebral cortex

Frontal lobe of cerebral cortex

Occipital lobe of cerebral cortex

Temporal lobe of cerebral cortex

Pons

Cerebellum

Pituitary gland

Hypothalamus

Medulla oblongata

The Human Brain

CHAPTER 20
The Nervous System

THE SPINAL CORD

A series of neurons form a cable, called the **spinal cord,** descending from the brain down along the backbone. All information traveling back and forth between the brain and the body passes through the neurons of the spinal cord.

The spinal cord is divided into two zones:

- The inner zone composed of interneurons, motor neurons, and the supporting cells **neuroglia**

- The outer zone composed of sensory axons along the dorsal (back) column of the spinal cord, and motor axons along the verntral (front) column

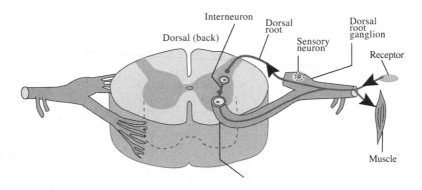

The Spinal Cord

Reflexes

The spinal cord also functions in producing **reflexes,** sudden, involuntary movements of muscles that do not require higher processing to be carried out. Reflexes are characterized as either simple or complex:

- **Simple reflexes:** Direct communication between a sensory neuron and a motor neuron in the spinal cord produces a motor response. Simple reflexes produce a jerking motion when the knee is tapped.

- **Complex reflexes:** An interneuron relays information between the sensory and motor neuron. In complex reflexes, interneurons may also send information regarding the stimulus from the sensory neuron to the brain. The involuntary blink of an eye to avoid contact with a foreign object, such as dust or an insect, is an example of a complex reflex.

The Peripheral Nervous System

The peripheral nervous system connects the body with the CNS and transfers information between the two. Three types of cells make up the nerve tissue in the PNS:

- **Cranial nerves** extend from the brain and stimulate the organs of the head and upper body.

- **Spinal nerves** extend from the spinal cord and stimulate activity in the rest of the body.

- **Ganglia,** the supporting cells of the PNS, surround the nerves and perform tasks such as removing wastes, supplying nutrients, and assisting the transport of nerve impulses along the axon.

THE SENSORY AND MOTOR DIVISION OF THE PNS

Cranial and spinal nerves can be separated into two divisions:

- The **sensory division** consists of sensory neurons that send information from the internal and external environments to the CNS.

- The **motor division** consists of motor neurons that relay impulses from the CNS to cells throughout the body, directing their responses to environmental stimuli.

THE SOMATIC AND AUTONOMIC NERVOUS SYSTEM

The motor division can be further divided by systems regulating either under voluntary or involuntary control:

- The **somatic (voluntary) nervous system** stimulates the contractions of the skeletal muscles. Although the majority of activity in the somatic nervous system results in voluntary reactions, such as the movement of limbs, automatic reflexes such as blinking are also the result of activity in the somatic nervous system.

- The **autonomic (involuntary) nervous system** controls smooth and cardiac muscle and organs of gastrointestinal, cardiovascular, excretory, and endocrine systems.

THE SYMPATHETIC AND PARASYMPATHETIC SYSTEM

The autonomic nervous system can be further divided into the sympathetic and parasympathetic divisions, both coordinated by the medulla oblongata, which is located in the hindbrain. Activation of the **sympathetic division** prepares the body for action by increasing energy consumption and blood flow. During times of stress, the sympathetic system coordinates those activities that increase the energy necessary for vital processes, such as heart rate and metabolic rate, and decrease energy expended on nonvital processes, such as digestion. Activation of the **parasympathetic division** relaxes the body by decreasing energy consumption and heart rate. The parasympathetic division promotes increases in bodily functions, such as digestion, that gain and conserve energy for the body. The antagonistic (opposing) relationship of the activities of these two divisions helps the body maintain stability (homeostasis) by regulating the cycle gaining and consuming energy.

THE SENSES

The body detects sensations, such as heat, pain, taste, and sound, through **sensory receptors** that respond to various stimuli. These receptors relay information about that stimulus to the brain. The brain then forms a perception, which is an interpretation of the meaning of the stimulus, on which a response is based.

The body uses two specific sensory receptors to detect internal and external stimuli:

- **Interoreceptors** detect internal stimuli and convey information about the internal environment. For example, interoreceptors in the arteries detect changes in blood pressure.

- **Exteroreceptors** detect external stimuli and convey information about the external environment. For example, exteroreceptors in the skin detect changes in air temperature.

Transmission of Sensory Information

Stimuli from a sensory receptor follow a four-step process on their way to the brain:

1. **Stimulation:** A sensory receptor detects a stimulus.

2. **Transduction:** Energy from the stimulus changes the membrane potential of the receptor cell, creating graded potentials that accumulate until an action potential forms.

3. **Transmission:** The action potential travels along nerve axons until it reaches the central nervous system.

4. **Interpretation:** The brain receives the action potential and creates a perception of the stimulus. The brain then forms a response based on the perception. For example, if the brain perceives that the body is cold, the nervous system might trigger shivering to increase body temperature.

EXAMPLE: Contact with a sharp object stimulates pain receptors in the bottom of the foot. The opening of ion channels in the nerve initiates transduction, leading to the transmission of the action potential. The action potential travels from the PNS to the brain, where it is interpreted as pain. The brain reacts by sending nerve signals to the muscles of the legs, which cause the foot to move away from the sharp object.

Senses and Sensory Receptors

Biologists categorize interoreceptors and exteroreceptors according to the type of stimuli they detect and use to create action potentials. The following table summarizes the types of

CHAPTER 20
The Nervous System

receptors and their use by the five senses to detect sensations in the environment.

Type of Receptor	Sense	Detection of Stimuli
Mechanoreceptors are stimulated by physical change such as pressure, touch, sound, and motion, which alter the receptor membrane.	Sound and touch	Mechanoreceptors in the ear respond to motion in sound waves. The brain perceives this motion as sound. Mechanoreceptors in the skin respond to changes in pressure, such as a breeze or more intense movements such as contact by another person, which the brain perceives as touch.
Electromagnetic receptors respond to forms of electromagnetic energy, including visible light, electricity, and magnetism.	Sight	**Photoreceptors** are electromagnetic receptors that respond to light. Photoreceptors in the eyes respond to changes in light, which the brain perceives as vision.
Chemoreceptors are stimulated by changes in chemical compositions, such as the concentration of solutes within a solution. In addition, these receptors can be stimulated by specific chemicals that bind to the receptor membrane.	Smell and taste	**Olfactory receptors** in the nose respond to chemicals that bind to their membranes. The brain perceives these chemicals as smells. **Gustatory receptors** in taste buds respond to chemicals that the brain perceives as tastes.

In addition to these three receptors, the body uses two other receptors: pain receptors and thermoreceptors. **Pain receptors** are stimulated by specialized sensations such as excess heat or pressure or certain chemicals released from damaged areas of the body. **Thermoreceptors** located in the skin are stimulated by changes in temperature, such as heat or cold.

Summary

Development of the Nervous System

- Organisms detect and respond to stimuli through the use of the nervous system.

- A nerve is a cell composed of a **neuron** that transmits a **nerve impulse** and supporting cells that support, nourish, and insulate the neuron.

- The **nerve net** is the simple nervous system found in all cnidarians, including corals, box jellies, jellyfish, and anemones.

- **Radial nerves,** a modified version of the nerve net, are found in echinoderms, such as starfish.

- As animals developed bilateral symmetry, the nervous system developed into central and peripheral components. Flatworms were the first animals to display this form of nervous system.

- **Cephalization,** also present in flatworms, is the formation of a head end where nervous tissue concentrates. The nervous system gradually became more complicated in organisms displaying cephalization.

Neurons, Nerve Signals, and Synapses

- The main components of the nervous system are neurons, or nerve cells.

- There are three types of neurons present in all vertebrates and most invertebrates: **sensory neurons, motor neurons,** and **interneurons.**

- Neurons have a cell body, **dendrites,** and an **axon.** They are accompanied by supporting cells known as **neuroglia.**

- Neurons communicate with each other via **actions potentials,** or nerve impulses.

- A nerve impulse passes from cell to cell through **gated ion channels** in the cell's membrane open that cause the membrane potential to depolarize. When the membrane reaches a threshold potential, the nerve impulse is transmitted.

- A **synapse** is a junction between two cells across which an action potential is transferred. Synapses can be either chemical or electrical.

- Nerve impulses are transferred across a **synaptic cleft** when the presynaptic neuron releases **neurotransmitters.** Neurotransmitters travel across the synaptic cleft and bind to neurotransmitters on the postsynaptic cell.

The Central Nervous System

- The central nervous system is made up of the brain and **spinal cord.**

- The human brain has three major regions: the forebrain, midbrain, and hindbrain.

- The brain has five major components: the brainstem, the cerebellum, the thalamus, the hypothalamus, and the cerebrum.

- The cerebrum is the largest component of the brain and performs most associative functions, including the interpretation of language and communication, conscious thought, memory, and emotion. The cerebrum also processes sensory input and coordinates the body's motor responses.

- The cerebrum is divided into two hemispheres separated by a layer of nerve tissue called the **corpus callosum.** The regions of the cerebrum include the **cerebral cortex,** the **basal ganglia,** the **hippocampus,** the **amygdala,** and the **reticular formation.**

- The spinal cord is a cable of neurons extending from the brain down the backbone.

- The spinal cord transmits reflexes, which are sudden, involuntary movements of muscles.

The Peripheral Nervous System

- The peripheral nervous system transfers information back and forth from the internal and external locations of the body to the CNS.

- The PNS is made of **cranial nerves, spinal nerves,** and **ganglia.**

- The PNS is also divided into the sensory division and the motor division.

- The sensory division consists of **sensory neurons** that send information from the internal and external environments to the CNS.

- The motor division sends signals from the CNS to cells throughout the body, directing their responses to their environment. The motor division

  is divided into the **somatic (voluntary) nervous system,** which controls skeletal muscles, and the **autonomic (involuntary) nervous system.**

- The autonomic nervous system is divided into the **sympathetic** and **para-sympathetic systems,** which perform antagonistic activities in the gaining and consuming of energy to maintain stability in the body.

- **Sensory receptors,** which can be either **interoreceptors** or **exteroreceptors,** respond to various stimuli, such as heat, pain, taste, and sound.

- Interoreceptors detect internal stimuli and convey information about the internal environment, while exteroreceptors detect external stimuli and convey information about the external environment.

- The transmission of information from sensory receptors to the brain occurs in four steps: **stimulation, transduction, transmission,** and **interpretation.**

- Sensory receptors are also categorized according to the type of stimuli they detect and use to create action potentials.

  » **Mechanoreceptors** are stimulated by physical change.

  » **Electromagnetic receptors** respond to forms of electromagnetic energy.

  » **Chemoreceptors** are stimulated by changes in chemical compositions.

  » **Pain receptors** are stimulated by pain.

  » **Thermoreceptors** are stimulated by changes in temperature.

Sample Test Questions

1. Describe the steps that occur when a neuron is stimulated from its resting membrane potential to send a nerve impulse.

2. Identify the four regions of the cerebral cortex and describe the major function of each.

3. Daniel touches a hot stove with his hand. Describe the four steps that occur as he formulates a response to the stimulus.

4. Which type of nervous system is present in a jellyfish?

   A. Nerve net
   B. Nerve plexus
   C. Radial nerves
   D. Nerve cords
   E. Central nerves

5. Which of the following terms describes the formation of a head end where nervous tissue is concentrated?

   A. Generalization
   B. Adaptation
   C. Cephalization
   D. Differentiation
   E. Interpretation

6. Which part of a neuron is an extension of the cell body that picks up information from other cells?

   A. Nucleus
   B. Axon
   C. Neuroglia
   D. Dendrite
   E. Synapse

7. Which of the following cells in the PNS wraps around the axon of a neuron to produce a myelin sheath?

    A. Schwann cells
    B. B cells
    C. Oligodendrocytes
    D. Receptor cells
    E. Mast cells

8. To transmit a nerve impulse, the cell membrane potential depolarizes until an action potential is produced. What point must the membrane potential reach for an action potential to be produced?

    A. Depolarization potential
    B. Transfer potential
    C. Resting membrane potential
    D. Target potential
    E. Threshold potential

9. Which part of the brain coordinates motor responses?

    A   Cerebrum
    B. Thalamus
    C. Hypothalamus
    D. Cerebellum
    E. Brainstem

10. Which of the following is NOT a region of the cerebral cortex?

    A. Parietal lobe
    B. Frontal lobe
    C. Posterior lobe
    D. Occipital lobe
    E. Temporal lobe

11. Which part of the cerebrum connects the two hemispheres of the cerebral cortex and allows information to pass between them?

    A. Corpus callosum
    B. Basal ganglia
    C. Hippocampus
    D. Amygdala
    E. Reticular formation

12. Which of the following is required to relay information between a sensory neuron and a motor neuron in a complex reflex?

A. Synaptic vesicle
B. Neurotransmitter
C. Interneuron
D. Myelin sheath
E. Action potential

13. Which of the following systems controls the skeletal muscles?

A. Somatic nervous system
B. Autonomic nervous system
C. Sympathetic nervous system
D. Parasympathetic nervous system
E. Involuntary nervous system

14. Which of the following activities would result from activating the parasympathetic nervous system?

A. An increase in energy consumption
B. An increase in heart rate
C. An increase in digestion
D. An increase in metabolic rate
E. An increase in breathing rate

15. Which type of receptor detects smells that enter the nose?

A. Mechanoreceptors
B. Photoreceptors
C. Thermoreceptors
D. Electromagnetic receptors
E. Chemoreceptors

ANSWERS

1. When a neuron is stimulated, gated channels open and allow ions to flow across the cell membrane, depolarizing the cell membrane. Depolarization continues until the cell reaches its threshold potential. An action potential is generated when a cell reaches its threshold potential. The action potential passes from the cell body, along the axon, and to another cell, where the process begins again.

2. The cerebral cortex is divided into the parietal, frontal, occipital, and temporal lobes. The parietal lobe integrates sensory functions; the frontal lobe coordinates motor functions; the occipital lobe controls vision; and the temporal lobe controls hearing and smell.

3. Sensory receptors in the skin of the hand are stimulated by the contact with the hot stove. This stimulus changes the membrane potential of the receptor cell, creating an action potential. The action potential travels along nerve axons to the central nervous system. Once the brain receives the action potential, it interprets the stimulus as heat and pain. The central nervous system stimulates the peripheral nervous system to respond, directing Daniel's muscles to move his hand away from the source of heat.

4. **A**   All cnidarians have nerve nets, the simplest type of nervous system.

5. **C**   Cephalization is the formation of a head end where nervous tissue concentrates.

6. **D**   Dendrites are long extensions from the cell's body that pick up information from other cells in the body.

7. **A**   Schwann cells are neuroglia that wrap around the axons of nerve cells in the PNS to produce a myelin sheath.

8. **E**   A cell membrane must reach its threshold potential for an action potential (nerve impulse) to be produced.

9. **D**   The primary function of the cerebellum is to coordinate movement via information received from joints, muscles, and the auditory and visual senses.

10. **C**   The cerebral cortex is divided into the parietal, frontal, occipital, and temporal lobes.

11. **A**   The corpus callosum is a thick band of fibers connecting the left and right hemispheres of the cerebral cortex.

12. **C**   Interneurons are required to relay information between the sensory and motor neuron in a compound reflex.

13. A The somatic (voluntary) nervous system is the part of the PNS controlling the skeletal muscles.

14. C Activation of the parasympathetic nervous system promotes an increase in functions, such as digestion, that result in the gaining and conserving of energy.

15. E Olfactory receptors in the nose are a type of chemoreceptor that responds to chemicals that bind to their membranes.

ANIMAL REPRODUCTION AND DEVELOPMENT

Asexual and Sexual Reproduction

The Human Reproductive System
and Process

Human Development

Asexual and Sexual Reproduction

Reproduction results in the creation of new individuals, or offspring, from either one or two parent organisms. Offspring can either be genetically identical to the parent, as is the case in **asexual reproduction,** or vary genetically from their parents, as is the case in **sexual reproduction.** In both cases, genetic material is passed on from one generation to the next. The processes involved in each form of reproduction, and consequently the resulting genetic variation in the offspring, differ greatly.

ASEXUAL REPRODUCTION

In asexual reproduction, the parent passes on an exact replica of its genetic material to the offspring, resulting in an offspring, also referred to as a clone, genetically identical to the parent organism. No genetic variation exists from one generation to the next in organisms that reproduce asexually, apart from variation introduced by random and uncontrollable mutations during gene replication.

Asexual reproduction allows the rapid reproduction of numerous offspring in the absence of a partner. An isolated organism reproducing asexually can colonize an ideal habitat very quickly. In addition, the relatively low rate of genetic variation is beneficial to organisms inhabiting marginal or harsh environments to which they are already ideally suited. The lack of genetic variation in asexual reproduction, however, inhibits species adaptation that results from advantageous traits being passed down to subsequent generations, as is the case in sexual reproduction.

Types of Asexual Reproduction

Three main types of asexual reproduction are exhibited in species of the animal kingdom:

1. **Fission:** An individual organism splits into two roughly equal-sized organisms, which then grow to the size of the original. Sea anemones reproduce through fission.

2. **Budding:** Small individuals split off from a parent organism and develop into full-sized adults. Cnidarians, which includes the jellyfish and corals, can reproduce through budding.

3. **Fragmentation/regeneration:** A parent individual splits into several parts, each of which develops into an adult. Some sponges and worms reproduce through fragmentation.

SEXUAL REPRODUCTION

Offspring created during sexual reproduction receive genetic material from two separate parent organisms, a male and a female, that are mixed together to form a fertilized cell, a **zygote.** Females package their genetic material in an immobile haploid gamete (sex cell) called an ovum. Males package their genetic material in a mobile haploid gamete called a sperm. Genetic material from the sperm and ovum combine in a fertilized cell, the zygote, resulting in an offspring that will vary genetically from its parents, as well as from other offspring of those parents.

FERTILIZATION

Animals employ one of two methods of fertilization, the process in which a sperm fuses with an ovum to form a zygote:

- **External fertilization,** in which both parents expel their gametes into another medium, such as water, without necessarily coming into contact with each other.

- **Internal fertilization,** in which the male deposits sperm inside the female reproductive tract.

EXAMPLE: The external fertilization of frogs requires a moist environment. A female frog deposits its eggs into water while the male frog clasped to her back discharges sperm over the eggs. During the mating season, male frogs of some species will clasp on to any female-sized object, whether it is a frog, a tree branch, or a stone, holding on for hours as they wait for it to eject eggs.

Development Following Internal Fertilization

After fertilization, a zygote will begin cell division and form an embryo, which is the multicellular developmental stage of an organism. Embryos resulting from internal fertilization undergo one of three forms of development, depending on where embryonic and fetal development takes place within the organism:

- **Oviparity:** The embryo formed inside the female is deposited outside her body as an egg. After development, offspring hatch out of the egg and directly into the environment. All birds and some reptiles are oviparous.

- **Ovoviviparity:** The embryo develops inside the female body, although it still obtains all nourishment from the egg yolk. The young hatches fully developed and is released from the female's body. Many reptiles and some fish undergo ovoviviparity.

- **Viviparity:** The embryo develops inside the female's body and the young obtain their nourishment from the female's blood, rather than egg yolk. The young emerges fully developed from the female body. Almost all mammals undergo viviparity.

The Human Reproductive System and Process

Humans reproduce through viviparous internal fertilization. Both the male and female possess gonads, called testes in males and ovaries in females, which are specialized reproductive glands for the production and storage of gametes. The female body also has specialized sites for fertilization and embryo development.

THE FEMALE REPRODUCTIVE SYSTEM

The female body contains two ovaries surrounded by round structures, called follicles, which house individual egg cells (ovum). Though women are born with anywhere from tens of thousands to hundreds of thousands of follicles, only a few hundred are released during their lifetimes.

Other important structures in female reproduction are the **vagina,** where sperm is deposited during copulation, the cervix, a tube connecting the vagina and the uterus, and the **uterus,** which provides a place for developing follicles.

The menstrual cycle, during which one follicle matures and releases its ovum, takes place in the female body over the course of approximately twenty-eight days. During the menstrual cycle, the ovum undergoes two phases separated by a period of ovulation:

1. **Follicular phase**

2. **Luteal phase**

THE FOLLICULAR PHASE

The follicular phase can be described in six stages:

1. The follicle-stimulating hormone (FSH) is released from the anterior pituitary gland and stimulates development in several follicles, only one of which will fully mature.

2. One of the stimulated follicles matures into a Graafian follicle.

3. A **primary oocyte,** an immature ovum housed within the Graafian follicle, undergoes its first meiosis division to produce two daughter cells: the **secondary oocyte** and the **polar body.** The secondary oocyte, or egg cell, is larger than the polar body and contains nearly all of the primary oocyte's cytoplasm, better equipping this cell for eventual fertilization. The smaller polar body disintegrates.

4. Stimulation by the luteinizing hormone (LH), which is also released from the anterior pituitary gland, induces the release of the secondary oocyte, beginning the process of **ovulation.** The secondary oocyte moves along one of the fallopian tubes (also called oviducts) with the assistance of cilia (tiny hairs) lining the tube.

5. Fertilization of the egg cell by a sperm cell takes place within the fallopian tube. Unfertilized egg cells disintegrate within a day following ovulation. Fertilized egg cells complete a second meiotic division, resulting in a fully mature ovum.

6. The nuclei of the ovum fuses with the nuclei of the sperm to produce a diploid zygote. Following fertilization, the zygote travels along the fallopian tube to the uterus, where it will develop. The zygote takes approximately three days to move from the fallopian tube to the uterus. Within two to three days of reaching the uterus, the zygote attaches to the **endometrium,** which is the wall surrounding the uterus.

Luteal Phase

The luteal phase can be described in five stages:

1. The release of LH during ovulation stimulates the Graafian follicle, which initially held the primary oocyte, to develop into the corpus luteum.

2. The corpus luteum releases two additional hormones: estradiol (also known as **estrogen**) and **progesterone.** These hormones exert negative feedback pressure on the anterior pituitary gland to signal the termination of LH and FSH secretion. This pressure inhibits further ovulation and the release of additional follicles.

3. Estrogen and progesterone encourage the growth of the endometrium, causing it to become more vascular, glandular, and enriched with glycogen deposits.

4. If the secondary oocyte is not fertilized, the decreased levels of LH and FSH cause the corpus luteum to degenerate, and the release of estrogen and progesterone ceases. The enlarged endometrium breaks down as a result of the decline in these two hormones. Endometrium cells are removed and exit the body along with accompanying bleeding, a process referred to as menstruation.

5. If the secondary oocyte is fertilized, the developing embryo releases the hormone human chorionic gonadotropin (hCG), which maintains the corpus luteum and prevents menstruation.

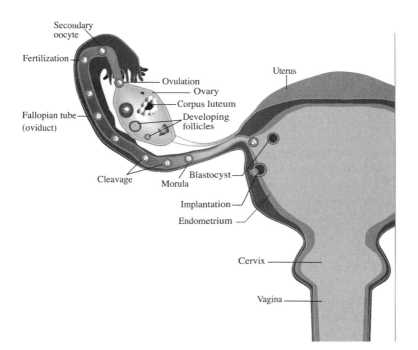

The Menstrual Cycle in the Human Female Reproductive System

THE MALE REPRODUCTIVE SYSTEM

Sperm, the male gametes, are produced in paired **testes,** which are reproductive glands situated inside sacs called the scrotum. Human sperm development takes place in temperatures three degrees lower than normal body temperature. The **scrotum,** which is located outside the abdominal cavity, keeps sperm at a lower temperature than the body.

Sperm Production

Sperm develop in the **seminiferous tubules,** which are highly coiled tubes located in the testes. Germ cells, called spermatogonia, covering the walls of the seminiferous tubules divide by mitosis to produce two diploid cells. One diploid cell, called the **primary spermatocyte,** will then undergo meiotic division to produce four haploid cells, called **secondary spermatocytes.** The other diploid cell remains a spermatogonia, thus ensuring a continuing supply for future sperm production.

Each secondary spermatocyte undergoes a second meiosis to produce sperm cells. Sperm leaving the male reproductive system first move into a temporary storage tube called the **epididymis.** During ejaculation, sperm exit this chamber and enter another tube called the **vas deferens,** which connects to two **ejaculatory ducts.** The ejaculatory ducts connect to the **urethra,** which moves the sperm out of the body.

A sperm cell consists of a head, body, and flagellum (tail). The head consists of a compact nucleus capped off by a vesicle, called an **acrosome.** Enzymes within the acrosome help the sperm dissolve the egg's protective layers. The tail of the sperm is used to propel the sperm as it moves along the female's reproductive tract.

Male Reproductive Glands

Sperm is transported out of the body in a mixture called **semen.** Sperm make up only 1 percent of the semen volume; the remaining volume is composed of fluids that nourish, protect, and help transport the sperm as it moves to the vagina.

Three glands in the male reproductive system produce the remaining fluids in semen.

Gland Name	Fluids Produced
Seminal vesicles	Connected to the vas deferens this gland secretes a fructose-rich fluid that nourishes and protects the sperm. This fluid makes up approximately 60 percent of the semen volume.
Prostate gland	Connected to the ejaculatory ducts, this gland secretes a milky fluid that is alkaline (low pH) and balances the acidity of the vagina and the male urethra. Secretions from this gland comprise approximately 30 percent of the semen volume.
Bulbourethral glands	Pea-sized glands located at the tip of the penis. These glands secrete small amounts of lubricating fluid into the urethra to help lubricate it for the passage of the sperm.

The Penis

Semen exits the body through the **penis.** The penis consists of the urethra surrounded by two columns of erectile tissue:

1. The **corpora cavernosa,** located along the dorsal side of the penis

2. The **corpus spongiosum,** located along the ventral side of the penis

Nitric oxide is released by the nervous system to produce an erection. The nitric oxide dilates blood vessels surrounding the penis, causing it to become engorged with blood. Veins in the penis are compressed, preventing blood that enters through the arteries from leaving. Erection and sexual stimulation leads to **ejaculation,** during which semen is released from the penis. During sexual intercourse, the semen is released into the vagina. Sperm travel through the reproductive tract of the female to the egg.

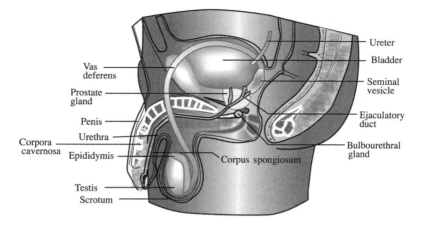

The Human Male Reproductive System

Human Development

Human development begins at **fertilization,** the process in which sperm and egg fuse to form a zygote. The initial development of the zygote is divided into two major stages:

1. Cleavage

2. Gastrulation

CLEAVAGE

In **cleavage,** the zygote divides multiple times to form a ball of cells. While DNA replication and cell division take place rapidly during cleavage, individual cells do not grow. Consequently, this initial ball of cells remains roughly the same size as the original fertilized egg.

The ball, consisting of thirty-two cells, is called a morula, and each cell in it is called a **blastomere.** As cleavage continues, a cavity forms at the center of the morula, into which the blasto-meres secrete fluid. Once the ball reaches a size of five hundred to two thousand cells, it is called a **blastocyst.** At this point in development, the fluid-filled cavity, called the **blastocoel,** concentrates at one end of the blastocyst. The outer layer of blastocyst cells, called the **trophoblast,** will eventually form the placenta, which nourishes the embryo throughout develop-ment and helps dispose of the embryo's waste.

GASTRULATION

Once cleavage is complete, the blastocyst enters the stage known as **gastrulation.** As cell division continues through gastrulation, the cells, now called a **gastula,** begin to organize into three distinct layers of tissue:

1. **Ectoderm**

2. **Endoderm**

3. **Mesoderm**

All tissues found in an adult animal are derived from these three tissue layers.

Gastrula Layer	Adult Tissue
Ectoderm	Found in the epidermis, nervous system, and sensory system
Endoderm	Forms the innermost lining of the digestive tract, as well as associated digestive organs, such as pancreas and liver; also found in parts of the reproductive system
Mesoderm	Found in most other organs and tissues, including the kidneys, heart, muscles, dermis, and the notochord or backbone

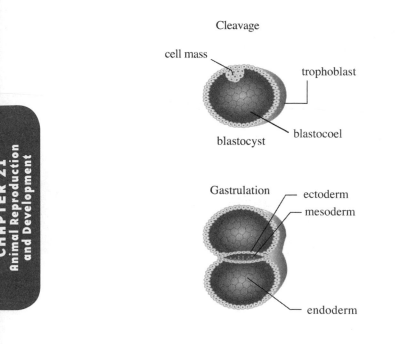

Cleavage

cell mass

trophoblast

blastocoel

blastocyst

Gastrulation

ectoderm

mesoderm

endoderm

Cleavage and Gastrulation

NEURULATION

Gastrulation is followed by a process known as neurulation, in which the dorsal nerve cord, which will eventually form the brain and spinal cord, develops. During neurulation in vertebrates, key architectural structures begin to form within the body that provide a frame for the development of organs (organogenesis):

Key Architectural Structure	Description
Neural tube	A crease along the long axis of the embryo, called the neural groove, begins to pinch together to form a long, hollow cylinder, called the neural tube. The neural tube will develop into the spinal cord and brain.
Notochord	A rodlike structure found only in chordates. The notochord forms from the mesoderm tissue and lies along the dorsal midline of the embryo. This structure eventually forms the backbone in vertebrates.
Neural crest	Prior to formation of the neural tube in vertebrates, a small strip of the neural groove pinches off and forms the neural crest. Cells move from the neural crest to other parts of the embryo during development, where they develop into a variety of body structures, such as organs and sensory neurons.
Nervous system	Some cells that migrate from the neural crest form the body's various sensory neurons.
Sensory organs and skull	Special cell formations called placodes form from ectodermal cells. Sensory organs and the skull then develop from these structures.

CHAPTER 21
Animal Reproduction and Development

EMBRYONIC TISSUE LAYERS

In addition to the ectoderm, endoderm, and mesoderm, four embryonic tissue layers assist the human embryo during development:

1. **Amnion:** Forms the **amniotic sac,** the pouch in which the embryo will float throughout development

2. **Chorion:** Develops into part of the placenta

3. **Yolk sac:** Provides nourishment to the growing fetus

4. **Allantois:** Forms into the umbilical cord

TRIMESTERS

The development of the human embryo is divided into three periods, known as trimesters, each lasting roughly three months.

Trimester	Development
First trimester	The zygote undergoes cleavage and forms an embryo. The placenta develops, allowing the exchange of materials, such as nutrients, gases, and waste products, between the mother and embryo. In the first month of the first trimester, gastrulation, neurulation, and organogenesis take place. In the second month, the embryo undergoes morphogenesis, during which the embryo takes shape and limbs begin to form. The nervous system develops in the third month, and the embryo begins to show reflexes and facial expressions.
Second trimester	The bones of the fetus enlarge. During this period, kicking by the fetus can be felt by the mother. Hair, called lanugo, is formed, although this hair is lost later in development. Toward the end of this trimester, the fetus begins to undergo rapid growth, increasing dramatically in size and weight.
Third trimester	Characterized by fetal growth and organ maturation, rather than development. The fetus increases in weight, as well as neural development, such as the brain and number of neurons. Neurological development continues several months after the birth of the baby.

BIRTH AND POSTNATAL DEVELOPMENT

The end of the third trimester signals the female body to enter **labor,** which is a series of rhythmic uterine contractions that bring about birth. Birth is stimulated in humans by the release of hormones, known as **prostaglandins,** which are released by the uterus. Uterine contractions begin, forcing the fetus downward. The pressure of the fetus stimulates an increased release of hormones, which in turn stimulates contractions and movement of the fetus. This positive feedback system continues until the child is born and the stimulus is removed. Once contractions are strong enough, the baby, followed by the placenta, is forced from the uterus.

Human babies, and other mammal babies, are nursed on milk produced and released from the mother's breast. Nursing usually continues for a year but can be longer or shorter. Following birth, babies grow very rapidly. Organs develop at differing rates in a trend known as **allometric growth.** For this reason, body proportions of babies differ greatly compared to adults.

Summary

Asexual and Sexual Reproduction

• In **asexual reproduction,** genes are passed directly from one parent to the offspring.

• Animals can exhibit three main forms of asexual reproduction: **fission, budding,** and **fragmentation/regeneration.**

• Offspring created through **sexual reproduction** share genetic material from two separate parent organisms.

• In sexual reproduction, the male gametes are **sperm,** which is small and has a flagellum used for locomotion, and the female gamete is an **ovum,** which is larger and stationary.

• **Fertilization** following sexual reproduction can be internal or external.

• Development following fertilization can be **oviparous, ovoviviparous,** or **viviparous.**

The Human Reproductive System and Process

• The female reproductive system consists of the **ovaries,** follicles, **fallopian tubes, uterus, vagina,** and cervix.

• An egg typically travels from follicles within the ovaries, along the fallopian tubes in which fertilization takes place, and into the uterus.

• The vagina and cervix are structures in which sperm are injected during intercourse and the fetus is pushed out during birth.

- Follicle-stimulating hormone (FSH) stimulates the development of follicles and release of egg cells.

- Luteinizing hormone (LH) stimulates the release of the egg from the ovaries into the fallopian tubes.

- Other hormones, mainly **progesterone** and **estrogen,** play major roles in preparing the uterus for embryo implantation.

- The two main phases during the female reproductive cycle, which typically takes thirty days, are the follicular phase, in which the egg is released, and the luteal phase, in which the uterus is prepared for the embryo.

- Menstruation occurs when an egg is not fertilized. The uterine wall breaks down and exits the body, accompanied by bleeding.

- The male reproductive system consists of **testes,** the **scrotum, semi-niferous tubules, vas deferens, ejaculatory ducts, urethra, seminal vesicles, prostate gland, bulbourethral glands,** and the **penis.**

- **Semen** is a fluid made up of sperm and a mixture of secretions from the seminal vesicles, prostate glands, and bulbourethral glands.

- Semen forms as sperm passes from the testes, which are inside the scrotum, through the seminiferous tubes, and along the vas deferens.

- Semen exits the body via the ejaculatory ducts that connect to the urethra.

- The penis comprises the urethra and two columns of erectile tissue. The penis becomes erect and ejaculates the semen during sexual stimulation. During sexual intercourse, the semen is ejaculated into the vagina.

- Sperm from the semen travel along the female reproductive tract to an egg.

Human Development

- Early embryo development takes place in two main stages: **cleavage** and **gastrulation.**

- During cleavage the zygote undergoes multiple divisions and forms a **blastocyst.**

- During gastrulation, the blastocyst undergoes further cell divisions, during which specific tissue layers form: the **ectoderm, endoderm,** and **mesoderm.** These layers give rise to the organs, tissues, and other structures of the body.

- In addition to the ectoderm, endoderm, and mesoderm, four embryonic tissue layers form and surround the embryo during development. These embryonic tissue layers are the **amnion, chorion, yolk sac,** and **allantois.**

- Human development is broken into three periods called trimesters, each of which lasts roughly three months.

- After birth, human babies undergo **allometric growth,** during which organs undergo different rates of development.

CHAPTER 21
Animal Reproduction
and Development

Sample Test Questions

1. Describe some of the major advantages and disadvantages of asexual reproduction.

2. Explain the main developmental stages that occur during the first trimester of human development.

3. How might a benign cyst blocking one of the fallopian tubes affect a woman's ability to have children?

4. Which of the following structures becomes the corpus luteum in the female reproductive system?

   A. The oviduct
   B. The ovary
   C. The follicle
   D. The vas deferens
   E. The uterus

5. Which of the following events is most directly stimulated by high levels of FSH and LH during the ovarian cycle?

A. The growth of one follicle and ovulation
B. The shedding of the uterine wall
C. The development of the corpus luteum
D. Menstruation
E. Prolonged estrus

6. The innermost lining of the digestive tract is derived from which of the following tissue?

A. Ectoderm
B. Both ectoderm and mesoderm
C. Mesoderm
D. Both endoderm and ectoderm
E. Endoderm

7. Which of the following is NOT a tissue layer that exists outside of the human embryo during development?

A. The amnion
B. The yolk sac
C. The corpus luteum
D. The chorion
E. The allantois

8. Which of the following processes involves the breakdown of the endometrium tissue lining the uterus walls?

A. Organogenesis
B. Gastrulation
C. Neurulation
D. Menstruation
E. Morphogenesis

9. Which of the following glands secretes an alkaline fluid present in semen?

    A. Pancreas
    B. Bulbourethral glands
    C. Seminal vesicles
    D. Prostate gland
    E. Epididymis

10. What is the name of the hollow ball of cells formed by the end of cleavage?

    A. Blastocyst
    B. Trophoblast
    C. Placenta
    D. Amnion
    E. Chorion

11. What type of mechanism describes the cycle that induces childbirth, in which hormones stimulate uterine contractions to force a fetus downward, creating further pressure on the uterine tissues stimulating the release of more hormones?

    A. Negative feedback
    B. Allometric growth
    C. Indeterminate growth
    D. Positive feedback
    E. Determinate growth

12. Which type of cell goes through a second phase of meiosis to become sperm during spermatogenesis?

    A. Primary spermatocyte
    B. Secondary spermatocyte
    C. Diploid spermatocyte
    D. Seminiferous cells
    E. Primary oocyte

13. The nervous system is derived from which tissue layer during development?

A. Protoderm
B. Mesoderm
C. Endoderm
D. Metaderm
E. Ectoderm

14. Sponges reproduce through which form of asexual reproduction?

A. Budding
B. Fission
C. Splicing
D. Fragmentation/regeneration
E. Partitioning

15. Which of the following structures is the typical site for fertilization in human females?

A. The ovaries
B. The uterus
C. The fallopian tubes
D. The cervix
E. The vagina

ANSWERS

1. Asexual reproduction provides a mechanism for reproduction in animals that are immobile. Also, asexual reproduction requires little energy, since parts simply fall off the parent organism in budding, fission, and fragmentation. Additionally, the animal is not required to expend energy and resources to produce and protect gametes. Finally, because no genetic recombination occurs, animals are able to colonize a suitable environmental niche and maintain it over numerous generations with little chance for detrimental variation. This final advantage also suggests a possible downside to asexual reproduction. Since clones do not allow for genetic variability in the offspring produced, the population will be very slow to adapt to any changes in the environment. Drastic environmental changes could result in offspring that would not be able to survive and reproduce, threatening the survival of the entire population.

2. The main developmental stages during the first trimester include cleavage, gastrulation, neurulation, and organogenesis. All three developmental stages occur within the first month of the first trimester. During cleavage and gastrulation, the zygote is transformed into a blastocyst, a ball of cells in which three tissue layers form. During neurulation and organogenesis, structures that form the frame of the body develop, allowing the formation of organs to begin.

3. A benign cyst that blocks a fallopian tube would decrease a woman's reproductive capacity by approximately one-half. Women have two ovaries, and a fallopian tube leads from each ovary to the uterus. During sexual reproduction, a mature follicle releases an ovum, which passes out of the ovary and into the fallopian tube. Sperm entering the oviduct may fertilize the ovum. Cilia in the tube move the ovum toward the uterus. Development begins after a fertilized embryo is implanted in the wall of the uterus. An egg prevented from moving from the ovary to the fallopian tube would not be able to be fertilized by sperm.

4. C   After the follicle matures and releases its ovum, the remaining follicular tissue develops into the corpus luteum.

5. A   Levels of follicle-stimulating hormone (FSH) and luteinizing hormone (LH) are relatively high at the beginning of the ovarian cycle. These hormones stimulate the growth of a follicle and ovulation.

6. E   The endoderm tissue forms the lining of the digestive and respiratory systems, as well as the liver, thyroid, pancreas, parathyroid, thymus, urinary system, and some parts of the reproductive system.

7. C   The corpus luteum is a structure in the female reproductive system. The four external embryonic tissue layers are the amnion, the chorion, the yolk sac, and the allantois.

8. D   Menstruation occurs when an egg fails to fertilize. The reduction in hormones stimulating endometrium enlargement causes the tissue to be sloughed away and exit the body, accompanied by bleeding.

9. D   The seminal vesicles secrete a fluid that nourishes and protects the sperm. This fluid makes up approximately 60 percent of the semen volume.

10. A   At the end of cleavage, the hollow ball of cells is known as a blastocyst. The outer layer of cells of the blastocyst is called the trophoblast. The trophoblast will eventually form the placenta, which nourishes the embryo throughout its development and also helps dispose of the embryo's waste.

11. D   The cycle that induces the body to give birth is an example of a positive feedback mechanism, in which a stimulus amplifies a response until the stimulus is stopped. During birth, the pressure of the fetus amplifies the release of hormones, which stimulates further pressure from the fetus. Once the fetus is born, the response subsides.

12. B   In sperm production, diploid cells in the seminiferous tubules undergo mitosis and then differentiate into primary spermatocytes. Each primary spermatocyte will undergo meiosis I to produce two haploid secondary spermatocytes. At this point, each secondary spermatocyte undergoes a second phase of meiosis to produce sperm.

13. E   The ectoderm becomes the epidermis, the lining of the mouth, rectum, tooth enamel, parts of the eye, and the nervous system.

14. D   Sponges are capable of reproducing via fragmentation and regeneration. If a sponge is split up into several parts, each part is capable of developing into an adult.

15. C   An ovum moves out of the ovary and into the fallopian tubes. Fertilization typically takes place in the fallopian tubes. The resulting zygote then undergoes division and continues to move toward the uterus.

ANIMAL BEHAVIOR

22

Nature and Nurture

Behavior Development

Communication

Behavioral Ecology

Nature and Nurture

An animal's **behavior** refers to the manner in which it responds to stimuli in its environment. Behavior influences the way animals reproduce, find food, and communicate with other animals. Behavior can be preprogrammed according to an organism's genetic makeup (instinct), or influenced by an animal's life experiences (learning). Nature, what biologists consider instinct, refers to genetic factors that determine behavior. Nurture refers to experiences after birth that influence an organism's behavior.

NATURE

Biologists initially believed all animal behavior to be **innate,** or instinctive. An animal was born with a set of predetermined responses, governed by genetics, to all manner of stimuli. Although recent research has shown that behavior is influenced by a mixture of nature and nurture, instinct does play a large role in determining behavior. Researchers have identified three models to illustrate the ways that innate factors can be applied to certain behaviors.

Model	Description	Example
Preset pathways	Animals are preprogrammed to perform tasks or elicit a response when triggered by a specific stimulus.	The belly of a male stickleback fish grows bright red when preparing to mate. During breeding season, males react aggressively as they approach each other. However, males will also react aggressively when approaching any bright red object. Biologists believe the male stickleback fish are preprogrammed to react aggressively to the color red.
Inherited responses	Behaviors are passed on through genes from one generation to the next.	Mice adept at learning to complete a maze are bred with other equally skilled mice. Offspring tend to be even faster learners than the parents. In the same way, slow-learning mice will produce slow-learning offspring.
Gene-controlled behavior	Different versions, or alleles, of a specific gene control specific behaviors.	Geneticists determined that the allele *fosB* determines whether a female mouse will be nurturing toward her offspring. Mice with the allele *fosB* nurtured their young, while mice without the *fosB* allele did not.

NURTURE

Learning refers to an animal's ability to change its behavior in response to past experience. Learning can take one of two main forms: **nonassociative learning** and **associative learning.**

Learning	Example
Non-associative learning: An animal develops a behavior without forming a connection between two stimuli or between a stimulus and a response.	**Habituation** is nonassociative learning in which an animal learns to ignore stimuli that have no positive or negative consequences. For example, a young bird may hide as a response to objects passing overhead. Over time, after objects, such as other birds, pass frequently overhead with no impact to the bird, the response lessens.
	Sensitization is the opposite of habituation. A particular behavior increases after repeated exposure to a stimulus. For example, a marine slug will withdraw its gill in response to a strong prodding. In experiments in which an electric shock is administered, the slug learns to withdraw its gill at even the slightest tap.
Associative learning: An animal changes its behavior by forming a connection between two stimuli or between a stimulus and a response.	**Classical conditioning** occurs when two stimuli are presented to an animal repeatedly. Eventually, the animal learns to respond identically to each stimulus when presented separately. For example, a dog will not naturally salivate when presented with a ringing bell but will salivate when presented with meat. After repeated exposure to both stimuli together, the dog associates salivating with the ringing bell. Eventually, the dog learns to salivate when presented with just the ringing bell.
	Operant conditioning occurs when a reward or punishment is presented following a certain response to stimuli. For example, a toad that is stung while attempting to eat a bee will associate the bee with pain. The toad learns not to eat anything that resembles the bee's shape and color patterns.

Because animals cannot form associations with every stimulus they encounter, instinct plays an important role in determining which stimuli an animal will form an association with. For example, pigeons have evolved to associate specific colors with individual types of seeds, allowing the pigeon to more easily find the seeds when foraging. However, pigeons do not associate particular sounds with seeds, because this association does not benefit them when foraging.

Behavior Development

Interactions between instinct and learning mold an animal's behavior during development. Biologists study many forms of behavior in an attempt to determine which factors shape these forms of behavior. Based on their studies, biologists have determined that some behaviors are purely instinctive, while others are a combination of instinct and learning.

EXAMPLE: The cuckoo bird lays its eggs in other birds' nests and abandons them. The young hatchling, therefore, grows up in the absence of its parents. Yet, despite never hearing the male cuckoo song, an adult cuckoo is capable of performing the tune perfectly. This phenomenon indicates that the cuckoo bird's song is purely instinctive.

EXAMPLE: Male birds of a particular species develop a specific courtship song while maturing to an adult. Young males who hear an older male sing the song can copy it and, with practice, perfect it. Birds who do not hear an adult sing are still capable of singing the song; however, they do so poorly. This phenomenon indicates that although the song is instinctive, learning is required to perfect it.

IMPRINTING

Imprinting refers to the process whereby social attachments developed as an animal matures influence behavior later in life. There are two main forms of imprinting:

- **Filial imprinting:** Social attachment formed between a parent and offspring influence behavior. In many animal species, social interactions between parents and offspring are necessary for normal behavioral development.

- **Sexual imprinting:** An animal learns to direct its sexual behavior toward members of the same species. Individuals that are raised by a different species will often attempt to mate with the adoptive species, rather than members of their own species.

COGNITIVE BEHAVIOR

Cognitive behavior refers to behavior influenced by conscious thought. Although the extent to which other animals are capable of thought is a subject of debate among biologists, examples from both nature and experiments suggest that many animals are capable of some degree of thinking. Observations of problem-solving behavior, where an animal demonstrates a novel response after processing information from the environment, provide the strongest support for cognitive behavior in animals.

EXAMPLE: In its natural habitat, a chimpanzee will strip a branch of all its leaves and use that branch to flush termites from a nest. The preparation displayed in stripping the branch suggests that the chimpanzee consciously intends to use the bare branch for a specific purpose.

Many experiments have been designed to test cognitive abilities in animals. In one classic experiment, a chimpanzee is placed in a room with several boxes and a banana dangling out of reach. The chimpanzee moves the boxes to a position that allows it to climb up and reach the banana. In this experiment, the chimpanzee appears to use cognitive thinking to devise a novel response to a problem.

DETERMINING MOVEMENT

An animal's movements can be determined by a combination of instinct and learning influenced by environmental stimuli. For example, some insects are drawn toward light; others are drawn away. Long-distance movements, called **migration,** are common among a variety of animals, including butterflies, birds, and turtles.

EXAMPLE: Starlings possess an instinctual internal compass that allows them to navigate using the Earth's magnetic field. However, starlings also navigate by way of learned migratory routes. As it grows older, a starling will adjust its route according to learned information.

Communication

Social groups commonly form among species of insects, fish, birds, and mammals that are capable of some form of sharing information, or communicating. **Communication** is the process by which animals share information, such as warnings about potential predators, methods of attack against prey, or the whereabouts of food sources, with other members of their social groups.

Animals can use three types of cues to communicate, as shown in the following table.

	Cue	Examples
1.	**Visual cues:** Behaviors or particular visual features, such as color, of other individuals.	Fireflies produce a flash of light that identifies males to females and vice versa. Stickleback fish communicate the willingness to mate with members of the opposite sex through various behavioral movements, such as a head bob.
2.	**Chemical cues:** The release of specific pheromones, which are chemical messengers recognizable to individuals of the same species.	Male silk moths have receptors on their antennae that detect sex pheromones released by female silk moths.
3.	**Auditory cues:** The production of sounds recognizable by other individuals.	Vocal sacs on a bullfrog produce a mating call as air enters and leaves. Females can distinguish the male bullfrog's call from calls of different species. Similarly, many male birds produce songs to attract females or to warn other males of their presence.

LEVELS OF SPECIFICITY

Levels of specificity measure the amount of information a specific signal communicates about the sender. Depending on their function, some signals will simply identify an individual's presence, while other signals provide information about an individual's identity. A signal can be either individually specific, species specific, or between species.

- **Individually specific:** The song of a male bird informs neighboring birds of its presence, either to warn other males away or to attract females.

- **Species specific:** The call of a bullfrog informs a female frog of its presence and willingness to mate.

- **Between species:** Potential prey communicates to a predator that it is aware of the predator's presence, impairing the predator's ability to hunt the prey.

PRIMATE COMMUNICATION

Primates have the most advanced communication abilities of all members of the animal kingdom. Many primate species use a complex vocabulary, where individual sounds communicate specific meanings. The velvet monkey, for example, can warn members of its social group of the presence of specific predators, using a distinct vocalization for different predators.

The human language incorporates numbers and symbols to communicate intricate meaning. Biologists believe humans are genetically predisposed to learn language. Human infants have been shown to recognize consonant sounds, and all infants also undergo a babbling phase, in which they learn to mimic the sounds of language. As development continues, children begin to learn the thousands of words that make up language and to create meaning by combining those words.

Behavioral Ecology

Behavioral ecology is an area of biology that seeks to determine how natural selection shapes those animal behaviors that are genetically based. Behavioral ecologists ask two basic questions when determining how natural selection influences a particular behavior:

1. **Is the behavior adaptive or adapted in response to the environment?** Not all behavioral changes are adaptive, and therefore not all behavior is shaped by natural selection. Genetic drift, selection of unrelated traits, and gene flow can influence changes in inheritable traits that affect behavior.

2. **How does an adaptive behavior increase an animal's fitness, or reproductive success?** Behaviors can increase fitness by attracting fitter mates, decreasing vulnerability to predators, or increasing energy intake that can be put toward producing more offspring. Behaviors that increase fitness have a greater likelihood of being passed on to later generations.

EXAMPLE: Ecologist Niko Tinbergen observed that gulls removed broken eggshells from their nests after eggs hatched. To determine if gulls are demonstrating an adaptive behavior, Tinbergen designed an experiment in which he left broken eggshells in a gull's nest. Tinbergen observed that the white of the eggshells attracted crows, which prey on gull hatchlings, to the nest. Tinbergen concluded that the gull's behavior is adaptive, since it increased the chance of survival of the gull hatchlings.

FITNESS TRADE-OFFS

While some behaviors may increase one aspect of an animal's fitness, that same behavior may be detrimental to another aspect of the animal's fitness. These trade-offs occur frequently in foraging and territorial behavior, where an increase in energy intake or protection of territory comes at the expense of exposure to predators. Although natural selection will generally favor behaviors that maximize energy gain, other factors that increase fitness, such as predator avoidance and mating opportunities, may also influence behavior selection.

> **EXAMPLES:** Shore crabs may benefit by gaining more energy from feeding on larger mussels. However, the increased exertion required to open the shell of the larger mussels would cost the shore crab in energy expenditure.
>
> A male bird defending a territory containing many breeding females may benefit from the increase in mating opportunities. However, that male bird may assume greater risk of injury from fighting with competing males.

REPRODUCTIVE STRATEGIES

Reproduction is necessary for animals to pass genetic material to the subsequent generations. The cost of reproduction, sexual selection, and a species' mating system all influence the reproductive behaviors, or strategies, used to increase reproductive success.

Cost of Reproduction

Reproduction requires the use of resources at the expense of the parent. Cost of reproduction refers to the amount of resources an animal invests into reproducing offspring. Species and individuals exhibit either low or high investment in the cost of reproduction.

- **Low investment,** such as small gametes or no parental care after mating, results in behavior in an individual, particularly the male of a species, that increases the frequency of mating opportunities.

- **High investment,** such as large gametes or parental care after mating, results in behavior in an individual that encourages greater selectivity when choosing a mate. High investment is most often required of females and results in **mate choice,** a behavior in which the female evaluates several potential males and chooses the one she determines to be most fit.

The roles of males and females are reversed in some species, where males demonstrate high investment and display mate choice. In other situations, males and females demonstrate equal levels of investment, leading to similar mating behavior in both sexes.

Sexual Selection

Sexual selection refers to the competition for mating opportunities. The limit to the reproductive opportunities for both males and females in any animal population influence the development of behaviors to enhance an individual's changes for reproductive success. The traits of successful individuals will be passed on to offspring and therefore have a higher likelihood of influencing the gene pool of the population.

Individuals in a population compete through two methods of sexual selection:

- **Intrasexual selection:** Individuals of the same sex, generally male, compete with each other for mating opportunities. Traits that allow one male to outcompete another will be favored.

   EXAMPLE: Deer with large antlers have an advantage in fights and will likely outcompete deer with smaller antlers. The trait for large antlers will be passed on, and over time large antlers will become more common in the population.

- **Intersexual selection:** Individuals of one sex attempt to entice individuals of the opposite sex to mate with them. In some cases, the benefits associated with a certain mate will be direct or obvious. In other cases, benefits that lead an individual to choose a certain mate are indirect and therefore not obvious.

EXAMPLE: A female that requires protection for her and her offspring will likely mate with the largest male. The benefit in this scenario is direct, since a larger male will provide better protection. Many female birds, however, demonstrate indirect mate selection by choosing the mate with the brightest coloring. Bright coloring does not provide an obvious or direct benefit; however, bright coloring may be an indication that the mate is healthy. This selection provides the indirect benefit of passing on genes for good health to the offspring.

Sexual selection often leads to sexual dimorphism, which refers to differences between the sexes. For example, in many species where intrasexual selection occurs, males tend to be larger than females. Over time, the size of males in a population will increase as larger-size males outcompete smaller males. Females in the same population will remain the same size.

Mating Systems

The potential number of individuals an animal mates with during a breeding season is characterized by different mating systems. Three mating systems, monogamy, polygyny, and polyandry, have evolved in each species based on their interactions among individuals and the environment.

	Mating System	Interaction
1.	**Monogamy,** in which one male mates with one female.	Both parents tend to provide care for altricial young, which are young that require long and extensive care, reducing the tendency for one individual to leave and mate with other individuals. Therefore, many animals, such as birds, that produce altricial young are **monogamous.**
2.	**Polygyny,** in which one male mates with more than one female.	Precocial young, which are young that require little parental care, are generally cared for by only one parent, allowing the other parent to leave and mate with other individuals in the population. Many animals that produce precocial young are polygamous. Polygyny is particularly beneficial to females in a population where a male defends a territory and those females that mate with the male, as is the case in elephant seal populations.
3.	**Polyandry,** in which one female mates with more than one male.	Animals that produce precocial young may also be polyandrous. Polyandrous females, such as the spotted sandpiper, will mate with more than one male, leaving several males to provide parental care for the offspring.

EXAMPLE: Recent research has revealed that many birds thought by biologists to be monogamous actually display polygamous and polyandrous behavior. Scientists using DNA testing to determine the identity of fathers of individual offspring in a nest have found that more than one male fathered the offspring in some cases. Scientists, therefore, conclude that the females must have been mating with more than one male. Scientists call this extra-pair copulation (EPC). Biologists theorize that males perform EPC to increase the number of offspring, while females perform EPC to either enhance the genes of their offspring with those of genetically superior males or to enlist more helpers to aid in chick-rearing.

ALTRUISM

Altruism, which is most common among animals that live in social groups, characterizes behavior performed by one individual that is beneficial to others in a population but detrimental to the individual. For example, an individual sounding a warning call after sighting a predator will allow others to hide but will draw the predator's attention. Biologists debate the evolution of altruism, which seems to contradict the expected behavior of individual survival. In theory, if an altruistic behavior is detrimental to an organism, it would not be favored by natural selection and the behavior would become diminished over time.

Biologists have proposed three major theories to explain the development of altruism.

	Theory	Example
1.	**Reciprocal altruism** suggests that partnerships exist in which an individual provides a benefit to others by performing an altruistic behavior but benefits from others who also perform that behavior. Reciprocal altruism also suggests nonreciprocators are cut off from these benefits.	Vampire bats that find a source of blood give a small amount of their blood to other vampire bats in the group. This behavior prevents other vampire bats from starving. This behavior is reciprocated when another bats finds a source of blood. Vampire bats do not give blood to individuals that have failed to reciprocate in the past.
2.	**Kin selection** asserts that the reproductive success of a relative is beneficial to an individual who shares some of the same genetic material. A percentage of a relative's genes will be present in the offspring, ensuring that at least some of the relative's genetic material is passed on to the next generation, even if they never reproduce themselves.	Studies have shown that individual ground squirrels that have relatives nearby are more likely to sound an alarm to alert others of a predator.
3.	**False altruistic acts** characterize acts that appear altruistic but are actually performed for the benefit of the individual.	An animal that issues a warning call to others appears to be sacrificing itself. However, the reaction of other individuals in the population may actually direct attention away from the caller.

CHAPTER 22
Animal Behavior

CHAPTER 22
Animal Behavior

SOCIAL SYSTEMS

A group of animals living together in a cooperative manner form a **society.** Social systems present several distinct benefits to the individuals living in them:

- Related individuals benefit from additional help provided by others in the group.

- The presence of more individuals decreases any individual's chances of being singled out by a predator.

- Individuals may learn of food sources from other individuals in their group.

Social behavior in societies is often based on altruism. Insects such as ants, bees, wasps, and termites tend to form very structured, altruistic societies based on **caste systems.** In caste systems, individuals of the same species perform different tasks, such as foraging or protecting, based on characteristics such as size. Many vertebrates also form social groups in which individuals differ based on characteristics and tasks performed. However, vertebrate groups tend to be less organized and less altruistic than those observed among species of insects.

Summary

Nature and Nurture

- Nature, or instinct, refers to genetic factors that determine **behavior.**

- There are three ways that genetics can determine behavior: preset pathways, inherited responses, and gene-controlled behavior.

- **Nurture** refers to experiences after birth that determine behavior.

- Learning, which refers to an animal's ability to change its behavior based on past experience, is divided into nonassociative learning and associative learning.

- **Nonassociative learning** refers to an animal's ability to develop a behavior without developing a connection between two stimuli or between a stimulus and a response. Nonassociative learning is divided into habituation and sensitization.

- **Associative learning** refers to an animal's ability to change its behavior by forming a connection between two stimuli or between a stimulus and a response. Associative learning is divided into classical conditioning and operant conditioning.

Behavior Development

- Interactions between nature and nurture mold an animal's behavior during development.

- Behavior can be either instinctive or a combination of instinct and learning.

- **Imprinting** refers to the process where social attachments developed as an animal matures influence behavior later in life. Imprinting is divided into **filial imprinting** and **sexual imprinting.**

- **Cognitive behavior** refers to a response that is determined through the processes of thought. The extent to which animals are capable of cognition is a subject of debate, although problem-solving behavior observed in some animals suggests that they do think.

- The movement of animals, including **migrating,** can be a combination of instinctual and learned behaviors.

Communication

- **Communication** is the process by which animals share information with other members of its social group.

- Communication is used to inform members about potential predators, organize attacks against prey, or to inform members of the whereabouts of food sources.

- Communication can be based on visual cues, chemical cues, or auditory cues.

- Communication signals can be individually specific, species specific, or between species.

- Primate communication is the most advanced form of communication and involves the use of a vocabulary to signal specific meaning.

- Communication is more complex in humans than in other primates. Biologists believe the human ability to learn language is genetically programmed.

Behavioral Ecology

- **Behavioral ecology** is an area of biology that seeks to determine how natural selection shapes animal behaviors.

- Behavioral ecologists consider whether a specific behavior is adaptive and how that adaptive behavior increases an animal's fitness.

- Many behaviors that increase fitness in one way reduce fitness in another way. This is known as a fitness trade-off.

- Reproduction is the only way an individual can pass on its genetic material to subsequent generations. Animals use many strategic behaviors to increase their reproductive success.

- The cost of reproduction is the amount of resources an animal invests in reproducing offspring.

- **Sexual selection** refers to the competition for mating opportunities. In **intrasexual selection,** same-sex individuals compete with each other for mating opportunities. In **intersexual selection,** individuals of one sex try to convince the opposite sex to choose them as mates.

- Mating systems can be monogamous, polygamous, or polyandrous.

- **Altruism** is an act performed by one individual that is detrimental to the individual performing the action but beneficial to others in the community. Behavioral ecologists debate the evolution of altruism since it seems to contradict what would be expected in nature.

- Reciprocal altruism, kin selection, and false altruistic acts are three theories that attempt to explain altruism.

- A **society** is a group of animals living together in a cooperative manner. Social behavior is often based on altruism, although the extent of altruism varies between societies.

Sample Test Questions

1. In an experiment, mice are trained to associate a ringing bell with food. Each time a bell rings, food is placed at one end of the cage. Eventually, mice run to the end of the cage after a bell is rung, even when food is not placed there. What form of conditioning is being performed, and how does it alter mice behavior?

2. Explain how intrasexual selection leads to sexual dimorphism.

3. Explain how the cost of reproduction influences reproductive behavior.

4. Which term describes experiences after birth that determine behavior?

   A. Nature
   B. Nurture
   C. Instinct
   D. Genetics
   E. Inheritance

5. Which form of learning describes an animal's tendency to ignore stimuli that have no positive or negative consequences?

   A. Associative learning
   B. Sensitization
   C. Habituation
   D. Classical conditioning
   E. Operant conditioning

6. A rabbit that attempts to eat the leaves of a rose bush is pricked by the thorns. The rabbit learns not to eat rose bushes. This an example of which type of learning?

   A. Nonassociative learning
   B. Sensitization
   C. Habituation
   D. Classical conditioning
   E. Operant conditioning

7. What type of imprinting is based on social attachment formed between a parent and offspring?

   A. Filial imprinting
   B. Genetic imprinting
   C. Social imprinting
   D. Sexual imprinting
   E. Instinctive imprinting

8. Problem solving is an example of which type of behavior?

   A. Genetic behavior
   B. Instinctive behavior
   C. Cognitive behavior
   D. Social behavior
   E. Altruistic behavior

9. Chemical messengers that allow communication between individuals of the same species are known as what?

   A. Hormones
   B. Receptors
   C. Lipids
   D. Alleles
   E. Pheromones

10. What is the term for a behavior that increases the fitness of an animal in one respect but decreases the fitness of that animal in another?

    A. A trade-off
    B. Cause and effect
    C. Altruism
    D. Foraging
    E. Genetic drift

11. Which form of selection describes the competition between male elephant seals for female mating partners?

    A. Heterosexual selection
    B. Intrasexual selection
    C. Homosexual selection
    D. Intersexual selection
    E. Extrasexual selection

12. Which term describes a female of a species that mates with more than one male?

A. Monogamous
B. Polygamous
C. Polyandrous
D. Altricial
E. Precocial

13. Which theory suggests that altruistic behaviors are based on partnerships?

A. Reciprocal altruism
B. Caste system
C. Kin selection
D. Cooperative living
E. False altruistic acts

14. Which theory states that altruistic behaviors are performed because the reproductive success of a relative is beneficial to an individual that shares some of the same genetic material?

A. Reciprocal altruism
B. Caste system
C. Kin selection
D. Cooperative living
E. False altruistic acts

15. Which type of animal tends to form very structured, altruistic societies?

A. Mammals
B. Insects
C. Birds
D. Fish
E. Reptiles

ANSWERS

1. This experiment involves classical conditioning. Mice are presented with two stimuli: the ringing bell and food. The stimuli are always presented together, causing the mice to form a connection, or association, between them. Eventually, the association becomes so strong that the mice react to just the ringing bell.

2. Intrasexual selection describes how same-sex individuals compete with each other for mating opportunities. In many species, the males generally compete with other males for the females. Any heritable trait that allows males to outcompete other males, such as size, is likely to be passed on to successive generations. The larger males reproduce more often, and the trait for large size is passed on to a large portion of the population. Future generations of males become universally larger, leading to sexual dimorphism in which the males of the species become gradually larger, while the females remain the same size.

3. Individuals that do not provide parental care after mating have a low cost of investment. These individuals usually try to mate as much as possible and are not selective in regards to the partner they choose. Animals with high investment, such as having to provide parental care after mating, tend to mate less often and are more selective in the male partner they choose.

4. B Nurture refers to experiences after birth that determine behavior.

5. C Habituation is a form of nonassociative learning in which an animal learns to ignore stimuli that have no positive or negative consequences.

6. E Operant conditioning describes how the rabbit's behavior changes because of the punishment that follows a certain response to stimuli.

7. A Filial imprinting is based on the social attachment formed between a parent and offspring.

8. C Cognitive behavior describes responses determined by thought.

9. E Pheromones are chemical messengers an individual releases as cues to communicate with other members of the same species.

10. A Fitness trade-off refers to behaviors that increase fitness in one respect but decrease fitness in another.

11. B   In intrasexual selection, same-sex individuals compete with each other for mating opportunities.

12. C   A female that mates with more than one male is polyandrous.

13. A   The theory of reciprocal altruism suggests that an individual that performs an altruistic act will do so to benefit from the altruistic behavior of others.

14. C   Kin selection states that the reproductive success of a relative is beneficial to an individual that shares some of the same genetic material.

15. B   Insects tend to form very structured, altruistic societies.

POPULATION DYNAMICS & COMMUNITY ECOLOGY

Populations

Population Change

Communities

Community
  Interactions

23

Populations

A **population** is a group of individuals of the same species living in the same geographic location. The properties of populations change over time as the population adapts or responds to changes. Three main characteristics define populations and the changes they undergo:

1. Range

2. Spacing

3. Size

POPULATION RANGE

Population **range** refers to the area that the population inhabits at any given time. The range of a population may fluctuate if the environment becomes more or less suitable for the population.

> *EXAMPLE:* Increases in temperature can allow populations to expand northward, as areas that were previously too cold for a population to inhabit become warmer and more tolerable. Decreases in temperature can have the opposite effect, confining populations to a smaller suitable area and reducing the range.

Range may also change if the population moves to a new, more hospitable area.

> *EXAMPLE:* A native African bird crossed nearly 2,000 miles of ocean to North America. The new habitat met the bird's requirements, so it was able to occupy the area and begin colonizing.

POPULATION SPACING

Population **spacing** refers to how the individuals in a population are distributed within that population. Individuals can be spaced in three ways:

1. **Random spacing:** Individuals do not interact a great deal with one another or with items in their environments, such as soil type or water sources. This type of spacing is uncommon in nature.

2. **Uniform spacing:** Individuals develop their own space, or territory, and will often exclude other individuals from it. This common type of spacing usually results from competition for resources in the area. For example, a lion will have a set territory that it considers its hunting space and will defend this territory from other lions.

3. **Clumped spacing:** Individuals group together. This kind of spacing may result from the uneven distribution of resources. For example, a species of monkey may clump together in a certain type of preferred tree. The formation of social groups also causes clumped spacing. For example, grazing animals, such as sheep, often travel together in large groups.

POPULATION SIZE

Population **size** refers to the number of individuals in a population. Population size is influenced by the rate at which the population is increasing, the availability of resources, and the maximum population size a habitat can support.

A **metapopulation** is a network of distinct populations that are able to interact with one another. Interactions occur when one population receives individuals, or **immigrants,** while another population sends out individuals, or **emigrants.** Population size can be affected by this exchange of individuals.

Population Change

Population dynamics refers to how a population changes over time and is affected by various factors such as the population's reproduction rate, proportion of different aged individuals, and mortality, or death rate. **Demography** is the analysis of statistics about these factors.

FACTORS AFFECTING GROWTH RATES

The main factors affecting the growth rate of a population are outlined in the table below.

Factor	Example
Sex ratio is the ratio of males to females in a population.	Consider a species in which each male mates with several females. In this species, a small number of males will not drastically affect the number of births. But in a monogamous species, in which one male mates with one female, a small number of males will decrease the number of births, because fewer matings can occur.
Generation time is the average time between the birth of an individual and the birth of its offspring.	Generation times vary across species. Smaller-sized animals tend to have shorter generation times than larger animals. When all other factors are equal, populations with short generation times can increase in size more rapidly than populations with long generation times.
Age structure refers to the proportion of different age groups within a population. A group of individuals of the same age is known as a cohort. Different age groups tend to have different birthrates, or fecundity, and different death rates, or mortality.	Young individuals are usually fecund and have low mortality rates; old individuals are not as fecund and have higher mortality rates. If a population has a high percentage of young members, the growth rate is likely to be fast. If a population is older, the growth rate is likely to be slow.

REPRODUCTION EVENTS

The number of reproductive events an organism undergoes during its lifetime also affects reproductive success. Some organisms undergo **semelparity,** or the production of as many offspring as possible in one reproductive event.

> *EXAMPLE:* Some animals are unlikely to survive between breeding seasons: species that live in harsh conditions, for example, or those that migrate long distances to spawn. Species of this type tend to marshall all their resources in the service of one big reproductive event, rather than in an attempt to survive.

Other organisms undergo **iteroparity,** producing offspring many times throughout their lifetimes.

> *EXAMPLE:* Humans are likely to survive each breeding season, so they reserve some resources for the survival effort. If they succeed in surviving, they give themselves more chances to reproduce.

Some long-lived animals put off the age of first reproduction, using their energy to grow and gain experience, rather than to reproduce. For short-lived animals, time is limited, and the benefits gained from experience are outweighed by the benefits gained by reproducing as early as possible.

COST OF REPRODUCTION

Environmental resources are not abundant enough to allow organisms to reproduce and care for offspring continuously. Therefore, organisms must make trade-offs between reproduction and survival. The cost of reproduction is the reduction in future reproductive potential caused by current reproduction events.

Effect	Factor	Explanation
Low cost of reproduction	Abundant resources	If there is plenty of food and water, for example, parents and offspring will not have to share scant resources.
	High mortality rates	If an organism is unlikely to survive until the next breeding season, reproducing will not affect its future survival.
High cost of reproduction	Scarce resources	More offspring will reduce the resources available, which will decrease the survival rate and future reproductive success rates.
	Danger involved in reproduction	The process of reproduction can imperil the survival of an organism. For example, some animals are likely to be targeted as prey while carrying offspring.

CHAPTER 23
Population Dynamics

THE RATE OF POPULATION INCREASE

Populations can increase either because of births in the population or because of the immigration of individuals into the population. Populations can decrease due to death or emigration. The rates for each of these four factors affect the rate of population increase, r, as described by the following equation:

$$r = (b - d) + (i - e)$$

where r is the rate of population increase, b is the birthrate, d is the death rate, i is the immigration rate, and e is the emigration rate.

This equation is used to make models that can predict the growth of a population. There are two main models: the **exponential growth model** and the **logistic growth model.**

The Exponential Growth Model

The exponential growth model is the simplest method of predicting growth. It calculates the biotic potential: the rate at which a population can grow when there are no factors limiting potential growth. The model uses the following equation:

$$dN/dt = r_i N$$

where dN/d_t is the rate of change in a population's numbers over time, r_i is the rate of natural increase for a population, and N is the number of individuals in a population.

The Logistic Growth Model

The logistic growth model assumes that growth is limited by environmental factors such as shortages of space, water, or nutrients. The maximum size of a population that a habitat can support is called the **carrying capacity,** K. Population growth slows as a population approaches carrying capacity. This is because each new individual in the population places more demands on the available resources. To model this scenario, the logistic growth model uses the following equation:

$$dN/dt = rN \, ((K - N)/K)$$

where dN/dt is the rate of change in a population's numbers over time, r is the rate of natural increase for a population, N is the number of individuals in a population, and K is the carrying capacity.

The second half of the equation, $K - N$, refers to the remaining carrying capacity of the population. When the population is below carrying capacity, this value is positive, and the population is increasing. When the population is above carrying capacity, this value is negative, and the population is decreasing. When the population is at carrying capacity, $K = N$, the population is stable. Populations tend to grow to the carrying capacity of their environment and then fluctuate around that value.

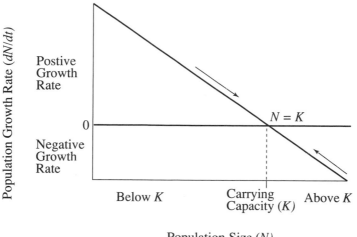

Logistical Growth Model

REGULATION OF GROWTH

The growth of a population can be regulated by factors such as population density, environmental conditions, food availability, and the presence of predators. These factors have three main effects on population growth.

	Effect on Population Growth	Description	Possible Cause
1.	**Density-dependant effect**	As the population size increases, the growth rate decreases.	Growth rates might suffer due to increased competition for resources, increased predation as predators focus on common prey, accumulation of toxic waste products from the large population size, or increased emigration as individuals leave the population in search of a better habitat.
2.	**Allee effect**	As the population size increases, the growth rate increases.	Populations may benefit if their large numbers help them ward off predators or breed more.

CONTINUED

	Effect on Population Growth	Description	Possible Cause
3.	**Density-independent effect**	The rate of growth of a population is limited due to factors unrelated to population size.	Limited growth is usually caused by changes to the environment such as cold winters, droughts, storms, or volcanic eruptions, which affect survival of individuals regardless of population size. Growth is rapid while the environment is calm, but growth declines sharply when the environment is hostile.

Communities

A number of species living together at the same location form a **community.** Communities can include many species or just a few species. The numerous interactions that occur between members of a community can affect **biotic factors,** which are the living parts of the community, and **abiotic factors,** which are the nonliving parts of the community.

Scientists studying the makeup and functioning of communities have two opposing theories:

- **The Individualistic Concept** suggests that a community is an aggregation of species that happen to coexist in the one location and that tend to respond independently to changes in the environment. This concept is the most accepted one today.

- **The Holistic Concept** suggests that the species in communities form an integrated unit. According to this theory, species in a community respond in a similar manner to changes in the environment.

NICHES

A **niche** is the biological role played by a particular species in its community or environment. The **fundamental niche,** which does not actually exist in nature, is the niche the organism is capable of occupying in theory.

The **realized niche** is the actual niche the organism occupies. Several interactions can affect the size of the realized niche.

- **The absence of another species:** If organisms require the presence of another organism, the absence of that other organism may cause the niche to decrease or disappear.

- **Predator presence:** Organisms may reduce their niche to avoid predators.

- **Interspecific competition:** When two organisms try to use a limited resource, one or both may have to limit their niche.

> *EXAMPLE:* Two species of clams live side by side on the seashore, competing for space and food. Because of their proximity to each other, both species have reduced realized niches.

COMPETITION

The principle of **competitive exclusion** states that in a community in which resources are limited, no two species can inhabit the same niche indefinitely. Species can temporarily coexist while competing for the same limited resource. However, over a long period, either one species will outcompete the other, causing its extinction, or natural selection will decrease the competition between them.

EXAMPLE: A number of different species of warbler feed on insects in the same tree. Each species has evolved to feed on insects at particular locations in the tree, so the warblers are not in competition with one another.

Determining Competition

It is not always easy to tell whether competition is causing population changes. If one population is winning out over another, it may not be the result of a struggle for resources; instead, one population may be thriving better than the other in the given environment. For example, one species of earthworm might do better in a warm climate than another species of earthworm in the same climate.

In addition, the presence of one organism can have negative effects on another organism—effects not related to competition. For example, the first organism may attract predators that prey on the second organism.

PREDATION

Predation, the consumption of one organism by another, is a common interaction between organisms. A fox may consume a rabbit, for example, and a rabbit may consume a plant. Interactions between predator populations and prey populations are dynamic, as a change in one can cause drastic changes in the other.

Evolution Prompted by Predation

To survive predation, prey populations have evolved traits that aid in defense. Defensive traits in plants and animals are divided into three categories: morphological defenses, physiological defenses, and behavioral defenses.

Name	Type of Defense	Examples in Animals	Examples in Plants
Morphological defenses	Physical features	The spines of an echidna and the thick skin of an alligator. Many animals have camouflage, which makes it harder for prey to spot them.	Spines, thorns, and thick leaves and stems.
Physiological defenses	Functions	Toxic chemicals. Invertebrates such as bees, wasps, and spiders inject venom into their prey.	Chemicals that disturb the functioning of herbivores, causing sterility or death, for example. Some plants also produce chemicals that make themselves foul tasting.
Behavioral defenses	Behaviors	Herding behavior. Sheep group themselves to reduce the danger from predators.	Plants do not exhibit behavioral defenses.

EXAMPLE: Batesian mimicry, a special type of morphological defense, occurs when organisms mimic the physical appearance of distasteful species. For example, nontoxic butterflies and moths take on the form of a toxic species, thereby tricking predators into avoiding them.

Community Interactions

All organisms experience competitive interactions and predator-prey interactions. Other, less common interactions include symbiosis and succession.

SYMBIOSIS

Symbiosis is an interaction between two or more species that benefits at least one of the species. There are three main forms of symbiosis: commensalism, mutualism, and parasitism.

1. **Commensalism** occurs when one species benefits and the other is neither benefited nor harmed. For example, anemone fish live in the toxic tentacles of the sea anemone. Immune to the tentacles' sting, the anemone fish receive protection and feed on leftovers from the sea anemone's meals. The sea anemone is not benefited or harmed by the interaction.

2. **Mutualism** occurs when both species benefit from an interaction. Animals and plants often enjoy mutually symbiotic relationships. A plant produces a fruit with seeds; the animal eats the fruit and then defecates, depositing the seeds in a new location. The animal benefits by gaining nourishment, while the plant benefits by having its seeds dispersed.

3. **Parasitism** occurs when one species benefits and the other is harmed. For example, a barnacle called *Sacculina* attaches to the underside of crabs. Female *Sacculina* take over the body and ready it for egg fertilization. Male *Sacculina* take over the body and use it to find female *Sacculina* and fertilize eggs. The crab can no longer reproduce its own eggs, and so it takes care of the parasite's eggs.

SUCCESSION

Succession occurs when a community changes from simple to complex. This is a natural progression that occurs even when the environment is stable. Succession is divided into primary succession and secondary succession.

- **Primary succession:** Life begins to form from bare, lifeless substrates. Usually an event such as a volcano wipes out a habitat. Eventually, soil builds up and life begins to develop and evolve.

- **Secondary succession:** Life continues to form from a substrate already containing soil. Secondary succession usually results from the destruction of a habitat, such as the clearing of a forest, which destroys most life forms but maintains the soil. Life begins to develop and evolve.

Succession generally occurs via the following three-step process:

1. **Tolerance:** Species tolerant to harsh conditions inhabit a new environment.

2. **Facilitation:** The first species cause changes to the environment that encourage new species to invade. These new species continue changing the environment, making it more tolerable to more species.

3. **Inhibition:** The changes to the environment may prevent old species from surviving.

Vegetation and animals influence each other during succession. As the habitat changes, the presence of certain vegetation favors colonization by certain animal species. These animals change the environment, thereby affecting which vegetation best survives in the habitat. Old plants die out as new ones begin to thrive. The change in vegetation affects which animals are able to thrive in the habitat, animal populations change, and the cycle continues.

Summary

Populations

- **Populations** are groups of individuals from the same species that inhabit the same area.

- Populations are defined by their range, spacing, and size.

- The **range** of a population is the area the population inhabits at any given time.

- Population spacing is a term describing how the individuals in a population are distributed within that population. Populations can have **random spacing, uniform spacing,** or **clumped spacing.**

- Population **size** is a measure of the number of individuals in a population.

Population Change

- **Population dynamics** refers to how a population changes over time. Dynamics are affected by a population's rate of reproduction, proportion of different aged individuals, and mortality, or death rates.

- The cost of reproduction is the reduction in future reproductive potential caused by current reproductive events.

- The number of reproductive events an organism undergoes during its lifetime affects reproductive success. **Semelparity** is one-time production of as many offspring as possible, while **iteroparity** is the repeated production of offspring throughout a lifetime.

- The rate of population growth can be predicted using the **exponential growth model** or the **logistic growth model.** The exponential growth model assumes that there are no factors limiting the potential growth of the population, while the logistic growth model takes into account factors that limit the potential growth of the population.

- The growth of a population is regulated by factors such as population density, environmental conditions, food availability, and the presence of predators.

Communities

- A **community** is a number of species living together in the same location.

- The **individualistic concept** conceives of communities as aggregations of species that happen to coexist in the one location and that tend to respond independently to changes in the environment.

- The **holistic concept** conceives of communities as integrated units of species. According to this theory, species in a community react and respond in a similar manner to changes in the environment.

- A **niche** is the biological role played by a particular species in its community or environment.

- The **fundamental niche** is the theoretical niche an organism would be capable of occupying if it did not face competition. The **realized niche** is the actual niche the organism occupies.

- The principle of **competitive exclusion** states that in a community in which resources are limited, no two species can inhabit the same niche indefinitely without one species being driven to extinction due to competition or forced to find a new niche.

- **Predation,** a common interaction between organisms, involves one organism consuming another.

- Predation has spurred evolution by favoring those plants and animals whose defensive traits help them survive. Defensive traits are divided into morphological defenses, physiological defenses, and behavioral defenses.

Community Interactions

- **Symbiosis** is an interaction between two or more species that benefits at least one of the organisms.

- **Commensalism** is a type of symbiosis that occurs when one species benefits while the other receives no benefit and no harm during an interaction.

- **Mutualism** is a type of symbiosis that occurs when both species benefit from an interaction.

- **Parasitism** is a type of symbiosis that occurs when one species benefits while the other is harmed during an interaction.

- **Succession** occurs when a community changes overtime, often from simple to complex. Succession is divided into **primary succession** and **secondary succession.**

- Succession occurs via a three-step process involving **tolerance, facilitation,** and **inhibition.**

Sample Test Questions

1. The range of a population can change either because of environmental changes or because of population changes. Explain how a population expands its range in both cases.

2. Describe two situations in which the cost of reproduction for an organism will be high.

3. Explain the difference between the exponential growth model and the logistic growth model.

4. The area a population inhabits at any given time is known as what?

   A. Size
   B. Range
   C. Spacing
   D. Demography
   E. Mortality

5. What is the term for population spacing in which individuals develop their own space, separate from other individuals' space?

   A. Random spacing
   B. Clumped spacing
   C. Irregular spacing
   D. Territorial spacing
   E. Uniform spacing

6. What is the term for a network of distinct populations that are able to interact with each other?

   A. Metapopulation
   B. Superpopulation
   C. Interpopulation
   D. Demipopulation
   E. Intrapopulation

7. The terms semelparity and iteroparity refer to what?

   A. The ratio of males to females in a population
   B. The average time between the birth of an individual and the birth of its offspring
   C. The proportion of different age groups within a population
   D. The number of reproductive events an organism undergoes during its lifetime
   E. The reduction in reproductive potential caused by current reproductive events

8. Which growth model assumes that there are no factors limiting the potential growth of a population?

   A. Exponential growth model
   B. Biotic growth model
   C. Maximum growth model
   D. Capacity growth model
   E. Logistic growth model

9. In which of the following situations would the cost of reproduction be low?

   A. When resources are abundant
   B. When resources are scarce
   C. When reproduction leads to sterility
   D. When mortality rates are high
   E. When reproduction increases susceptibility to predators

10. What is the term for the maximum size of a population that a habitat can support?

   A. Emigration rate
   B. Threshold potential
   C. Carrying capacity
   D. Mortality limit
   E. Biotic maximum

11. The Allee effect refers to populations that benefit from what?

A. High reproductive rates
B. Large size
C. A hostile environment
D. Low population density
E. An abundance of food

12. The view that a community is an integrated unit of all its species is known as what?

A. Individualistic concept
B. Superficial concept
C. Metapopulation concept
D. Holistic concept
E. Communal concept

13. Which factor is most likely to limit the actual niche that an organism occupies?

A. Lack of shelter
B. Limited food supply
C. Immigration and emigration
D. The presence of competitors
E. Mortality and fecundity

14. The spines of an echidna and the thorns of a rose bush are examples of what type of defense?

A. Physiological
B. Anatomical
C. Morphological
D. Chemical
E. Behavioral

15. What is the first step in succession?

A. Facilitation
B. Inhibition
C. Predation
D. Tolerance
E. Speciation

ANSWERS

1. Environmental changes are changes that make the conditions in an environment more or less suitable for a population. When an environment becomes more suitable for a population, the population is able to expand its range. Population changes result from a population's move to a new, more suitable area that allows the population to expand its range.

2. The cost of reproduction is high for an organism when resources are scarce. Producing offspring will further reduce the available resources, thereby reducing the organism's chances of survival. The cost of reproduction is also high for those organisms that are weakened when they carry offspring and therefore more likely to be targeted by predators.

3. The exponential growth model is a simple model used to predict growth. It assumes that there are no factors limiting the potential growth of the population. The growth predicted by this model is known as the biotic potential. The logistic growth model takes into account environmental factors that limit growth. It is a more complicated growth model and gives a more accurate prediction of actual population growth.

4. **B** The range of a population is the area that the population inhabits at any given time.

5. **E** Uniform spacing is what happens when individuals develop their own space, separate from others' space.

6. **A** A metapopulation is a network of distinct populations that are able to interact with each other by exchanging individuals.

7. **D** Iteroparity describes an organism that reproduces many times throughout its lifetime. Semelparity describes an organism that has one reproductive event.

8. **A** The exponential growth model assumes that there are no factors limiting the potential growth of a population.

9. **A** The cost of reproduction is low when resources are abundant and high when resources are scarce.

10. **C** The maximum size of a population that a habitat can support is called the carrying capacity.

11. **B** The Allee effect refers to populations that benefit from a large population size, increasing their growth rate as their size increases.

12. **D** The holistic concept conceives of a community as an integrated unit of all its species.

13. **D** Competitors are most likely to limit an organism's niche.

14. **C** Morphological defenses are physical features such as the spines of an echidna or the thorns of a rose bush.

15. **D** Tolerance is the first step of succession. It is followed by facilitation and inhibition.

ECOSYSTEMS AND BIOMES

The Flow of Matter

The Flow of Energy

Land Ecosystems

Marine Ecosystems

24

The Flow of Matter

An environment or **habitat** is an area with physical boundaries. All things living in a particular environment are known as a **community.** Together, the habitat and community make up the **ecosystem.** The living parts of an ecosystem are known as the **biotic factors.** The nonliving portions of the ecosystem, such as atmosphere, water, and rocks, are known as the **abiotic factors. Ecology** is the study of the interactions between living things and the nonliving things in an environment.

> *EXAMPLE:* One ecosystem can contain many smaller ecosystems. For example, a forest habitat and all the living things in the forest constitute an ecosystem. A single tree in that forest and all the living things in the tree also constitute an ecosystem. So do a single piece of bark and all the living things that live on the piece of bark.

To survive, living things need matter, including water, carbon, nitrogen, and phosphorus, which provide the raw materials necessary to produce all the molecules that support life. These materials continually cycle through the ecosystem, moving through its abiotic and biotic parts.

> *EXAMPLE:* Nitrogen is contained in the soil, which is an abiotic part of the ecosystem. When nitrogen is taken up by plants, it enters the biotic part of the ecosystem. Animals eat the plants, consuming the nitrogen as they do so; when the animals die, their bodies decay, and the nitrogen is returned to the soil.

There are four main cycles that occur in ecosystems:

1. **The water cycle:** Liquid water from the ocean evaporates, condenses to form clouds, falls to the Earth as precipitation, and returns to the ocean as runoff.

2. **The carbon cycle:** Plants convert the carbon dioxide in the atmosphere into carbon compounds, which are used by animals that consume the plants.

3. **The nitrogen cycle:** Nitrogen from the atmosphere is fixed by bacteria and made available to plants.

4. **The phosphorus cycle:** Phosphate dissolves in water, making phosphate ions available to plants. Plants convert the phosphorus into biological molecules, which are then passed through the ecosystem's food chain.

THE WATER CYCLE

Water is essential for the survival of all living organisms. It is the main component of all living things. For example, it composes about two-thirds of the human body. The amount of water that an ecosystem receives is a major determinant of the living things that can survive there. This explains why rainforests are able to support a diverse range of living things, while deserts are only able to support living things specially adapted to surviving in a dry climate.

The water cycle begins with the ocean, the largest source of water, which cycles through ecosystems as follows:

1. Water in the ocean evaporates and becomes part of the atmosphere in the form of clouds.

2. Water in clouds condenses and falls to Earth as precipitation, such as rain, snow, or hail, or condenses as dew.

3. Runoff moves water from the ground to surface water bodies including lakes, ponds, and rivers.

4. Water flows from lakes, ponds, and rivers back to the ocean.

Groundwater

Water exists in surface water bodies such as lakes, ponds, and rivers. Water also exists in **groundwater,** which is water located beneath the surface of the ground. Surface and ground water are equally important to the system. Groundwater is divided into two layers:

- The **water table** is the upper section of groundwater. Water that seeps through the soil from ponds, lakes, streams, and rainfall is added to the water table. The water table is accessible to plants, which absorb water through their roots. Through this exchange, the water table is recycled relatively quickly and regularly.

- **Aquifers** make up the lower section of groundwater. Aquifers are deeper than the water table and are not accessible to plants. They are replenished slowly, as water flows down from the water table.

The Organismic Water Cycle

In addition to cycling through the environment through evaporation and precipitation, water cycles through plants. The organismic water cycle occurs as follows:

1. The roots of plants take up water in the water table.

2. Water passes through the plants, exiting via pores on the surface of leaves.

3. The water evaporates from the surface of the plant's leaves in a process known as **transpiration.**

> *EXAMPLE:* Transpiration is a natural process that occurs when a plant opens its stomata to allow in carbon dioxide. Water flows continually through the plant, allowing nutrients to move through the plant.

The organismic water cycle is the main source of water in rainforest ecosystems. Around 90 percent of the water taken in by plants is transpired back into the atmosphere. Because the cool conditions in rainforests allow water to condense, it remains in the ecosystem. The condensation gives plants a constant source of water, which means they don't need to rely on rainfall.

THE CARBON CYCLE

Carbon compounds are the basis for the molecules necessary for life, including lipids, proteins, nucleic acids, and carbohydrates. Carbon dioxide in the atmosphere provides carbon for living things, delivering it to animals via plants. Photosynthetic plants, bacteria, and protists capture carbon from the atmosphere and convert it to carbon compounds such as glucose. Through this process, carbon enters the biotic part of the ecosystem. Carbon is then transferred through the biosphere as living things consume the photosynthetic organisms and each other. When plants and animals die, the carbon compounds (detritus) are broken down by organisms such as bacteria and fungi (detritivores), releasing the carbon dioxide back into the atmosphere or into the water.

There are three primary ways that carbon dioxide is returned to the atmosphere:

1. **Respiration** occurs when living things, including herbivores and carnivores, break down the food molecules produced by plants. They return carbon dioxide to the atmosphere by releasing it as a waste product.

2. **Combustion** occurs when material containing high amounts of stored carbon atoms, such as wood, coal, or oil, is burned. When the material is burned, carbon dioxide is released into the atmosphere.

3. **Limestone erosion** occurs when limestone erodes. Limestone forms from the remains of the calcium carbonate shells of marine organisms. The erosion of the calcium carbonate returns carbon to the oceans, where it can be returned to the atmosphere as carbon dioxide.

EXAMPLE: The burning of fossil fuels for energy has greatly increased the amount of carbon released into the atmosphere via combustion. The increased levels of carbon dioxide in the atmosphere are thought to be a major contributor to global warming.

THE NITROGEN CYCLE

Nitrogen is essential to living things because it is a main component of proteins and nucleic acids. In the form of nitrogen gas (N_2), nitrogen is a large component of the atmosphere. N_2 must be converted to ammonia by bacteria in the soil before plants can use it. The nitrogen cycle describes how nitrogen from the atmosphere is cycled through the system.

1. In a process called **nitrogen fixation,** bacteria in the soil fix nitrogen from the air, converting it to ammonia (NH_3), a compound that can be used to synthesize organic compounds such as amino acids.

2. Plants take up the ammonia from the soil through their roots, and the nitrogen enters the biosphere.

3. After the plant dies, in a process called **ammonification,** the nitrogen-containing compounds are broken down by bacteria and fungi. The nitrogen is released in the form of ammonium ions (NH_4^+).

4. The ammonium ions are converted to nitrate (NO_3^-) by **nitrifying bacteria.** The nitrates can then be taken up by plants.

5. The nitrates can also be converted to nitrogen gas by certain types of bacteria known as **denitrifying bacteria.** The nitrogen gas is returned to the atmosphere.

EXAMPLE: The majority of fertilizers used by humans to encourage plant growth contain nitrogen. These fertilizers are used to such an extent that human activity is often considered another important part of the nitrogen cycle. In many cases, nitrogen is added to the soil via fertilizer in much greater quantities than through nitrogen fixation.

THE PHOSPHORUS CYCLE

Phosphorus is an essential component in many biological molecules, such as ATP, phospholipids, and DNA. Phosphorus is often a limiting factor in plant growth because it is not available from the atmosphere, as carbon and nitrogen are. Phosphorus does not exist as a gas and is primarily found in the solid form of calcium phosphate ($Ca_2(PO_4)_2$) in soil and rocks. When calcium phosphate dissolves in water, phosphate ions become available to plants, which take them up through their roots. Plants convert the phosphorus into biological molecules, which are then passed through the ecosystem's food chain in much the same way as carbon. Phosphorus is returned to the soil via excretions and the death of plants and animals. Bacteria break down the dead organism and return phosphorus to the soil. In contrast to atmospheric elements, such as carbon and nitrogen, phosphorous, because it usually can be found in solid form, tends to recycle locally through soil, plants, and other organisms.

The Flow of Energy

Unlike matter that is recycled, energy enters the system, is used by organisms, and is lost as heat. All of the Earth's energy is initially provided in the form of solar energy. The amount of solar energy that reaches each individual ecosystem varies, and ecosystems are limited by the amount of energy they receive.

Energy first enters an ecosystem when photosynthetic organisms use sunlight to produce food. The flow of energy continues through the ecosystem as photosynthetic organisms are consumed by other organisms. As these organisms break down the food molecules to obtain nutrients, the majority of the energy contained in them is converted to heat and lost to the atmosphere. However, the organisms use approximately 10 percent of the energy to produce new molecules. As organisms are consumed by other organisms, the breakdown and loss of energy continues.

TROPHIC LEVELS

Organisms in an ecosystem are divided into two major trophic levels based on their source of energy:

*   **Autotrophs** manufacture their own food. Autotrophs, which include plants, algae, and some bacteria, generally use light energy to manufacture food via photosynthesis. Autotrophs are also known as **primary producers.** **Chemoautotrophs** can manufacture their food without the use of solar energy. They live in extremely hostile deep-sea hydrothermal vents and use carbon dioxide to oxidize sulfur and other inorganic compounds for the production of food, in a process known as **chemosynthesis.**

- **Heterotrophs** are unable to produce their own food. They gain their energy by consuming autotrophs or other heterotrophs that feed on autotrophs. All animals and fungi, as well as most protists and prokaryotes, are heterotrophs. Heterotrophs are also known as consumers.

Heterotrophs

Heterotrophs are divided into several trophic levels depending on their source of energy.

- **Primary consumers,** or herbivores, feed on plants. Examples include rabbits and caterpillars.

- **Secondary consumers,** or carnivores, feed on herbivores. Examples include tigers, foxes, and sharks.

- **Tertiary consumers** feed on secondary consumers.

- **Quaternary consumers** feed on tertiary consumers.

- **Detritivores** feed on the waste of an ecosystem. Detritivores include **scavengers,** which feed on the remains of dead animals. Vultures and jackals are scavengers. Detritivores also include **decomposers,** which break down dead plant and animal remains. Decomposers include bacteria and fungi.

The Food Chain

A **food chain** is a diagram showing how organisms from each trophic level feed on each other. A **food web** is a more complicated diagram that recognizes that one organism rarely feeds only on one type of organism. A food web shows all the relationships between organisms in the ecosystem.

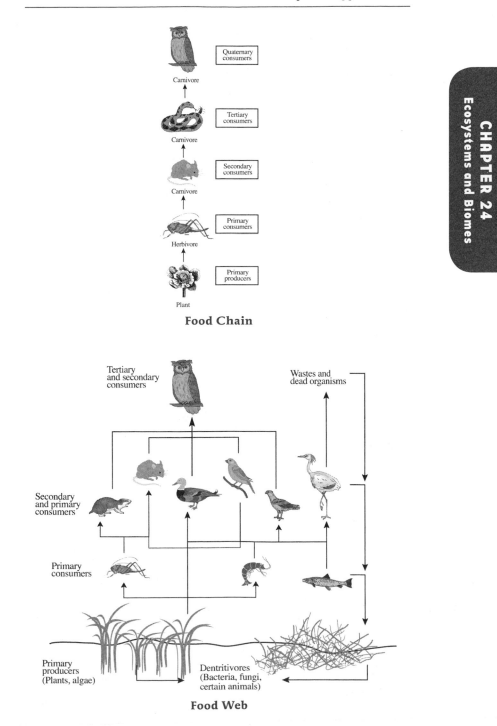

Food Chain

Food Web

PRODUCTIVITY IN ECOSYSTEMS

The flow of energy through an ecosystem is not an efficient process. Only around 1 percent of the solar energy that reaches a plant is used to produce food. From there, at each level that organisms consume each other for food, energy is lost as heat. Scientists use three calculations for measuring plant productivity in an ecosystem:

- **Primary productivity** measures how much energy is produced by the photosynthetic organisms in an ecosystem.

- **Gross primary productivity** measures the total amount of organic matter produced by the photosynthetic organisms in the ecosystem. Plants also use some of the matter they produce for respiration. The remainder is available for heterotrophs.

- **Net primary productivity** (NPP) measures the amount of matter produced that is available for heterotrophs. NPP for an ecosystem is generally measured in grams per square meter (g/m^2).

EXAMPLE: In an environment such as a rainforest or wetland, NPP can be as high as 3,000 g/m^2. In an environment such as a desert, it may be less than 100 g/m^2.

Biomass is the total weight of all an ecosystem's organisms. Biomass increases as net primary productivity increases. **Secondary productivity** is a measure of the rate of biomass production by an ecosystem's heterotrophs. Only around 10 percent of the energy available at one trophic level is available at the next. Energy may be lost as fecal matter or converted to heat. As energy passes from consumer to consumer down the food chain, the amount of available energy is continually diminished, limiting the number of steps a given ecosystem's food chain can sustain.

ECOLOGICAL PYRAMIDS

Ecological pyramids are diagrams that show the relationships between the trophic levels of an ecosystem. The pyramid shape reflects the loss of energy that occurs from one trophic level to the next. There will generally be many individuals at the lowest trophic level, where energy levels are more abundant. Since only 10 percent of the energy is passed on to the next level, there will be fewer individuals at the second level, and even fewer at the third.

Ecological pyramids can be represented in three ways:

- A **pyramid of numbers** shows the number of individuals at each level.

- A **pyramid of biomass** shows the amount of biomass produced at each level.

- A **pyramid of energy** shows the amount of energy available at each level.

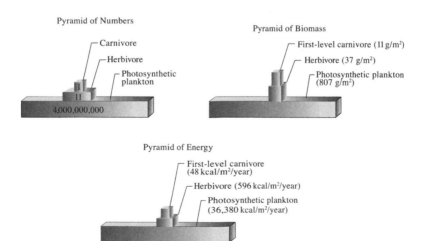

Ecological Pyramids

EXAMPLE: Photosynthetic algae fix 1,000 calories of potential energy. Of these, 850 calories are used by the algae, and 150 calories are transferred to a heterotroph that eats algae. Of the 150 calories, around 120 calories are used by the heterotroph, and 30 calories are transferred to a consumer of the heterotroph.

Land Ecosystems

An **ecosystem** is the environment itself and all of the organisms that live in it. Organisms depend on their environment to provide resources and a means of survival. Environments with similar characteristics are likely to have similar groups of organisms that exploit their environments in similar ways.

Biomes are ecosystems that have similar climates and organisms. There are eight major biomes: tropical rainforest, savanna, desert, temperate grassland, temperate deciduous forest, temperate evergreen forest, taiga, and tundra. The map below shows the distribution of the biomes.

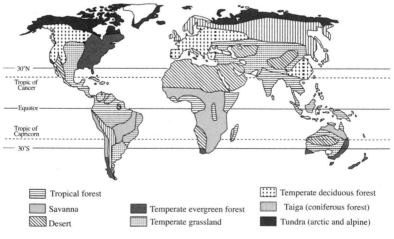

Tropical forest		Temperate deciduous forest
Savanna	Temperate evergreen forest	Taiga (coniferous forest)
Desert	Temperate grassland	Tundra (arctic and alpine)

Distribution of Terrestrial Biomes

The biomes and their characteristics are summarized in the table on the following page.

Biome	Climate and Characteristics	Example
Tropical rainforest	Tropical rainforests receive between 140 to 450 mm of rain per year. They are characterized by warm climates, high humidity, and year-round rainfall.	The Amazon rainforest of South America
Savanna	Savannas are dry tropical grasslands that border the tropics. Landscapes are usually open, and rainfall is seasonal.	The savannas of Africa
Desert	Deserts are dry environments that receive less than 25 cm of rain per year.	The Sahara and Gobi deserts
Temperate grassland	Temperate grasslands lie halfway between the equator and the poles. They typically have fertile soil, and many have been converted to agricultural land because of this characteristic.	The prairies of Canada
Temperate deciduous forest	Temperate deciduous forests have mild climates with warm summers and cool winters.	The hardwood forests of Canada
Temperate evergreen forest	Temperate evergreen forests have a cold winter and a seasonal dry period.	The evergreen forests of California
Taiga (coniferous forest)	The taiga experience long, cold winters. There is low rainfall in the summer, and the growing season is short.	Most of Alaska
Tundra	The tundra is characterized by low temperatures and low precipitation. Permafrost, permanent ice, exists within one meter of the surface.	The upper areas of Canada

CHAPTER 24 Ecosystems and Biomes

Marine Ecosystems

While ecosystems on land are based on climate, marine ecosystems are based on water depth. There are three major marine habitats, called zones.

- The **neritic zone** is the relatively shallow region along the edges of continents and islands. It is less than 300 meters below the surface. It is the smallest marine area in terms of size but contains the most species. The neritic zone is the source of fish for almost all the world's fisheries.

• The **pelagic zone** is the area of water above the ocean floor, stretching from the surface to depths approaching 1,000 meters. Plankton are the primary producers of the pelagic zone and live in the top 100 meters. Fish, whales, jelly-fish, and other sea life also live in the pelagic zone. These animals either feed on the plankton or on animals that consume the plankton.

• The **benthic zone** is the ocean floor. It exists at depths greater than 1,000 meters, and animals that live there have adapted to darkness, lack of food, and cold temperatures. Most organisms are small detritivores that feed on the limited leftovers that float down from animals living in the pelagic zone. Some animals in the benthic zone get their energy from compounds produced at hydrothermal vent systems, which are areas where water circulates through porous rock. These organisms use chemosynthesis in place of photosynthesis to produce energy.

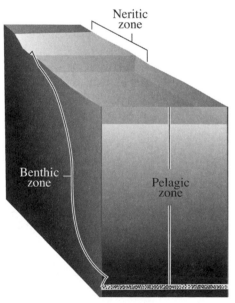

Marine Zones

Summary

The Flow of Matter

- An **ecosystem** is made up of an environment and all the living things in that environment.

- To survive in their environment, living things need the presence of matter including water, carbon, nitrogen, and phosphorus.

- Matter cycles through the ecosystem, moving through its **abiotic** and **biotic factors.**

- The **water cycle** explains how water makes its way through the system via the processes of evaporation, condensation, precipitation, and runoff.

- The **carbon cycle** explains how carbon from the atmosphere is used by plants and then transferred to the animals that consume the plants and the animals that consume those animals.

- The **nitrogen cycle** explains how nitrogen from the atmosphere is fixed by bacteria and made available to plants.

- The **phosphorus cycle** explains how calcium phosphate dissolves in water, making phosphate ions available to plants. Plants convert the phosphorus into biological molecules, which are then passed through the ecosystem's food chain.

The Flow of Energy

- All of Earth's energy is initially provided by the sun.

- Energy first enters an ecosystem when photosynthetic organisms use sunlight to produce food.

- The flow of energy continues through the ecosystem as photosynthetic organisms are consumed by other organisms.

- The organisms in an ecosystem are divided into **autotrophs,** which manufacture their own food, and **heterotrophs,** which are unable to manufacture their own food.

- Heterotrophs are divided into **primary consumers, secondary consumers, tertiary consumers,** and **quaternary consumers.**

- **Detritivores** feed on the waste of an ecosystem.

- Detritivores include **scavengers,** which feed on the remains of dead animals, and **decomposers,** which break down dead plant and animal remains.

- A **food chain** is a diagram showing how organisms from each trophic level feed on each other. A **food web** shows all the relationships between organisms in an ecosystem.

- As energy passes from consumer to consumer down the food chain, the amount of available energy is continually diminished, limiting the number of steps the food chain of a given ecosystem can sustain.

- Plant productivity in an ecosystem is measured by **primary productivity, gross primary productivity,** and **net primary productivity.**

- **Ecological pyramids** are diagrams that show the relationships between the trophic levels of an ecosystem.

Land Ecosystems

- An ecosystem consists of all of the organisms that live in a certain environment, as well as the abiotic components of the environment itself.

- Organisms depend on the characteristics of their environment to provide them with resources and the means of survival.

- **Biomes** are ecosystems with similar climates and similar organisms.

- There are eight major biomes: tropical rainforest, savanna, desert, temperate grassland, temperate deciduous forest, temperate evergreen forest, taiga, and tundra.

Marine Ecosystems

- Marine ecosystems are characterized based on water depth and light levels.

- The **neritic zone** is the relatively shallow region along the edges of continents and islands. It stretches to a depth of 300 meters.

- The **pelagic zone** is the area of water above the ocean floor, stretching from the surface to depths approaching 1,000 meters.

- The **benthic zone** is the ocean floor.

Sample Test Questions

1. Explain the difference between gross primary productivity and net primary productivity.

2. Explain the process of nitrogen fixation and why it is necessary to make nitrogen available to plants.

3. Explain why a food chain can have only a limited number of steps.

4. Which of the following two parts combine to form an ecosystem?

    A. Habitat and community
    B. Matter and energy
    C. Community and abiotic factors
    D. Autotrophs and heterotrophs
    E. Organisms and biotic factors

5. Which of the following is NOT one of the processes of the water cycle?

    A. Evaporation
    B. Condensation
    C. Runoff
    D. Ammonification
    E. Precipitation

6. Which of the following processes involves living things returning carbon dioxide to the atmosphere?

    A. Nitrogen fixation
    B. Limestone erosion
    C. Combustion
    D. Respiration
    E. Transpiration

7. During which cycle is ammonification an important process?

    A. Carbon cycle
    B. Water cycle
    C. Nitrogen cycle
    D. Oxygen cycle
    E. Phosphorus cycle

8. Primary consumers are also known as what?

    A. Detritivores
    B. Scavengers
    C. Chemoautotrophs
    D. Herbivores
    E. Decomposers

9. The amount of matter produced that is available for heterotrophs is known as what?

    A. Net primary productivity
    B. Secondary productivity
    C. Gross primary productivity
    D. Primary productivity
    E. Gross secondary productivity

10. Which of the following is a measure of the rate of biomass production by an ecosystem's heterotrophs?

    A. Net primary productivity
    B. Secondary productivity
    C. Gross primary productivity
    D. Primary productivity
    E. Gross secondary productivity

11. Which type of biome has between 140 and 450 mm of precipitation per year?

  A. Tropical rainforest
  B. Temperate grassland
  C. Taiga
  D. Tundra
  E. Savanna

12. Which type of biome is often converted to agricultural land because of its fertile soil?

  A. Tropical rainforest
  B. Temperate grassland
  C. Taiga
  D. Tundra
  E. Savanna

13. Which of the following describes the characteristics of tundra?

  A. Warm climate and high humidity
  B. Open landscape and seasonal rainfall
  C. Mild climate and warm summers
  D. Low temperature and low precipitation
  E. High temperature and high rainfall

14. What is the shallowest marine habitat?

  A. Pelagic zone
  B. Surface zone
  C. Neritic zone
  D. Plankton zone
  E. Benthic zone

15. What is the main type of organism found in the benthic zone?

  A. Producers
  B. Herbivores
  C. Autotrophs
  D. Carnivores
  E. Detritivores

ANSWERS

1. Gross primary productivity measures the total amount of organic matter produced by the photosynthetic organisms in the ecosystem. This includes the matter that plants use for respiration. Net primary productivity measures the total amount of matter produced that is available to heterotrophs. The gross primary productivity minus the matter used by the plant is the net primary productivity.

2. Nitrogen fixation is the process by which bacteria convert atmospheric nitrogen to ammonia. Plants are not able to use nitrogen gas from the atmosphere. However, they are able to use nitrogen in the form of ammonia. Once bacteria convert nitrogen from the atmosphere into ammonia, plants can take it up from the soil.

3. Animals gain energy by eating food. A portion of the food eaten is used by the animal. The remainder is stored in the animal and is available to the next trophic level. This means that only a portion of the food eaten is available to be used as energy for the next trophic level. This limits the number of steps there can be in a food chain.

4. **C** An ecosystem consists of the community and its abiotic factors.

5. **D** Evaporation, condensation, precipitation, and runoff are all parts of the water cycle.

6. **D** When organisms break down food molecules via the process of respiration, carbon dioxide is released to the atmosphere.

7. **C** Ammonification is part of the nitrogen cycle. It involves the breakdown of nitrogen-containing compounds to form ammonium ions.

8. **D** Primary consumers, or herbivores, feed on green plants.

9. **A** Net primary productivity (NPP) measures the amount of matter produced that is available for heterotrophs.

10. **B** Secondary productivity is a measure of the rate of biomass production by an ecosystem's heterotrophs.

11. **A** Tropical rainforests are characterized by warm climates, high humidity, and between 140 and 450 mm of rain per year.

12. B   Temperate grasslands are often converted to agricultural land because of their fertile soil.

13. D   The tundra is characterized by low temperatures and low precipitation.

14. C   The neritic zone is the relatively shallow region along the edges of continents and islands.

15. E   Most organisms in the benthic zone are small detritivores that feed on the limited leftovers that float down from animals living in the pelagic zone.

CHAPTER 24
Ecosystems and Biomes

CONSERVATION

Species Extinction

Causes of Species Extinction

Population Recovery and
   Extinction Prevention

25

Species Extinction

Extinctions are a normal part of nature; over 99 percent of the species known to science are extinct. However, these extinctions have occurred over the millions of years that living things have existed on Earth and have been caused by relatively rare natural events such as meteorite impacts and massive volcanic eruptions. In recent times, the pace of extinction has sped up greatly.

Humans' arrival to habitats is often followed by the extinction of native species. When humans arrived in North America roughly 12,000 years ago, they hunted and deforested; the result was an 80 percent reduction in **megafauna,** animals weighing more than 100 pounds, such as mammoths, lions, saber-toothed cats, horses, and camels. Similar patterns of extinction occurred in Australia, New Zealand, and Madagascar, as humans migrated to these previously uninhabited areas.

CHANGING EXTINCTION RATES

While extinction is a normal process, its rate has been increasing over the last few hundred years. Between 1600 and 1700, an average of one species of bird and one species of mammal became extinct per decade. From 1850 to 1950, this rate increased to one species of bird and one species of mammal per year. Between 1986 and 1990, the figure had increased to four species of bird and four species of mammal per year. The rate of extinction is expected to increase in future years.

Human-influenced extinction is generally triggered by the overconsumption of resources or the complete destruction of habitats, which makes it impossible for new species to develop and replace those that have been lost. The long-term effect is the permanent reduction of the planet's biodiversity.

CHAPTER 25
Conservation

EXAMPLE: Human colonization of Madagascar has resulted in the destruction of 90 percent of the island's forests and the extinction of sixteen primate species. Humans continue to utilize the land and its resources, making them unavailable to other species and decreasing the likelihood that new species will develop to increase the lost biodiversity.

Extinction triggered by human actions is generally considered more damaging than extinction triggered by natural causes such as meteorite impacts or climatic change.

EXAMPLE: The mass extinction that occurred 65 million years ago at the end of the Cretaceous period caused the extinction of dinosaurs. As a result, biodiversity decreased. Without the dinosaurs present, more resources were available for mammals, which grew and diversified. After a period of around 10 million years, biodiversity had rebounded.

SPECIES ENDEMISM

An **endemic** species is one that is found naturally in only one geographic area, whether it is an entire continent or a small place (the Komodo dragon, for example, is found on only a few Indonesian islands). Endemic species are of particular concern to **conservation biologists.**

A **hotspot** is an area that is home to many endemic plant and animal species. Biologists have identified twenty-five hotspots, including Madagascar, California, and the eastern Himalayas. These hotspots contain around half of all the terrestrial species in the world. Hotspots also tend to be sites where human populations are growing and threatening species. Because hotspots have so many species and are associated with a high number of extinctions, they are often the focus of conservation efforts.

> *EXAMPLE:* The California Floristic Province, a hotspot, contains over 2,000 endemic plant species and many threatened endemic species, including the giant kangaroo rat, California condor, island grey fox, and stellar sea lion. Urban expansion and road construction put the area at risk. To protect the Province, the California state government and various conservation societies have designated 37 percent of it protected land.

THE IMPORTANCE OF CONSERVING BIODIVERSITY

Conservationists identify three key benefits of conserving biodiversity:

1. **Direct economic value:** Living things are sources of food, medicine, clothing, and biomass. Reducing biodiversity reduces possible economic gains.

> *EXAMPLE:* Around 40 percent of medications have active ingredients that are sourced from either plants or animals. Reducing the diversity of plants and animals reduces the opportunities to discover new medications.

2. **Indirect economic value:** Diverse biological communities maintain the quality of natural water, buffer ecosystems against droughts, preserve soils, and absorb pollution, all of which have economic benefits. Reducing biodiversity can lead to problems including waterlogging, desertification, and mineralization. The cost of managing these problems can be significant.

> *EXAMPLE:* In the 1990s, New York City had a problem with the declining quality of drinking water. One option was to spend $6 billion installing filtration plants to clean the water. The second option was to spend $1 billion preserving the biodiversity in the areas where the water came from, such as the Catskill Mountains. With the ecosystems in the area preserved, the water was naturally purified.

CHAPTER 25
Conservation

3. **Ethical and aesthetic values:** Many people view conservation as an ethical issue. They believe that humans have a responsibility to protect species rather than exploit them.

Causes of Species Extinction

While the extinction of species has many causes, including pollution, natural disasters, and loss of genetic variation, human action is responsible for four main causes of the current rate of extinction:

1. Habitat loss

2. Overexploitation

3. Introduced species

4. Ecosystem disruption

HABITAT LOSS

Habitat loss occurs when an environment is altered so much that certain organisms can no longer survive there. It is currently the most common cause of extinctions.

Method of Habitat Loss	Description	Examples
Destruction	**Destruction** occurs when a habitat is destroyed.	Destruction can be caused by clearing forests, burning forests to provide grazing lands for livestock, or urban development.
Pollution	**Pollution** can make it impossible for species to survive in their habitat. It is most common for aquatic ecosystems.	Aquatic ecosystems can be compromised by acid rain caused by fossil fuel combustion or by pesticides, herbicides, and other chemicals that enter aquatic systems via runoff, atmospheric pollution, or direct pollution.
Disruption	**Disruption** occurs when human activity alters the natural activities or processes that occur in an ecosystem.	Human use of insecticides reduces the number of insects in an area, which may disrupt the natural process of pollination that occurs for plants and lead to the loss of plant life.
Habitat fragmentation	**Habitat fragmentation** occurs when habitats are physically divided into smaller segments.	Roads cutting through a forest divide the populations in the forest into smaller segments. Habitat fragmentation increases the amount of a habitat's edges, which can lead to the edge effect: the decreased chances of survival for individuals on the edge of ecosystems, who are more accessible to predators.

OVEREXPLOITATION

Overexploitation occurs when living things are hunted or harvested by humans to the extent that the population is not able to maintain itself, even if it was initially abundant. Some overexploited species include:

- Passenger pigeons: Once numerous across North America, they were hunted for food and are now extinct.

- Cod: Only 1 percent of the once-numerous original population now remains in the North Atlantic.

- Otter: Hunted for their fur, they have rapidly declined in North America.

- Whales: Many species of whales are nearly extinct because of whaling.

INTRODUCED SPECIES

Colonization is the expansion of a species into a new habitat. While it is a natural process, colonization is uncommon in nature, because the boundaries separating species rarely undergo significant, rapid change. Humans have made the introduction of new species a relatively common event.

Colonization can have a significant impact on endemic species, as the new species may prey on, or compete for resources with, existing species.

The introduction of new species can occur intentionally, when new crops are added, for example, or when animals are brought in as pest control agents. It can also occur unintentionally when, for example, foreign organisms are accidentally transported from one place to another on ships.

EXAMPLE: The introduction of new species can have drastic, unforeseen consequences. Three examples: The brown tree snake was introduced in Guam and eliminated almost all species of forest birds. Mosquitoes with avian malaria were introduced to Hawaii, and the result was the extinction of around 100 different species of birds. Leafy spurge was introduced into North America, where it competed with native grasses, thus ruining cattle grazing lands.

CHAPTER 25
Conservation

Introduced species can alter factors in the environment, such as the availability of nutrients, water, or sunlight. This can make the environment more suitable to the introduced species, less suitable to the original species, or more suitable to other species.

EXAMPLE: *Myrica faya,* a tree native to the Canary Islands, can fix nitrogen in the soil at high rates. When *Myrica faya* was introduced to Hawaii, it changed the soil, which now contains roughly 90 times the nitrogen it used to. As a result, other plant species that require large amounts of nitrogen have invaded.

ECOSYSTEM DISRUPTION

The living things in an ecosystem are often linked together. This link might be a food chain; the pollination of plants by insects, bats, and birds; or the shelter plants provide for certain animals. Because of these links, the reduction of the population of one species can affect the populations of other species.

EXAMPLE: Whales eat zooplankton. Because whales were overharvested, the zooplankton population increased. The pollack, a fish that feeds on zooplankton, began to increase in number. The pollock outcompeted other fish such as sea perch and herring, and these fish populations declined. Sea lions and harbor seals, which feed on sea perch and herring, did not get the same nourishment from the pollock, and their numbers declined. The orca whale, which feeds on sea lions and harbor deals, began to consume more sea otters because of the decline in their original food source. The number of sea otters declined. This led to an increase in sea urchins, which were the food source for the sea otters. The sea urchins consumed more kelp, which caused a decline in the kelp populations. The fish that relied on the kelp forests for protection from predators then began to decline.

Keystone species promote diversity in ecosystems. The keystone species might be a predator that limits population growth, which prevents one species from outcompeting others, or it might be an environment changer whose activities make a habitat suitable for other species. Changes that affect keystone species can be devastating for many other species.

EXAMPLE: The flying fox is a keystone species on the island of Guam because it pollinates plants and disperses seeds. Flying foxes are being reduced, which is having a dramatic impact on plant life, forcing many plant species toward extinction.

CHAPTER 25
Conservation

Population Recovery and Extinction Prevention

Once it is known that an ecosystem is declining or that a species is becoming endangered, action can be taken to restore the ecosystem and to prevent the species from declining further and becoming extinct. Two common methods for preventing extinction are habitat restoration and captive breeding.

HABITAT RESTORATION

Conservation is action aimed at preserving species and preventing extinction. This is only possible when the ecosystem still exists to be preserved, meaning that it has not declined to the point of complete loss or become an entirely new ecosystem. If the ecosystem does not exist to be preserved, habitat restoration, action taken to rebuild an ecosystem or return it to its original state, is required.

EXAMPLE: If tourist activity in a forest is affecting the ecosystem, conservation can reduce the human impact, because the ecosystem still exists. However, if a forest has been cleared and converted to farmland, conservation is not possible, because the ecosystem no longer exists. In this case, habitat restoration is required.

There are three ways that habitats can be restored:

- **Pristine restoration:** Plants and animals that previously inhabited the area are restored.

- **Removal of introduced species:** If introduced species caused the habitat loss, they are removed.

- **Cleanup and rehabilitation:** This method is used in habitats that have been affected by human actions such as pollution.

CAPTIVE BREEDING

Captive breeding programs involve breeding species in captivity and then releasing them into the wild. These programs are used to restore populations of animals under immediate threat of extinction.

EXAMPLE: In 1985, there were only nine wild condors in existence and twenty-one in captivity. The wild condors were captured and made part of a captive breeding program. By 2002, there were over 120 condors, and over thirty had been successfully reintroduced into the wild.

CONSERVATION OF ECOSYSTEMS

Efforts to protect species and prevent their extinction do not always focus on a single species. Protecting entire ecosystems is often the most effective way to promote biodiversity.

Megareserves, large areas of land containing a **core zone** (undisturbed habitat), are one means of protecting ecosystems. On these reserves, the core zone is surrounded by a **buffer zone,** where human activities such as education and tourism, which do not threaten the habitat, are allowed. In other parts of the reserve, economic activities such as building and manufacturing are often allowed.

Summary

Species Extinction

- Extinctions are a normal part of nature, but human activity is greatly increasing the rates of species extinction.

- The arrival of humans in North America around 12,000 years ago resulted in the extinction of around 80 percent of the **megafauna.**

- Between 1600 and 1700, an average of one species of bird and one species of mammal became extinct per decade. From 1850 to 1950, there was a tenfold increase in species extinction. Between 1986 and 1990, the figure had increased to four species of bird and four species of mammal every year.

- Extinction triggered by human actions is generally considered more damaging than extinction caused by natural events.

- An **endemic species** is one that is found naturally in only one geographic area.

- Species that are endemic to an area are of particular concern to **conservation biologists.**

- A **hotspot** is an area in which many endemic species of plants and animals live. The world's twenty-five hotspots contain around half of all the terrestrial species on Earth and are the focus of conservation efforts.

- Conservationists identify three main benefits of biodiversity: direct economic value, indirect economic value, and ethical and aesthetic values.

Causes of Species Extinction

- There are four main causes leading to the current extinction of species: habitat loss, overexploitation, introduced species, and ecosystem disruption.

- **Habitat loss** is the most common cause of extinctions at the current time.

- There are four ways that habitats are lost: destruction, pollution, disruption, and habitat fragmentation.

- **Overexploitation** occurs when living things are hunted or harvested by humans to the extent that the population is not able to maintain itself.

- **Colonization** is a process by which a species expands the area it lives in and enters a new habitat. Colonization is uncommon in nature. However, humans have made the introduction of new species a relatively common event.

- Colonization can be either intentional or unintentional.

- Introduced species can alter factors in the environment, making the environment more suitable to the introduced species, less suitable to the original species, or more suitable to other species.

- Ecosystem disruption occurs when a change that effects one population leads to changes in other populations.

- The changes are most significant when they affect a **keystone species,** which is a species that promotes diversity in an ecosystem.

Population Recovery and Extinction Prevention

- Conservation is action aimed at preserving species and preventing extinction.

- Habitat restoration aims to return an ecosystem to its original state.

- There are three ways that habitats can be restored: pristine restoration, removal of introduced species, and cleanup and rehabilitation.

- **Captive breeding** programs involve breeding species in captivity and then releasing them into the wild.

- Species can be protected when their entire ecosystem is the focus of protection efforts.

- **Megareserves,** large areas of land containing undisturbed habitats, have been developed to promote diversity.

- Megareserves are designed to contain a large, undisturbed habitat but also allow for human economic activity.

Sample Test Questions

1. Explain why endemic species are at particular risk of extinction due to habitat loss.

2. Explain how biodiversity would rebound after a large catastrophic event such as a major volcanic eruption and how this differs from extinction caused by human activity.

3. Explain how human activity has altered the introduction of new species as compared to how colonization occurs naturally.

4. Which term describes a species that is found in only one geographic area?

   A. Extinct
   B. Endemic
   C. Endangered
   D. Conservative
   E. Threatened

5. Around 40 percent of medications have active ingredients that are sourced from either plants or animals. What benefit does this describe?

   A. Ethical value
   B. Aesthetic value
   C. Direct economic value
   D. Indirect economic value
   E. Prevention value

6. Which term describes an alteration of the environment so extensive that certain organisms can no longer survive there?

   A. Habitat loss
   B. Defragmentation
   C. Overexploitation
   D. Ecosystem disruption
   E. Introduced species

7. Acid rain has altered some lakes and made them unable to support living things. This is an example of what type of habitat loss?

   A. Pollution
   B. Disruption
   C. Colonization
   D. Exploitation
   E. Fragmentation

8. Which of the following would be most likely to result in habitat fragmentation?

   A. A freeway cutting through a forest
   B. A farmer using insecticides to control pests
   C. A factory releasing chemicals into a lake
   D. A complete forest being burned and converted to grazing land
   E. A whaling company hunting large numbers of blue whales

9. Individuals on the borders of ecosystems are less likely to survive as conditions are altered at the borders. This situation is known as what?

    A. Fragment effects
    B. Physical effects
    C. Predation effects
    D. Change effects
    E. Edge effects

10. The hunting of passenger pigeons for food led to their extinction. This an example of what?

    A. Pollution
    B. Disruption
    C. Colonization
    D. Exploitation
    E. Fragmentation

11. A species that promotes diversity in an ecosystem is known as what?

    A. A keystone species
    B. A dependent species
    C. A linking species
    D. A growth species
    E. A limiting species

12. Which term describes actions aimed at preserving species and preventing extinction?

    A. Rehabilitation
    B. Regeneration
    C. Conservation
    D. Restoration
    E. Minimization

13. What does pristine restoration involve?

    A. Cleaning up an area to remove pollution introduced by human activity
    B. Restoring the plants and animals that previously inhabited an area
    C. Rehabilitating an area to undo physical environmental damage
    D. Removing species unintentionally introduced by humans
    E. Altering the environment to improve on original conditions

14. Captive breeding programs are commonly used for which category of animals?

A. Keystone animals
B. Endemic animals
C. Introduced animals
D. Endangered animals
E. Overpopulated animals

15. Which region of a megareserve contains an undisturbed habitat?

A. The buffer zone
B. The core zone
C. The transition zone
D. The research zone
E. The economic zone

ANSWERS

1. An endemic species is one that is found naturally in only one geographic area. If a nonendemic species declines in number due to habitat loss, it may not become extinct, because the species still exists in other areas. However, if habitat loss occurs in an area where a species is endemic, the extinction of the species might be the result.

2. A large catastrophic event such as a major volcanic eruption could lead to the extinction of many plant and animal species. However, the environment would return to normal over time. Without the species that became extinct, there would be opportunities for other animals to use the resources available. A biodiversity rebound would occur, and new species would develop to take the place of extinct ones. In contrast, after extinctions caused by human activity, humans remain to use the resources, and there are no opportunities for new species to develop or for biodiversity to rebound.

3. Colonization is a process that introduces species to new areas. In nature, it is rare, as there are usually geographic barriers that prevent species expanding into new areas. However, humans have overcome these barriers and are able to travel between geographic locations and introduce species to new areas. As a result, the introduction of new species has become a relatively common event.

4. B   An endemic species is one that is found naturally in only one geographic area.

5. C   Direct economic value refers to the fact that living things are
sources of food, medicine, clothing, and biomass.

6. A   Habitat loss occurs when an environment is altered to the extent
that certain organisms can no longer survive there.

7. A   Acid rain is a type of pollution. It can alter lakes and make them unable to support plant and animal life.

8. A   Habitat fragmentation occurs when habitats are physically divided
into smaller segments.

9. E   Edge effects describe how individuals on the edge of ecosystems are
less likely to survive as the conditions are altered at the borders.

10. D   Overexploitation occurs when living things are hunted or harvested
by humans to the extent that the population is not able to maintain
itself.

11. A   A keystone species is a species that promotes diversity in an
ecosystem.

12. C   Conservation refers to actions aimed at preserving species and
preventing extinction.

13. B   Pristine restoration is a method of habitat restoration that involves
restoring the plants and animals that previously inhabited an area.

14. D   Captive breeding programs involve breeding species in captivity and
then releasing them into the wild. These programs are usually used
for endangered animals, especially those under immediate threat of
extinction.

15. B   Megareserves are designed to include a core zone, which is an
undisturbed habitat.

Glossary

A

abiotic factors: Nonliving elements of an ecosystem, such as atmosphere, water, and rocks

absorption: Movement of small molecules into the mucous membranes that line the digestive tract

acetyl CoA: Chemical used to start the Krebs cycle in plant cellular respiration

acetylcholine: A common neurotransmitter that crosses the gap between motor neurons and muscle fiber in animals

acidic solutions: Solutions with a pH of less than 7

acid: Substance that raises the pH of a solution

acoelomates: Bilaterally symmetrical animal that do not have a body cavity

acrosome: Vesicle capping the head of the compact nucleus of a sperm and aiding in the penetration of the egg

actin: Protein filaments that, along with myosin, allow muscles to contract

action potential: Reversal of electric potential across a neuron membrane, allowing the transmission of a nerve impulse

activation energy: Initial input of energy required for chemical reactions to take place

activator: Protein that increases the rate of transcription by improving the ability of RNA polymerase to bind to the promoter

active immunity: Immunity developed after the body's exposure to an infection

active site: Part of the enzyme that binds with a substrate

active transport: Movement of materials powered by the expenditure of a cell's energy

adaptations: Inherited characteristics, physical or behavioral, that enhance an organism's ability to survive and reproduce in an environment

adenoids: Lymphatic tissue found at the back of the nasal passage

adenosine triphosphate (ATP): The energy currency of the cell; a nucleotide involved in intracellular energy transfer in all organisms

adrenal cortex: Part of the adrenal gland that controls the body's responses to stress through the production of specific hormones

adrenal medulla: Part of the adrenal gland that produces a group of hormones called catecholamines

adrenocorticotropic hormone (ACTH): Hormone that stimulates production of corticosteroids in the adrenal cortex

adventitious plantlets: Notches on plant leaves that give rise to new plants in some forms of asexual reproduction

alcoholic fermentation: Method by which pyruvic acid is converted to ethanol

aldosterone: Hormone that stimulates the distal tubes of the kidney to reabsorb sodium and water

allantois: Embryonic membrane that forms the umbilical cord

allele: Each variation of a given gene

allergies: Abnormal or excessive immune defense to an antigen

allometric growth: Period of development when organs in an animal body develop at rates that vary from each other

allopatric speciation: Mechanism by which two geographically isolated populations evolve into separate species

allosteric site: Any site on an enzyme that is not an active site

alternation of generations: Plant reproductive cycle in which a haploid phase, during which a gametophyte is formed, leads to a diploid phase, in which a sporophyte is formed. Spores produced by the meiotic division of the sporophyte then lead to new gametophytes.

altruism: Acts performed by one individual in a population that are beneficial to others but detrimental to that individual

alveoli: Tiny sacs in the lungs where gases are exchanged with the blood

amines: Functional group containing nitrogen and three substituents. The functional groups within a given molecule are those atoms responsible for the characteristic result in a chemical reaction.

amino acid: Small molecule composed of a central carbon atom, an amino group, a carboxyl group, a hydrogen atom, and a functional group labeled "R"

ammonification: Process during which nitrogen-containing compounds are broken down by bacteria and fungi and converted to ammonium ions

amnion: Membrane enclosing the sac containing amniotic fluid and the embryo in amniotic eggs

amniotic eggs: Shelled eggs containing a nutrient-filled amniotic sac that prevents dehydration of the embryo and provides it with protection and nourishment

amniotic sac: Two thin, transparent membranes that hold the developing embryo until just prior to birth

amygdala: Region of the brain within which sensory data converges from other sources, such as the thalamus and brainstem

analogous trait: Similar traits found in two species as a result of convergent evolution

anchoring junction: Junction that connects the cytoskeletons of two or more cells

androgen: Hormone produced in high levels in males and small levels in females. In males, the testes produce and secrete androgens to stimulate the development of the male sexual structures and secondary sex characteristics.

angiosperm: Flowering plant whose seeds are enclosed in fruit

Animalia: The kingdom of animals. Organisms in the kingdom Animalia are multicellular and heterophic, and are composed of eukaryotic cells that lack cell walls.

anion: Atom that has gained electrons to take on a negative charge

annual: Plant that completes its life cycle within a calendar year or growing season

anterior pituitary gland: Gland that communicates with both the hypothalamus and target cells via an endocrine pathway

anthers: Site at the end of the stamen where meiosis takes place to produce pollen

anthropoid: Diurnal organism possessing excellent color vision, an extended juvenile period, and a relatively large brain. Anthropoids include monkeys, apes, and humans.

antibody: Immunoglobulin protein produced by lymphocytes and released into the bloodstream in response to the presence of an antigen

anticodon: Base triplet that recognizes the complementary codon on an mRNA molecule

antidiuretic hormone (ADH): Hormone that stimulates water retention in the kidneys in response to increased levels of solutes within the blood

antigen: Any substance that elicits an immune response from the body

antigen receptor: Site on an antigen to which an antibody binds

antigen-presenting cell (APC): A cell, such as a phagocyte or B cell, that protects an antibody against an antigen

antiparallel: Opposing alignment of the two nucleotide chains that compose a molecule of DNA

anus: External opening of the rectum

aorta: Artery that carries oxygen-rich blood from the heart to the rest of the body

apical dominance: Process in plant growth during which growth of the main stem is dominant over growth of side stems

apical meristem: Plant tissue that produces new growth in the tips of roots and shoots

apomixis: Process during which seeds form in asexual reproduction

appendicular skeleton: Portion of the vertebrate skeleton that includes the appendages, pelvic girdle, and shoulder blades

aquifers: Layers of rock, sand, and gravel where groundwater can be found

archaebacteria: Type of bacteria in the kingdom Monera

arteries: Blood vessels that transport blood away from the heart

asexual reproduction: Reproduction in which a single parent organism passes on an exact replica of its genetic material to offspring

associative learning: Form of learning in which an animal changes its behavior after forming a connection between two stimuli or between a stimulus and a response

atom: Fundamental unit of matter; the smallest unit that has all the characteristics of an element

atomic nucleus: Core of an atom composed of protons and neutrons

atomic number: Number of protons in an atom

atrioventricular (AV) node: Node in the heart through which an electrical signal travels

atrioventricular (AV) valves: Valves in the heart that cover the openings between the atria and ventricles

autoimmune diseases: Diseases that result from the immune system's inability to recognize the body's own cells and therefore attacks them

autonomic (involuntary) nervous system: Portion of the nervous system that controls the involuntary actions of the smooth and cardiac muscles and the organs of the gastrointestinal, cardiovascular, excretory, and endocrine systems

autotrophs: Organisms that manufacture their own food; also known as primary producers

axial skeleton: Portion of the skeleton composed of the head and torso, including the skull, spinal column, and ribcage

axillary buds: Undeveloped shoots located where a petiole meets the stem

B

B cells: See B lymphocytes.

B lymphocytes: White blood cells produced when immature lymphocytes continue development in the bone marrow

bacteria: Single-celled organisms that contain their genome in one double-stranded, circular piece of DNA

bacteriophage: Virus that infects bacteria

basal ganglia: Portion of the brain that receives sensory information from the body and motor commands from other regions of the brain

basal metabolic rate: Amount of energy required to perform basic body functions

base: Substance that lowers the pH of a solution

basic solution: Solution with a pH greater than 7

behavior: An animal's response to stimuli in its environment

behavioral ecology: Field of biology that focuses on determining how and why animal behaviors have evolved

benthic zone: Bottom surface of an aquatic environment

biennial: Plant that completes its life cycle within two years

bilateria (bilateral): Animals that are symmetrical along a single, central axis; examples include worms, insects, and all vertebrates

bile: Bodily fluid composed of pigments from broken-down red blood cells. Bile is excreted along with digestive waste.

binary fission: Process used by prokaryotes to reproduce new cells. The parental chromosome is replicated and packaged into a new cell.

binomial system of nomenclature: Two-word system consisting of the genus and species-specific epithet, used to name species—for example, Homo sapiens.

biodiversity crisis: Area of science concerned with the rate of extinction over the last five centuries

biological clock: Internal control mechanism that determines certain biological processes dependent on environmental cues such as day and night

biology: The scientific study of life

biomass: Total dry-weight of all organic matter constituting a group of organisms or habitat

biome: Ecosystems characterized by similar climates and organisms. The Earth's eight major biomes are tropical rainforest, savanna, desert, temperate grassland, temperate deciduous forest, temperate evergreen forest, taiga, and tundra.

biotechnology: Technology based on biology or biological processes

biotic factors: All organisms living in an ecosystem

bird: Feathered animal that produces amniotic eggs, has scaled legs, and is endothermic

bladder: Hollow muscular organ that stores urine prior to excretion

blade: Flat part of a leaf

blastocoel: Central, fluid-filled cavity of a vertebrate blastocyst. The blastocoel forms when a zygote divides into many cells

blastocyst: Early stage of embryonic development in mammals

blastomere: Cell of the blastula produced after zygote division

blastopore: Opening in the gastrula during early embryonic development. The blastopore develops into the mouth in protostomes and the anus in deuterostomes.

blastula: Hollow ball of cells formed when a zygote divides

blood: Bodily fluid composed of plasma, platelets, and many different cells, such as red and white blood cells

body cavity: Fluid-filled compartment in some animals. In coelomates, this cavity develops into the coelom; in pseudocoelomates, this cavity develops into a pseudocoel.

bone: Main structural tissue in most vertebrates and the site of blood cell production

bottleneck: Change in the genetic representation of a population occurring when a population undergoes a sudden and drastic reduction in size, such as during a natural disaster

Bowman's capsule: Cup-shaped sac on the kidney, through which filtrate enters the nephron from the blood

brain hormone (BH): Hormone, produced by neurosecretory cells in the brain, that controls the release of ecdysone

breathing: Process of gas exchange in some animals in which air is inhaled as the chest cavity expands and the diaphragm contracts (moves downward) and air is exhaled as the chest cavity decreases and the diaphragm relaxes

bronchi: Tubes leading from the trachea to the lungs

budding: Form of asexual reproduction in which new individuals split off from a parent and develop into adults

buffer zone: Region within a protected habitat in which nonthreatening human activities, such as education and tourism, are allowed

buffer: Substance that reduces the effect of acids and bases on the pH of a solution

bulbourethral glands: Pea-sized glands located near the tip of the penis. Bulbourethral glands secrete small amounts of lubricating fluid into the urethra to assist the passage of sperm.

C

C$_3$ plant: Plant that relies on the Calvin cycle to form a three-carbon compound during the initial steps during incorporation of CO$_2$ into organic material

C$_4$ plant: Plant that produces four-carbon compounds using CO$_2$ prior to entering the Calvin cycle to incorporate CO$_2$ into organic material

Calvin cycle: Light-independent phase of photosynthesis, where carbon dioxide is fixed to a three-carbon compound used to form glucose. ATP and NADH are consumed in the Calvin cycle.

CAM (crassulacean acid metabolism): Method by which plants reduce photorespiration and conserve water by performing light reactions during the day and the Calvin cycle at night

capillaries: Small vessels in animal bodies that move blood between veins and arteries and through which substances are exchanged with body tissues

captive breeding: Conservation program in which species are bred in captivity and released into the wild

carbohydrate: Group of organic compounds that include sugars (monosaccharides and polysaccharides), starch, glycogen, and cellulose

carbon cycle: Process during which carbon is exchanged between the biosphere, geosphere, hydrosphere, and atmosphere

carbon fixation: Incorporation of CO$_2$ into organic molecules; performed during photosynthesis in plants

cardiac muscle: Muscle tissue in animal bodies that pumps the heart and circulates blood through the body

carnivore: Animal that eats other animals to gain energy. Examples include cats, eagles, wolves, and frogs.

carpel: Female organ of a flower made up of the stigma, the ovary, and the ovule

carrier protein: Protein that binds specifically to a substance and shuttles it into a cell via pores in the cell membrane

carrying capacity: Maximum population size that a habitat can support

cartilage: Padding for joints in most vertebrates and the main structural tissue for some boneless aquatic organisms, such as sharks

caste system: Hierarchy in which individuals of the same species perform different tasks, such as foraging or protecting, based on differing characteristics, such as size

cation: Atom that has lost electrons and takes a positive charge

cell: Smallest biological unit that can carry out all of life's processes

cell cycle: Series of events occurring during the life of a dividing eukaryotic cell

cell differentiation: Developmental process of a multicelled animal during which cells separate and specialize

cell division: The splitting of one cell into two

cell membrane: Outer surrounding of a cell that separates it from the environment and aids in internal regulation by controlling the entrance and exit of materials

cell plate: Double-layered membrane lining a dividing plant cell. The cell wall develops between the cell plate during cytokinesis.

cell wall: Supportive and protective layer that surrounds the cell membrane of plant, fungi, bacteria, and some protist cells

cell-mediated immunity: Immunity involving specialized cells that circulate in the lymph and blood. Cell-mediated immunity is used by hosts to fight infection against invading organisms.

cell theory: The doctrine that every living organism is composed of cells and that all cells come only from other preexisting cells

cellular respiration: Process that converts energy found in food molecules, in particular glucose, to the more useable form of ATP

central nervous system (CNS): The brain and spinal cord. The CNS, mostly made up of interneurons, acts as the central command center of the body.

centrioles: Tubelike structures that aid in cell division in animal cells

centromere: Region within the chromosome at which sister chromatids join during cell division

cephalization: Formation of the head end of an animal where nervous tissue is concentrated

cerebral cortex: Portion of the cerebrum containing sensory and motor nerve cells. In mammals, the cerebral cortex is the largest and most complex region of the brain.

Chargaff's rule: Principle governing the balance of quantities of the four nucleic bases present in the DNA of any cell. Chargaff's rule states that the amounts of cytosine and guanine are equal, and the amounts of thymine and adenine are equal.

charophyceans: Water-bound green algae existing in abundance in oceans 500 million years ago. Biologists believe that the first organisms to colonize land evolved from charophyceans.

chemical bond: Attraction between atoms resulting from the interactions of their electrons

chemical digestion: Form of digestion during which stomach acids aid the breakdown of the chemical bonds of food, turning polymers into their smaller constituent parts

chemical synapse: Region across which a nerve impulse is transmitted through the use of chemicals

chemoautotrophs: Group of organisms that use carbon dioxide to oxidize sulfur and other inorganic compounds for the production of food. Chemoautotrophs live in extremely hostile environments such as deep-sea hydrothermal vents.

chemoreceptors: Receptors in the body that are stimulated by changes in chemical compositions, such as changes to solute concentration, or are stimulated directly by chemicals that bind to the receptor membrane, such as olfactory receptors in the nose

chemosynthesis: Process in which chemical energy is converted into food, similar to the conversion of solar energy through photosynthesis

chiasma (plural **chiasmata**): X-shaped region representing the exchange of genetic material between homologous chromatids during meiosis

chief cells: Cells in the stomach that secrete pepsinogen used to convert the enzyme pepsin to a form usable for digestion

chloroplast: Structure found in the cells of some organisms where light energy is converted into ATP during photosynthesis

Chordata: Group of animals, including mammals and birds, that have at some time in their development had a notochord, a muscular tail, an endostyle, pharyngeal pouches, and a hollow dorsal nerve

chorion: Membrane surrounding the embryonic membrane in many animals. In mammals, the chorion contributes to development of the placenta.

chromatids: Two identical strands of genetic material that join at the centromere to create a chromosome during cell reproduction

chromatin: Combination of DNA and proteins that form a eukaryotic chromosome

chromosome: Structure composed of pieces of DNA tightly folded up by proteins. Chromosomes are present in the cell nucleus and carry all of an organism's genes.

chyme: Mixture of gastric acid and partially digested food

circadian rhythm: Twenty-four-hour cycle that drives certain processes in all eukaryotic organisms

circulatory system: Network of vessels within most animal bodies in which blood is transported throughout the body. The circulatory system is responsible for the exchange of substances between cells and the blood.

class: One of eight ranks of classification used by taxonomists to classify organisms

cleavage: Stage of animal development during which the zygote divides multiple times to form a ball of cells

cleavage furrow: Region of constriction in an animal cell resulting from the contraction of microfilaments in the center of an elongated cell during division

cloaca: Common opening, present in all vertebrates except most mammals, through which genital, digestive, and urinary products are expelled from the body

clonal selection: Immune system process in which activated B and T cells grow and divide repeatedly to form a clone specialized to defend against a specific antigen

clone: Exact genetic copy of an original

closed circulatory system: Circulatory system in which blood is contained within closed tubes called vessels and kept separate from extracellular fluid

clumped spacing: Population spacing in which individuals are grouped together, sometimes resulting in the uneven distribution of resources

cnidocytes: Stinging cells present on the surface of tentacles that assist in capturing prey

codon: Three bases of DNA or mRNA that are used to code for a specific amino acid

coelomate: Animal that contains a body cavity, known as a coelom, formed entirely within the mesoderm

cognitive behavior: Actions performed by an animal after information is processed in the brain and an appropriate response is determined

collenchyma cell: Plant cell, composed of thick, single outer-cell walls, that elongates during plant growth and provides support for a plant without restraining growth

colonization: Natural process during which a species expands the area in which it lives and enters a new habitat

combustion: Process during which materials containing high amounts of stored carbon atoms, such as wood, coal, or oil, are burned and release carbon dioxide into the atmosphere

commensalism: Symbiotic relationship in which interactions provide a benefit to one species, while the other species receives neither benefit nor harm

communication: Process by which individuals share information with other members of a social group

community: All of the species inhabiting a specific location

comparative genomics: Field of science that compares the genomes of various organisms to determine relationships between species

competitive exclusion: Theory that states in a community with limited resources, no two species can inhabit the same niche indefinitely without one species being driven to extinction because of competition

competitive inhibitor: A substance that mimics a substrate that normally binds to an enzyme. Activity of the enzyme is reduced when the inhibitor binds with the enzyme in place of the true substrate.

complement protein: Protein used in defense against microbes. Complement proteins create an opening in a microbe's cell membrane through which water enters, causing the cell to expand and eventually burst.

complex reflex: Reflex requiring an interneuron to relay information between the sensory and motor neuron, such as the reflex causing the eye to blink involuntarily to avoid contact with an object

complex structures: Structures, such as the nucleus and endoplasmic reticulum, that biologists believe developed from more simple structures present in prokaryotic cells

concentration gradient: Difference in the concentration of substances on either side of a membrane

connective tissue: Tissue used to bind and support other tissue in an animal's body

conservation biology: Field of science that studies the preservation of species and ecosystems

convergent evolution: Form of adaptations that occurs when distantly related organisms facing similar environmental challenges evolve functionally similar traits

core zone: Undisturbed habitat of an organism

cork cambium: Meristematic plant tissue that forms a ring of cork and parenchyma

corpora cavernosa: Erectile tissue along the dorsal side of the penis

corpus callosum: Thick band of fibers connecting the left and right hemispheres of the cerebral cortex. Information is passed between the two hemispheres via the corpus callosum.

corpus spongiosum: Erectile tissue along the ventral side of the penis

corticosteroids: Steroid-like hormones produced in the adrenal cortex

covalent bond: Bond that forms between atoms that share valence electrons

cranial nerves: Nerves extending down from the brain that stimulate organs within the head and upper body

crossing over: Exchange of genetic material between homologous chromosomes during meiosis

crystal: Highly regular and ordered formation in which atoms are arranged in repeating units

cuticle: Waxy coating on above-ground plant structures that helps prevent water loss and protects the plants from physical damage or contaminants, such as bacteria, viruses, and dust

cytokines: Chemicals present in the vertebrate immune system that regulate cells, proteins, and peptides

cytokinesis: Division of a cell's cytoplasm after division of the nucleus

cytoplasm: Cellular material existing between the plasma membrane and the nucleus

cytosol: Fluid within cells and cellular components, such as organelles, macromolecules, ions, and filaments

cytotoxic T cells: Lymphocytes that attack body cells infected with viruses or bacteria

D

decomposer: Organism, such as bacteria and fungi, that breaks down dead plant and animal remains

demography: The scientific study of birth and death rates in a population

dendrite: Structure on a neuron responsible for transmitting nerve impulses. Dendrites are usually short and highly branched.

denitrifying bacteria: Bacteria that convert nitrates into nitrogen gas

deoxyribonucleic acid (DNA): Genetic material that determines the genetic characteristics of an individual. Identical copies of DNA are housed in every cell of an organism.

dermis: Layer of the skin beneath the epidermis. The dermis is primarily composed of connective tissue and a well-developed sensory nerve network.

desiccation: Process of completely drying out, especially in plants

detritivore: Organism that feeds on the waste products of other organisms in an ecosystem

deuterostome: Class of animals distinguished by the development of the anus from the blastopore; radial, indeterminate cleavage; and enterocoelous formation of the coelom; examples include starfish, sea urchins, and sea cucumbers

diaphragm: Sheet of muscle that provides structure to the organs in the lower abdomen and assists in breathing

diastole: Period of the cardiac cycle during which the heart is at rest and blood fills the chambers

dicot (dicotyledon): Type of flowering plant that contains two cotyledons

diffusion: Spontaneous movement of particles from an area of higher concentration to an area of lower concentration

digestion: Breakdown of food into particles small enough to be absorbed by the lining of the digestive tract

digestive system: Bodily system in animals that aids in the breakdown of food particles and the absorption of these particles by the body's cells

digestive tract: Tube extending from an animal's mouth to its anus through which food passes and is broken down via mechanical and chemical processes into simpler chemical compounds that can be readily used by the body

dihybrid cross: Experiment in which parents that differ in two traits are mated, so that dispersal of those traits among offspring can be studied

diploid cells: Cells containing two sets of every chromosome

directional selection: Selection in which individuals at one phenotypic extreme are selected for and those at the other extreme are selected against

disruptive selection: Selection in which individuals at both phenotypic extremes are favored, and those with intermediate phenotypes are selected against

distal convoluted tubule: Portion of the kidney that regulates potassium, sodium, calcium, and acidity

DNA fingerprinting: Use of an individual's genome for purposes of identification

DNA ligase: Enzyme used during DNA replication. DNA ligase links together fragments of DNA.

DNA polymerase: Enzyme that assists in the replication of DNA by stimulating elongation of a DNA sequence

DNA replication: Process by which a cell replicates its DNA and packages it into a new cell, resulting in two cells with a complete copy of the cell's genome

domain: One of eight ranks of classification used by taxonomists to classify organisms

dominant allele: Characteristics that, if present on a chromosome, will be expressed in an organism's phenotype

dormancy: Period during which an organism delays growth or development until conditions are favorable

dorsal nerve: Single, hollow nerve cord running under the dorsal (or back) surface of an animal

double fertilization: Reproductive process, performed by angiosperms, in which two sperm cells are released and combine with two cells in the embryo sac. One set develops into the zygote and the other develops into the endosperm, which will form seed tissue.

duodenum: Hollow tube connecting the stomach to the small intestine and ileum

E

ecological pyramid: Pyramid-shaped diagram showing the loss of energy between trophic levels of an ecosystem

ecology: Study of the interactions between livings things and their environment

ecosystem: All organic and nonorganic components of an environment

ectoderm: Specialized embryonic tissue that is later found in the epidermis, nervous system, and sensory system of adults

egg: The female genetic in sexual reproduction, also called an ovum

ejaculation: Release of semen from the penis occurring when the penis is erect and sexually stimulated

ejaculatory ducts: Muscular tubes in the penis that cause ejaculation in some animals

electrical synapse: Region across which action potentials are transmitted directly from one neuron to the next during the transfer of a nerve impulse

electromagnetic receptor: Receptor that recognizes and responds to electro-magnetic energy, such as visible light, electricity, and magnetism

electron: Negatively charged particle that exists in a cloud around an atomic nucleus

electronegativity: Property that describes the strength of the attraction an atomic nucleus has for electrons

electron transport chain: Series of molecules that carry the electrons in a redox reaction, during which the energy required to create ATP is released

elements: Substances that cannot be broken down into other substances in chemical reactions

elimination: The removal of waste products from the body

embryo: Before birth, the maturing cells that will grow into a fully formed organism

embryonic stem cells: Cells that have the potential to develop into any body tissue cell

ectotherm: Organism that is unable to retain heat produced by metabolic activities

emigrant: Individual that leaves one population and is received into another population

endemic species: Species found naturally in only one geographic area

endergonic reaction: Reaction requiring energy input from an outside source in order to proceed

endocrine glands: Ductless glands that secrete hormones into the blood-stream

endocytosis: Absorption of substances into a cell via vacuoles or vesicles

endoderm: Specialized embryonic tissue that gives rise to the epithelium, which lines internal structures

endometrium: Surrounding wall of the uterus

endoplasmic reticulum (ER): Eukaryotic cell structure composed of a network of rough and smooth membranes connected to organelles and the outer cell membrane

endoskeleton: Rigid internal skeleton composed of calcium-rich structures, such as the bones of vertebrates and the plates of echinoderms, which are attached to muscles

endosperm: Nutrient-rich tissue in a plant's seed that provides nourishment to the embryo

endotherm: Organism capable of generating heat internally to maintain a relatively high body temperature

energy: Capacity to perform work

enhancer: Protein that activates transcription by binding to DNA. Enhancers do not need to be close to the DNA transcription site to be effective.

enzyme: Protein that lowers the activation energy of a reaction by binding with the reactants (or substrates) and either changing the reactants in some way or bringing them in close proximity with each other

epidermis: Outer layer of animal skin; or dermal tissue system of plants

epidermal tissue: Layer of flattened epidermal cells that constitutes the outer skin of a plant

epididymis: Tightly coiled tube near the testes that connects the efferent ducts and the vas deferens in male animals

epiglottis: Flap of skin covering the windpipe to prevent food from traveling to the lungs

epinephrine: Hormone released by the adrenal gland to prepare the body for emergency action when an animal is threatened

epithelial tissue: Tightly packed cells lining the internal and external surfaces of the body, including the skin, digestive tract, and body cavities

erythrocytes: Cells that transport oxygen around the body through the bloodstream; also known as red blood cells

esophagus: Muscular tube that performs peristaltic contractions to move food along the digestive tract from the pharynx to the stomach

estrogens: Group of hormones that maintain the female reproductive system and are responsible for the development of female secondary sex characteristics

eukaryotic cell: Cell composed of a membrane-bound nucleus and membrane-enclosed organelles. Eukaryotic cells are present in all eukaryotes, such as animals, plants, fungi, and protists

Eumetazoa: Group consisting of all animals, excluding sponges

eutherians: Animals whose young develop within the womb and are nourished by a placenta

evolution: Genetic change in a population over time

exergonic reaction: Reaction in which energy is released into the surroundings

exhale: Release of air from the lungs occurring when the diaphragm relaxes and the chest cavity decreases in volume

exocrine glands: Glands that secrete substances through tubelike ducts

exocytosis: Removal of substances from a cell via vacuoles or vesicles

exons: Coding regions of mRNA

exoskeleton: Hard structure composed of chitin that attaches to the muscles and protects the body of certain animals, such as mollusks, crustaceans, and arthropods

exponential growth model: Simplest method used to predict growth of a population

external fertilization: Fertilization of egg and sperm that occurs when parents expel their gametes into another medium, such as water, without necessarily coming into contact with each other

exteroreceptor: Receptor on an animal that detects external stimuli and conveys information about the external environment to the brain

extracellular matrices: Material secreted by animal cells that surrounds them and binds them together to form tissues

extreme halophile: Organism that thrives in extremely salty habitats

extreme thermophile: Organism that thrives in extremely hot habitats

F

F_1 generation: First generation of offspring from two opposing parental varieties

F_2 generation: Offspring produced by the interbreeding of individuals from an F_1 generation

facilitated diffusion: Process in which particles move spontaneously across a cell membrane from an area of higher concentration to an area of lower concentration with the aid of special transport proteins

facilitation: The process by which a species causes changes to the environment that allow new species to invade

Fallopian tubes: Tubes leading from the female ovaries into the uterus in mammals

family: One of eight ranks of classification used by taxonomists to classify organisms

fat (triacylglycerol): Compound composed of three fatty acids and one glycerol molecule

fatty acids: Long carbon chain carboxylic acid

feathers: Defining feature of birds that are derived from reptilian scales

fermentation: Method of ATP production that does not require an electron transport chain. Fermentation produces a limited amount of ATP, in addition to the end products ethyl alcohol or lactic acid.

fertilization: Union of sperm and egg cells to form a zygote

fibrous root system: Vascular system comprising an extensive network of roots and root hairs extending away from a plant. The large surface area of a fibrous root system allows plants to absorb great amounts of water and nutrients.

filial imprinting: Social attachment formed between a parent and offspring that influences behavior in the offspring

fission: Asexual reproduction in single-celled organisms in which one individual divides into two roughly equal-sized individuals, which then grow to the size of the original individual

fitness: Measure of an organism's contribution to the gene pool of the next generation in comparison to other members of its population

food chain: Flow of food from one trophic level to the next, starting with producers

food web: Diagram of feeding interactions of all organisms within an ecosystem

founder effects: Genetic drift that occurs when a new population is colonized by only a few individuals from an original group

fragmentation: Asexual reproduction in which part of a parent plant breaks off and gives rise to a new individual

free energy: The sum of potential and kinetic energy within a system

fruit: Structure developed from the ovary of a flower and used to protect and disperse dormant seeds

functional genomics: Study of gene functions and their results

functional group: Those atoms in a molecule most likely to be involved in a chemical reaction

fundamental niche: Theoretical environmental niche that an organism is capable of occupying

G

gallbladder: Organ in the animal body that stores and concentrates bile

gamete: Haploid sperm or egg cell that combines with an opposing gamete during sexual reproduction to produce a diploid zygote

ganglia: Support cells that exist around nerves and perform tasks such as removing wastes, supplying nutrients, and assisting the transport of nerve impulses along the axon

gap junctions: Molecular channels that exist between animal cells

gastric acid: Mostly hydrochloric acid (HCl) secretion produced during digestion in the stomach to aid in the chemical breakdown of food

gastric juice: Acidic mixture of acid, enzymes, and mucus secreted by gastric glands located in the lining of the stomach

gastrovascular cavity: Cavity present in some invertebrates in which digestion occurs. Food enters and leaves via the same hole of a gastrovascular cavity.

gastrula: Three-layered embryonic stage that forms from the blastula

gastrulation: Embryonic development process in vertebrates during which a three-layered gastrula is produced from the blastula

gated ion channel: Channel that opens and closes, allowing the cell to regulate its membrane potential

gene: DNA sequence that carries instructions for making specific proteins. Genes are the fundamental unit of heredity in all organisms.

gene expression: The synthesis of proteins according to information enclosed in DNA

gene flow: Change in genetic makeup of a population resulting from the movement of organisms in and out of a population

gene pool: The total of all variations of all genes that exist in a given population

genetic divergence: Separation of genetic similarities between individuals within a population that can lead to the separation of one species into two

genetic drift: Unpredictable change in allele frequency within a population resulting from chance occurrences

genetic variation: Presence of a number of different alleles of a specific gene, resulting in variation from parent to offspring

genetically modified organism: Organism whose genome has been altered by either the insertion of foreign genes or the manipulation of genes already present in the organism

genome: Organism's complete genetic material, or DNA sequence

genomics: Study of entire species genomes, as opposed to individual genes, to determine functionality of genes, relatedness between species, and genetic requirements for life

genotype: Collection of alleles within an individual

genus: One of eight ranks of classification used by taxonomists to classify organisms

germination: Point at which a seed begins growth and development, usually occurring after the seed reaches suitable soil and water conditions

gill: Respiratory organ projecting from the external surface of many aquatic animals, such as fish, which allows extraction of oxygen from water and regulation of osmolarity

gland: Any organ in an animal body that produces and secretes a substance

glomerulus: Collection of capillaries in the nephron of the mammalian kidney, used to filter blood entering the nephron

glucagon: Type of hormone secreted by the pancreas to stimulate an increase in blood glucose levels

glucocorticoids: Hormone secreted by the adrenal cortex to assist in glucose metabolism and immune functions

glycogen: Altered form of glucose that is stored in the muscle and liver cells for use as food

glycolysis: First stage of cellular respiration, during which glucose is split into pyruvate

glycoprotein: Polypeptide chain longer than 100 amino acids attached to a carbohydrate

Golgi apparatus: Organelle, present in eukaryotic cells, that is used to modify, store, and transport products of the endoplasmic reticulum. The Golgi apparatus is composed of membranous sacs that are stacked together.

gradualist model: Theoretical model of evolution that suggests a slow, continuous process of species change

gravitropism: Movement of the shoots or roots of a plant in response to gravity

gross primary productivity (GPP): Measure of the total amount of organic matter produced by all photosynthetic organisms in an ecosystem

ground substance: Proteins secreted by cells and fibroblasts that compose part of the plant tissue structure

ground tissue: Tissue between a plant's epidermal and vascular tissue where food and water are stored and the functions of photosynthesis take place

groundwater: Water located beneath the surface of the ground

growth hormone: Hormone that stimulates the growth of muscle, bones, and other tissues and regulates the body's metabolism

guard cells: Paired cells, found in the epidermal plant layer, that surround the stomata

gustatory receptor: Type of chemoreceptor that detects chemicals in the mouth

gymnosperm: Type of plant that produces seeds that lack a protective fruit coating

H

habitat: An organism's environment

habitat loss: Most common cause of extinction, resulting when an environment is altered to the extent that certain organisms can no longer survive in it

hair: Integumentary structure composed of keratin protein growing outward from follicles embedded in the dermis

half-life: Amount of time it takes for half of the atoms in a radioactive substance to decay; used by scientists to measure the approximate age of fossils

haploid cell: Cell containing only a single set of chromosomes

Hardy-Weinberg equation: Equation used to estimate allele or genotype frequencies in a population. The two-allele Hardy-Weinberg equation is $p^2 + 2pq + q^2 = 1$

Hardy-Weinberg theorem: Theorem stating genetic mixing owing to the sexual shuffling of genes is not capable of altering the genetic makeup of a population

heart: Muscular organ that pumps blood through an animal's body

helper T cells: Cells in the immune system that activate cytotoxic T cells, stimulate B cells, and activate macrophages

hemolymph: Invertebrate body fluid that moves through an open circulatory system and bathes the body's tissues

herbivore: Animal such as cows, horses, and nearly all rodents that eats only plants

heritability: Genetic transmission of traits from one generation to the next

heterotrophs: Organisms that are unable to produce their own food and therefore must consume autotrophs or other heterotrophs to gain energy

heterozygous: Presence of two different alleles for the one genetic trait

hippocampus: Gland present in many vertebrates that is responsible for emotional responses and memory

holistic concept: Theory that a community is an integrated unit of all the species found in it. The holistic concept suggests that all species within a community react and respond in a similar manner to changes in the environment.

homeostasis: Maintenance of stable internal conditions, such as body temperature, water level, and osmolarity in the body

homologous chromosomes: Pair of chromosomes that are similar in length and centromere position and possess genes for the same characteristics at corresponding loci. Offspring receive one homologous chromosome from each parent.

homologous structure: Feature that is similar in structure in different species that share a common ancestry. For example, the wing of a bat and the hand of a human are homologous structures.

homologous trait: Similar trait found in two species as a result of common ancestry

homoplasy: Character states shared between two species that are not inherited from a common ancestor

homozygous: Presence of two of the same alleles for the one genetic trait

hormone: Regulatory chemical secreted into the blood from an endocrine gland

hotspot: Area that is host to many endemic species of plants and animals

humoral immunity: Form of immunity in which antibodies within the blood plasma and lymph fight bacteria and viruses

hybrid: Offspring of two different species

hydraulic skeleton: Fluid skeleton found in many soft-bodied invertebrates that allows an organism to change shape but not volume; also called a hydrostatic skeleton

hydrostatic skeleton: Internal system of hydrostatic pressure that provides structure to some soft-bodied invertebrates, such as starfish

hydrogen ion: A hydrogen atom that has one proton and no electron and takes a charge of +1

hydrogen bond: Weak attractions between molecules that form when a hydrogen atom covalently bonded to an electronegative atom is attracted to another electronegatve atom

hydrophilic (polar) molecule: A molecule that is attracted to water

hydrophobic (nonpolar) molecule: A molecule that is repelled from water

hydroxide ion: Diatomic ion composed of oxygen and hydrogen atoms. A hydroxide ion takes a charge of –1.

hyperosmotic solution: Solution that has a higher osmolality compared to the outside environment or another solution

hypertonic solution: Solution with a higher solute concentration than the cell

hypodermis: Connective tissue attaching the skin to internal organs, also known as the subcutaneous (sub-skin) layer

hypoosmotic solution: Solution that has a lower osmolality compared to the outside environment or another solution

hypotonic solution: Solution with a lower solute concentration than the cell

I

ileum: Portion of the small intestine that contains a large surface area and is instrumental in the absorption of nutrients from food

immigrant: Individual that moves into a population

immune system: Animal body system that protects the body and defends against foreign invaders

imprinting: Process in which behaviors learned by a newborn individual during the period of early development are influenced by social attachments

individualistic concept: Theory that a community is an aggregation of species that happen to coexist in a single location because of similarities in abiotic needs

inferior vena cava: Major vein in the circulatory system that collects blood from the lower body

inflammation: Immune response to infection or cell damage during which small blood vessels dilate, allowing an increase in the number of leukocytes to an area, and resulting in swelling and reddening of the area

ingestion: Consumption of food via a mouth or mouthlike structure

inhale: Intake of air into the lungs that occurs when the diaphragm contracts and the chest cavity increases in volume

inhibition: Changes to an environment that prevent an old species from surviving

inhibiting hormone: Hormone secreted by the hypothalmus that causes the anterior pituitary to stop secretion

inhibitor: Neurotransmitter that prevents a signal from being transmitted and an action from being stimulated

innate: Quality of traits, such as behaviors, present in newborn individuals and believed by biologists to be influenced by genetics

insulin: Hormone secreted by the pancreas in vertebrates that stimulates the uptake of glucose by liver, muscle, and adipose cells after an increase in blood glucose levels

integumentary system: Outer features of an animal, such as the skin, hair, nails, scales, or feathers of a vertebrate. The integumentary system forms the first line of defense against the outside world.

interferon: Chemical messenger produced and released by a virus-infected cell to inhibit viral replication

internal fertilization: Fusing of a sperm and egg that occurs inside the body following the deposition of sperm by a male into the female reproductive tract

interneuron: Neuron located between motor and sensory neurons that helps transfer signals between these neurons

internode: Stretch of stem between a plant's leaves

interoreceptor: Receptor that detects internal stimuli and conveys information about the internal environment throughout the body. For example, interorecepters in the arteries detect changes in blood pressure.

interphase: The phase of the cell cycle when the cell is not in the process of dividing

interpretation: Creation of a perception and response based on stimulus received by the brain

intersexual selection: Process in which individuals of one sex attempt to entice individuals of the opposite sex to mate with them

intrasexual selection: Process in which individuals of thc same sex compete with each other for mating opportunities

intron: Noncoding region of an mRNA strand

invertebrate: Animal that lacks a backbone

ion: Atom that has acquired a positive or negative electric charge by gaining or losing electrons

ionic bond: Bond formed when an atom of low electronegativity donates an electron to an atom of high electronegativity

isoosmotic solution: Solution that has the samc osmolality as the environment or another solution

isotonic solution: Solution with equal solute concentration as the cell

isotopes: Atoms of the same element that have different numbers of neutrons and therefore different atomic masses

iteroparity: Process by which individuals in some species will tend to undergo many reproductive events throughout their lives

J

jejunum: Portion of the small intestine that contains a large surface area and is instrumental in the absorption of nutrients from food

joint: Point at which two or more bones connect

juvenile hormone (JH): Hormone that promotes the retention of larval characteristics. JH is a key hormone in arthropods.

K

keratin: Protein found in nails and hair

keystone species: Species whose significance in influencing diversity in an ecosystem or community is greater than their relative abundance

kidney: Bean-shaped organ that plays a major role in filtering the blood and producing urine in many vertebrates

kinetic energy: Energy of objects in motion

kinetochore: Region on a centromere where sister chromatids join to mitotic spindles

kingdom: One of eight ranks of classification used by taxonomists to classify organisms

Krebs cycle: A stage of cellular respiration in which glucose molecules are broken down to produce carbon dioxide

L

labor: Series of rhythmic uterine contractions that bring about the birth of offspring in placental mammals

lactic acid fermentation: Mehtod by which pyruvic acid is converted to lactate

large intestine: Digestive organ responsible for waste compaction; also known as the colon

larynx: Structure in the trachea where the vocal cords are located

lateral roots: Roots that grow as offshoots of a primary root

leaves: Projections from the plant stem that are the primary site of photosynthesis. A leaf is made up of a flat blade and a petiole.

leukocyte: Type of white blood cell that acts to defend against infection in the body

ligament: Tissue that connects bones to one another at joints

light microscope (LM): Instrument that bends and magnifies visible light, causing images to be magnified

light-dependent reaction: Part of the photosynthesis process during which solar energy is converted into ATP and NADPH

lipid: Nonpolar organic compound that includes fats, phospholipids, and steroids

liver: Organ that produces bile, a secretion that breaks down fat in the small intestine

locomotion: An animal's ability to use its body to move through its environment

logistic growth model: Model used to predict when growth will level off as a population reaches carrying capacity

long-day plant: Plant that flowers in the summer or early spring, when the duration of daylight is long

loop of Henle: Structure consisting of an ascending and descending loop that is located in the nephron of the vertebrate kidney and assists in the reabsorption of salt and water

lungs: Respiratory surface found in some arthropods as well as all amphibians, reptiles, birds, and mammals

lymph: Fluid found within the lymphatic system

lymphatic system: Bodily system composed of vessels and lymph nodes that help to return water and proteins to the bloodstream

lymphocyte: Type of white blood cell used by the immune system. Lymphocytes develop into either B cells or T cells.

lysogenic cycle: Process by which viral genetic material combines with the DNA of a bacterium and becomes a phage

lysosome: Sac containing digestive enzymes used to break down food molecules or damaged parts of a cell

lysozyme: Enzyme that destroys bacterial membranes

lytic cycle: Process by which a virally infected bacterial cell replicates the virus's material and packages up new virus cells. The bacterial cell then bursts, releasing the new virus cells.

M

macroevolution: Large-scale evolutionary changes that occur over long stretches of time, such as the splitting of one species into two

macrophages: Large white blood cells found in extracellular fluid that engulf invading viruses and bacteria

Malpighian tubule: Excretory organ found in insects

mammals: Vertebrate group characterized by the presence of hair, a middle ear, and mammary glands

mantle: Heavy fold of tissue present in mollusks that protects the internal components. The shell of some mollusks may be secreted from the mantle.

marker proteins: Proteins that facilitate cell-to-cell recognition by identifying a cell's function or origin to other cells

marsupial: Mammal in which the young are born early and complete their development in a pouch called a marsupium

mass number: Sum of the protons and neutrons in an atom

mate choice: Behavior performed by animals in which one sex evaluates potential mating choices and chooses the fittest individual

matrix: Part of connective tissue composed of fibers embedded in a homogenous intracellular substance

maximum likelihood method: Statistical process in which an evolutionary tree is designed to weight evolutionary events according to their likelihood of occurrence

maximum parsimony method: Statistical method that seeks to determine phylogenic relatedness among species by assuming the least number of evolutionary branches on a phylogenic tree

mechanoreceptor: Receptor that is stimulated by physical changes, such as pressure, touch, stretching, sound, and motion

mechanical digestion: Form of digestion in which methods such as chewing and grinding are used to divide food into smaller pieces without disrupting chemical bonds

megafauna: Animals weighing more than 100 pounds, such as mammoths, lions, saber-toothed cats, horses, and camels

megareserves: Large area of land containing undisturbed habitats

meiosis: Process of cell division used in sexual reproduction to create gametes, which are sperm and eggs. Meiosis involves the replication and division of genetic material from each parent, such that new cells contain half the chromosome number of a parent cell.

meristem: Tissue composed of undifferentiated plant cells that divide, allowing indeterminate growth throughout the plant's life

membrane proteins: Proteins that interact with the cytoskeleton of the cell to stabilize the cell membrane, transport molecules into and out of the cell, relay messages between cells, and facilitate chemical reactions

memory B cell: Long-lived cell that remains in the system after an infection has been neutralized in order to initiate rapid immune response when the same antigen reenters the body

mesoderm: Specialized embryonic tissue that gives rise to muscle, bone, and connective tissues and to organs in the circulatory, excretory, and reproductive systems

messenger RNA (mRNA): RNA that is synthesized from DNA and attached to ribosomes in the cytoplasm, where it conveys information used in the formation of proteins

metabolism: Sum of all of an organism's chemical processes

metapopulation: Network of distinct populations that can interact with each other by exchanging individuals

methanogen: Organism that inhabits oxygen-free environments high in toxic gases

microevolution: Small-scale changes to a population's gene pool over a few generations

microfilaments: Contractible protein fibers, found in some cells throughout the vertebrate body, that are responsible for structure and movement

microscope: Instrument used to magnify and observe very small things, such as cells

microtubule: Cylindrical protein fiber found in all eukaryotic cells that gives the cell its shape and assists in cell division

microvillus: Outfoldings of epithelial cells that increase the surface area of the small intestine

migration: The act of moving from one region to another

mineralocorticoid: Corticosteroid responsible for regulating salt and water balance in the body

mitochondria: Organelle that is the site of energy production in eukaryotic cells

mitotic phase: The phase of the cell cycle during which the cell is dividing

mitosis: Method by which eukaryotic cells divide and produce two genetically identical cells

molecule: Collection of atoms joined together by chemical bonds

monocot (monocotyledon): Type of flowering plant that consists of only one cotyledon

monocyte: Immune system cell that is sent to an infected region, where it transforms into a phage

monohybrid cross: Experiment in which parents differing in only one trait are mated to study the distribution of that trait in the offspring

monomer: Smallest subunit of a polymer

monophyletic group: Group that encompasses all potential descendants of a single common ancestor

monosaccharide: Simplest form of carbohydrate molecule. Monosaccharide subunits form disaccharides and polysaccharides.

monotremes: Egg-laying mammal, such as the platypus and echidna

motor division: Part of the nervous system that consists of motor neurons, which send instructions from the central nervous system to cells throughout the body

motor neuron: Nerve cell that carries impulses from the central nervous system to parts of the body, such as muscles and glands, that operate in response to a particular stimulus

mouth: Oral cavity located at the beginning of the digestive tract

mucosa: Innermost layer of the digestive tract, which comes into contact with food as it is digested. The mucosa produces digestive enzymes and protective mucus.

multicellular organism: Organism composed of more than one cell

muscle tissue: Tissue that can be contracted to allow movement

muscularis: Muscle layer surrounding the submucosa and responsible for peristalsis contractions that mechanically break down the food and move it along the alimentary canal

mutation: Random change in the DNA sequence of a gene. Mutation is the only process capable of producing new alleles.

mutualism: Mutually beneficial interaction between two organisms

myelin sheath: Layer of insulating cell membrane covering a neuron

myofibrils: Bundles of filament composed of actin and myosin found in muscle fiber

myosin: Protein filaments that, along with actin, allow muscles to contract

N

nails: Integumentary structures composed of keratin and found at the end of toes and fingers

natural selection: Process in which environmental factors produce differences in reproductive success of certain phenotypic traits within a population

negative feedback: Mechanism of homeostatic regulation in which an end, or product, of a process inhibits an increase in that process

nephridia: Tubular structures found in many invertebrates that filter waste from the body cavity and expel it externally

nephron: Tiny tubule structure responsible for the filtering of blood in the kidneys of vertebrates

neritic zone: Relatively shallow regions of water along the edges of continents and islands. A large percentage of the oceans' fish species inhabit the neritic zone.

nerve cord: Thick bundle of nerves running down the length of the body, just beneath the dorsal (back) surface

nerve impulse: Electric or chemical signal, passed along nerve fibers to cells in the body, that either stimulates or inhibits the region receiving the signal

nerve net: Primitive nervous system, found in some invertebrates, that is composed of branching nerves

nervous system: Bodily system composed of nerves, the brain, spinal cord, and ganglia and involved in the transfer of electrical and chemical signals to cells throughout the body

nervous tissue: Tissue that produces and conducts electrochemical signals between organs of the body and the brain

net primary productivity (NPP): Amount of energy produced within an ecosystem that is available to heterotrophs

neural crest: Embryonic cells that migrate to various regions of a vertebrate's body and develop into structures such as organs and sensory neurons

neural tube: Embryonic tube that later forms into the spinal cord and brain

neuroglia: Supporting cells found in the central nervous system that assist neuron cells by removing waste products, supplying nutrients, and guiding impulses along the axon

neuron: Nerve cell across which nerve impulses, or signals, travel, carrying information throughout cells in the body

neurotransmitter: Chemical released by an axon that transmits information across chemical synapses

neutral solution: Solution with a pH of 7

neutron: Electrically neutral particle located at the atomic nucleus

niche: Specific role an organism fulfills in interactions with its environment

nitrifying bacteria: Bacteria that convert nitrogen compound ions to nitrates

nitrogen cycle: Process by which nitrogen from the atmosphere is fixed by bacteria and cycled through an ecosystem

nitrogen fixation: Process by which nitrogen from the atmosphere is converted into compounds that can be used for growth by plants

node: Region at which leaves are connected to the plant stem

nonassociative learning: Mechanism by which an animal learns a behavior without developing a connection between two stimuli or between a stimulus and a response

noncompetitive inhibition: Process by which an inhibitor binds to a region on an enzyme and changes the shape of that enzyme so that it no longer bonds with a substrate

nonrandom mating: Selection of mate based on evaluation and choice, where certain characteristics are favored over others, resulting in an unequal distribution of mating opportunities among members of the population

norepinephrine: Type of neurotransmitter released by the adrenergic system to constrict blood vessels

notochord: Flexible rod that runs along the dorsal side (back) of the body and serves as an attachment for muscles. Present in the embryos of all chordates, the notochord eventually becomes the vertebrate backbone

nucleoid: Unbound region on a prokaryotic cell where DNA is located

nucleotide: Molecule composed of a phosphate group, a five-carbon sugar, and a nitrogenous base

nucleus: Membrane-bound command center of all eukaryotic cells where the cell's DNA is located

nutrients: Chemical substance necessary for metabolic and physiological functions that an animal cannot produce by itself

nutrition: Study of the intake and use of food by the body

O

omnivore: Animal that eats both plants and other animals

open circulatory system: Circulatory system in which the blood mixes with interstitial body fluids, rather than being enclosed within vessels, and directly bathes the body organs

order: One of eight ranks of classification used by taxonomists to classify organisms

organ: Group of tissues organized to carry out a particular bodily function

organ system: Group of organs designed to carry out a particular bodily task

organelle: Group of molecules organized to perform a specific cellular function

organism: Any living thing, including everything from unicellular bacteria to highly complex multicellular plants and animals

osmolality: Concentration of an osmotic solution

osmolarity: Concentration of solutes within a solution measured in moles per liter

osmoregulation: Process by which an animal's body manipulates the internal movement of water according to its needs

osmosis: Process in which water moves from dilute solutions with a low concentration of solutes to concentrated solutions with high concentrations of solutes

osteoblasts: Bone cells that secrete calcium and collagen used to create new bone

osteoclasts: Cells formed from a type of white blood cell and used in the breakdown and reabsorption of bone cells as bones are reformed during growth

osteocytes: Bone cells produced as osteoblasts become covered in the calcium and collagen secretion and mature

osteons: Tubes composed of cylindrical layers of bone surrounding blood and nerve vessels

outgroup: Group of organisms that exhibits characteristics similar to another group being studied

ovary: In animals, site in which the female gametes and hormones are produced; in plants, site where the ovules develop

overexploitation: Over-hunting or over-harvesting of a population resulting in that population not being able to maintain itself

oviparity: Developmental condition in which eggs developed inside the female's body are deposited outside the body, and offspring hatch in the external environment

ovoviviparity: Developmental condition in which growth is completed inside the female's body and offspring hatch fully developed and are released from the female's body

ovulation: Process resulting in an egg being released from the female ovaries in animals

ovule: Female plant structure that houses the developing gametophyte

ovum: The female gamete, also called an egg

oxidation-reduction reactions: Reaction in which energy is transferred between molecules in the form of electrons; also known as a redox reaction

oxidation: Loss of one or more electrons from a molecule during a redox reaction

oxytocin: Female hormone that stimulates contraction of the uterine walls during childbirth and the production of milk from the nipples

P

pacemaker: Node on the right atrium of the mammalian heart that sets the rate of heart contraction; also known as the sinoatrial (SA) node

pain receptor: Receptor stimulated by pain, such as excess heat, excess pressure, and certain chemicals released from damaged areas of the body

pancreas: Organ that secretes pancreatic juice into the small intestine

pancreatic juice: Fluid secreted by the pancreas and composed of bicarbonate that neutralizes the acid in the chyme. Pancreatic juice also contains a host of enzymes that help digest molecules in the chyme.

paracrine regulator: Chemical messenger secreted by and used specifically within an organ

paraphyletic group: Group that includes some, but not all, of the descendants of a single common ancestor

parasitism: Interaction that benefits one organism while harming another

parasympathetic system: Part of the vertebrate nervous system that, when activated, decreases energy consumption and heart rate as the body relaxes

parathyroid glands: Four glands of the thyroid gland that secrete a parathyroid hormone, which plays a role in controlling blood calcium levels

Parazoa: Group consisting of primitive animals lacking any true tissues; each parazoan cell carries out all of the body's functions

parenchyma cell: Cells that make up the majority of cells in most plants and perform many functions including differentiation into various cell types, metabolism, and synthesis and storage of organic products

parietal cell: Cell that secretes hydrochloric acid or gastric acid during digestion

passive immunity: Immunity in which antibodies have been obtained from a source outside the body. For example, in humans, antibodies from a baby's mother are transferred to the baby via the placenta.

passive transport: Process by which materials are transferred between membranes in the body without the expenditure of energy

pelagic zone: Area of water above the ocean floor, stretching between the surface to depths approaching 1,000 meters

penis: Male organ from which sperm is deposited into the female reproductive system

pepsin: Enzyme used to break down proteins in food during digestion

pepsinogen: Precursor to the enzyme pepsin

perennial: Plant that completes its life cycle over the course of many years

perforin: Protein, produced by cytotoxic cells in the immune system, that act to destroy antigen-bearing targets

peripheral nervous system (PNS): Part of the central nervous system responsible for the transfer of information between internal and external locations of the body and the central nervous system

peristalsis: Rhythmic and stepwise muscle contractions that force food to move along the alimentary canal

permeability: Degree to which substances can pass through a cell membrane

petals: Colorful structures that attract pollinators to a flower

petiole: Plant structure that joins the stem to the leaf

pH scale: Hydrogen ion concentration of a solution ranging from 0 to 14. Solutions with pH values below 7 are acidic and those above are basic.

phage: A virus that infects bacteria

phagocytosis: Consumption of solid matter via endocytosis

pharyngeal slits: Openings that connect the pharynx and esophagus with the outside environment

pharynx: Part of the throat that connects the oral cavity to the esophagus

phenotype: Outward expression of an organism's genotype, typically displayed in physical characteristics and behavior

phloem: Portion of a plant's vascular system that conveys materials throughout the plant

phospholipid: Molecule constituting the cell's membrane and composed of a nonpolar tail and a polar head

phospholipid bilayer: Main component of a cell membrane composed of two layers of phospholipids arranged with the nonpolar tails facing inward and the polar heads facing outward

phosphorus cycle: Naturally occurring cycle during which calcium phosphate dissolves in water, making phosphate ions available to plants. Plants convert the phosphorus into biological molecules, which are then passed through an ecosystem's food chain.

photoperiodism: An organism's physiological response to changes in the length of day throughout the course of a year

photoreceptor: Type of electromagnetic receptor that responds to light

photorespiration: Plant respiratory process during which oxygen is consumed, carbon dioxide is released, and photosynthesis output is decreased. Photorespiration is performed when a plant's stomata are closed, reducing water loss in hot, dry, or bright environments.

photosynthesis: Process in which light energy is used to power chemical reactions that produce glucose in plants

photosystems: Clusters of molecules found in photosynthesizing organisms that are composed of light-absorbing pigments and a reaction center

phototropism: Process by which plants grow, or bend, toward light

phylogeny: Study of the evolutionary history of groups of species

phylum: One of eight ranks of classification used by taxonomists to classify organisms

physiology: Branch of biology that studies basic biological tasks and the processes by which they are carried out in animals

phytochromes: Class of photoreceptor pigments involved in detecting daylight and setting the biological clocks of plants

pinocytosis: Consumption of extracellular fluid and its dissolved solutes by a cell

pituitary gland: Gland that communicates with the hypothalamus via nerve pathways and with target cells via an endocrine pathway

Plantae: The kingdom of plants. Members of the kingom Plantae are composed of eukaryotic cells that contain cell walls. Almost all members are multicellular and have cholorplasts used in photosynthesis.

plasma: Component of blood in which blood cells are suspended

plasma membrane: Membrane that covers every cell and regulates the internal environment by controlling the entry and exit of materials

plasmid: Fragment of bacteria or yeast DNA that is independent of the genetic material and can be replicated separately from the main chromosome

plasmodesma: Molecular channels between adjacent plant cells

platelets: Cells present in the blood to facilitate clotting, thereby preventing blood leakage when vessels break or tear

pollen: The male gametophyte of gymnosperms and angiosperms

pollination: Process in fertilization during which pollen grains are transferred from the anther to the stigma

polymer: Molecule composed of similarly structured subunits bonded together

polypeptides: Polymers of amino acids that are folded into complex three-dimensional shapes

polyphyletic group: Group that does not include the most recent ancestor common to all members

polysaccharide: Carbohydrate used by the body to store energy and provide structural support

population: Group of individuals of the same species living in the same geographic area and capable of interbreeding

population dynamics: Changes in a population over time that are influenced by various factors, such as the rate at which a population reproduces, the proportion of different aged individuals in the population, and death rates within the population

positive feedback: Homeostatic mechanism in which changes in a variable increase the frequency or intensity of a process

postanal tail: Tail extending beyond the anus

posterior pituitary gland: Gland that communicates with the hypothalamus via nerve pathways and with target cells in the body via an endocrine pathway

postsynaptic membrane: Region on a neuron where a synapse is formed

postzygotic isolating mechanisms: Traits that prevent hybrids, the offspring of different species, from developing into healthy, fertile adults

posttranscriptional control: Regulation of gene expression occuring at any point following transcription

potential energy: Energy available for use, such as the chemical energy stored in the glucose molecule

predation: The consumption of one organism by another

prezygotic isolating mechanisms: Traits that prevent mating and fertilization between different species

primary consumer: Organism that feeds on plants; also known as herbivores

primary growth: Growth that has been initiated by the apical meristem of a plant root or shoot

primary immune response: Immune system's initial response to an antigen, during which memory cells are produced for use in subsequent attacks of the same antigen

primary oocyte: An immature ova, or animal egg cell

primary producer: Autotrophic organism at the first trophic level of an ecosystem that support all other trophic levels

primary productivity: Measure of the rate at which autotrophs convert light or chemical energy into organic compounds

primary root: Central root of the plant

primary structure: Sequence of amino acids composing the polypeptide of a protein

primary succession: Colonization of bare, lifeless substrates, such as rock or open water, where life did not previously exist

Primates: Group of animals, including monkeys and apes, that evolved from small, arboreal mammals, known as Archonta, and are characterized by grasping hands and eyes located at the front of the head

primer: Chain consisting of RNA and DNA to which DNA nucleotides are added during DNA synthesis

primitive character: Trait that can be traced back to the ancestor of a species being studied

progesterone: Female sex hormone secreted by the ovaries

prokaryote: Single-celled organism characterized by cells that lack a membrane-enclosed nucleus and membrane-enclosed organelles; found in only the domains Bacteria or Archaea

prokaryotic cell: Cell that lacks membrane-bound organelles and a membrane-bound nucleus

promoter: Sequence of DNA nucleotides responsible for binding with RNA polymerase and determining where to begin RNA transcription

prostaglandins: Diverse group of paracrine regulators that are produced in almost every organ and function as chemical messengers

prostate gland: Gland, connected to the ejaculatory ducts, that secretes a milky alkaline (low pH) fluid that acts to balance the acidity of the vagina and the male urethra

protein: Organic compound composed of amino acids that perform a variety of biological functions

protonephridium: Externally opening excretory tubules that are branched throughout the body. Protonephridium form the excretory system in some invertebrates.

protons: Positively charged particles located at the atomic nucleus

protostome: Group of coelomates characterized by the development of the mouth from the blastopore

pseudocoelomate: An animal that contains a body cavity, called a pseudo-coel, formed between the mesoderm and the andoderm

pulmonary arteries: Vessels that carry blood to the lungs for oxygenation

pulmonary circuit: Route of blood circulation to and from the lungs

pulmonary veins: Vessels that carry oxygenated blood from the lungs to the heart

punctuated equilibrium model: Model stating that evolution occurs in rapid spurts interrupted by long periods during which little evolutionary change takes place

Punnett square: Diagram illustrating the ratio of dominant to recessive phenotypes for any given genetic cross

R

radial nerve: Nerve that spreads out from a central cluster of nerves, known as a nerve plexus

radiata (radial symmetry): Body symmetry around a center point

radioactive isotope: Unstable atom that spontaneously decays, releasing energy

random spacing: Population distribution in which individuals do not interact strongly with one another or with other factors in their environments, such as soil type or water sources

range: Area that a population inhabits at any given time

reabsorption: Process by which useful ions and nutrients are taken from the filtrate passing through the kidneys and put back into the circulatory system for use by the body

realized niche: Niche occupied by an organism owing to competition for resources in the environment

receptor protein: Protein that binds only to specific signaling molecules and relays chemical messages between cells

recessive allele: Trait that is only fully expressed when homozygous, rather than paired with a dominant partner

recombination: Bacterial process resulting in genetic variation via the exchange of genetic material

red blood cell: Cell that transports oxygen throughout the body; also known as an erythrocyte

GLOSSARY

red marrow: Marrow found within spongy bone that is the major site of blood cell production

reduced molecule: Molecule that gains an electron during an oxidation-reduction reaction

reducing atmosphere: Atmosphere that lacks oxygen and therefore allows small molecules to remain stable and react with each other to form larger molecules

reflex: Involuntary reaction to stimuli

releasing hormone: Type of hormone that stimulates the anterior pituitary to secrete other hormones

renal artery: Vessel that carries oxygenated blood to the kidney

renal cortex: Outermost layer of the kidney

renal medulla: The middle region of the layers of the kidney

renal tissue: Tissue making up the kidneys

repeated segments: Numerous individual segments making up the body of an annelid

repressor protein: Protein that provides negative control of gene activity by reducing the ability of RNA polymerase to bind to DNA

reproduction: Process resulting in the creation of new individuals from either one or two parent organisms

reproductive isolating mechanism: Trait that prevents different groups from successfully interbreeding

reproductive system: Organ system that regulates reproductive functions in an organism

reprogramming: Process in which the activity of a gene is programmed to occur at specific times during development

respiration: Process of gas exchange between an animal and the environment. During respiration, oxygen is consumed and carbon dioxide is released.

respiratory surface: Surface of the lungs across which gases can diffuse

respiratory system: Organ system that controls gas exchange in animals

resting membrane potential: Difference in charge that exists across a neuron at rest

reticular activating system (RAS): Part of the reticular formation that controls both sleeping and waking states and is responsible for the arousal of the body and the stimulation of brain activity

reticular formation: Collection of neurons within the brain stem that governs the activities of the nervous system as a whole

rhizome: Horizontal stems formed underground near a parent plant that can give rise to a new plant

ribonucleic acid (RNA): Nucleic acids responsible for carrying out the DNA instructions for protein synthesis

ribosomal RNA (rRNA): Constituent of ribosomes that assist in protein synthesis

ribosomes: Organelles within the cell's cytoplasm where protein synthesis occurs

RNA polymerase: Enzyme involved in catalyzing the assembly of nucleotides to form a growing RNA strand

RNA splicing: Process of removing the noncoding regions of mRNA, called introns

root: Downward growth of vascular plants to obtain minerals and water from the soil

root cap: The cap that covers and protects the cells that make up the apical meristem of a plant root

root hair: Extension of an epidermal root cell that increases the surface area for absorption of minerals and water

rough ER: Organelle that transports proteins formed by attached ribosomes to the cell's Golgi bodies for further processing

rubisco: Enzyme that catalyzes the molecules used in the first step of the Calvin cycle

runner: Stem that runs beneath the surface of the soil and contains nodes from which new plants can form

S

saliva: Secretion of the salivary glands in the mouth that serves to clean and moisten the mouth and performs the first stage of the chemical digestion of starches

salivary amylase: The active enzyme in saliva

salivary gland: Any gland that secretes saliva into the mouth cavity

scales: Rigid plates that grow out of the skin and provide protection for an animal

scanning electron microscope (SEM): Microscope that uses electrons to scan the surface of a specimen that has been coated with metal

scavenger: Animal that feeds on the waste left by other organisms in an ecosystem

Schwann cells: Supporting cells of neurons that assist in peripheral nerve fiber regeneration and insulate nerves by forming myelin sheaths

sclerenchyma cell: Cell that forms the simple tissue of flowering plants

scrotum: External sac containing the testes in male animals

sebaceous gland: Gland located in the epidermis that produces sebum

sebum: Oily secretion that lubricates hair and skin and can prevent bacterial growth

second messenger: Molecule within the cell that mediates the signal arriving at the receptor and amplifies the response to a hormone

secondary consumer: Carnivore that feeds on herbivores and primary consumers; the second level of consumers in the food chain

secondary growth: Period of growth during which a plant expands in girth, or bulk

secondary immune response: Rapid and prolonged response by white blood cells to a previously encountered antigen that has been recognized by memory cells

secondary oocyte: Daughter cell of the primary oocyte formed after meiosis I

secondary spermatocyte: Haploid spermatocyte produced after the first meiotic division, prior to the second division

secondary structure: Folding of a single region of a polypeptide chain in a protein

secondary succession: Process in which a community develops and evolves in a habitat that is recovering from disturbance

seed: Plant ovule containing the embryo surrounded by a protective coating that is produced by flowering plants and gymnosperms

self proteins: Markers within an organism's cells that identify those cells as not being foreign

semelparity: Process by which individuals in some species will tend to produce as many offspring as possible in one reproductive event before dying

semen: Fluid containing sperm that is expelled during ejaculation

semilunar valve: Valve in the heart that controls the opening between each of the ventricles and the arterial system

seminal vesicle: Gland connected to the vas deferens that secretes a fructose-rich fluid constituting about 60 percent of the semen volume to nourish and protect the sperm

seminiferous tubule: Long, convoluted duct within the testes that is the site of sperm production and maturation

semipermeable: Allowing some but not all matter to pass through the membrane

sensory neuron: Neuron that relays impulses from sensory receptors to the spinal cord and brain

sensory receptor: Structure that detects specific stimuli in the internal or external environment

sepal: Green, leaflike structure that encloses the flower and protects it prior to opening

serosa: Thin tissue layer covering the body cavities that secretes fluid to reduce friction from muscle movement

sessile: Nonmoving organisms that are anchored to their environment

sexual imprinting: Mechanism by which animals learn to direct sexual advances only toward members of the same species

sexual reproduction: Creation of offspring from the fusion of gametes of two parents

sexual selection: Natural selection favoring traits that give an organism a competitive edge in reproductive success

shared derived character: Evolutionarily unique trait shared among a group of organisms and used by taxonomists in classification

shoot: The above-ground portion of a plant that includes the stem

short-day plant: Plant only able to flower in the late summer, fall, or winter, when the duration of daylight is relatively short

sieve cell: Slender cells found in the secondary phloem of plants

sieve-tube member: Elongated cells of the conducting tubes in phloem tissue of flowering plants

simple epithelial tissue: Epithelial tissue comprising a single layer of cells

simple reflex: Stereotyped movement caused by direct stimulation of the motor neuron owing to the arrival of a sensory message at the spinal cord

sinoatrial (SA) node: Structure located in the right atrium in most vertebrate hearts that sets the rate of contractions through the generation of an electrical impulse

size: The number of individuals in a given population

skeletal muscle: Striated muscle that attaches to the bones by tendons and contracts to create movement

skin: Outermost protective layer of the integumentary system

small intestine: Region of the digestive system where the majority of digestion and absorption takes place

smooth ER: Eukaryotic intracellular organelle responsible for lipid synthesis, carbohydrate metabolism, calcium storage, and attachment of proteins to cell membrane receptors

smooth muscle: Nonstriated muscle lining the walls of hollow organs, the bladder, and the abdominal cavity

society: Group of organisms of the same species living together in a cooperative manner

solute: A substance that is dissolved in a solvent

solution: Homogeneous mixture of molecules

solvent: A solution that is used to dissolve other molecules

somatic nervous system: Nerves connecting the central nervous system to the muscles

spacing: Distribution of individuals within a population

speciation: Evolutionary process by which new species arise from the transformation of an existing species or the branching off of one species into two descendant species

species: Group of interbreeding populations reproductively isolated from other groups

sperm: The male gamete in sexual reproduction

sphincter: Muscle ring that constricts to close passages between regions of a tubular system

spinal cord: Bundle of neurons enclosed in the vertebral column that allow reflex connections between sensory and motor neurons and transmit messages to and from the brain and the rest of the body

spinal nerve: Nerve extending through the vertebrae in the spinal cord

spongy bone: Highly porous, inner cavity of long bones containing the blood vessels and marrow

spore: Haploid reproductive cell released by plants between meiosis and fertilization

stabilizing selection: Favoring of individuals with average phenotypes, as opposed to those with phenotypic extremes

stamen: Male reproductive organ of a flower

stem: Above-ground vascular tissue structure of plants that supports leaves and flowers

steroid: Lipid derived from cholesterol that acts as a chemical messenger in metabolic processes, intercellular communication, and in cell membranes

stigma: Upper surface of the ovary in flowering plants where pollen grains are captured

stimulant: Neurotransmitter that causes an increase in nervous system activity

stimulation: Activation of a sensory receptor through the detection of a stimulus

stomach: Muscular organ in the gastrointestinal tract in which food is digested by enzymes and gastric acid

stomata: Pores, located between guard cells on the surfaces of leaves, that allow the absorption of carbon dioxide and the release of water vapor during photosynthesis

stratified epithelial tissue: Epithelial tissue comprising multiple cell layers

submucosa: Connective tissue layer of the gastrointestinal tract that is located beneath the mucosa and contains blood vessels and nervous tissue

succession: Natural progression of changes in the composition of a community over time

sucker: Structure produced by the roots of a plant that can give rise to new plants

superior vena cava: Major vein that carries blood from the upper body to the right side of the heart

sweat gland: Gland that produces sweat to aid body temperature regulation

symbiosis: Interaction between two or more species that benefits at least one of the species

sympathetic division: Components of the autonomic nervous system responsible for regulating homeostatic mechanisms and involved in the mediation of the body's responses during times of stress

sympatric speciation: Splitting of one species into two as a result of ecological, behavioral, or genetic barriers without geographical isolation

synapse: See electrical synapse and chemical synapse

synapsis: Pairing of homologous chromosomes during meiosis

synaptic cleft: Junction between two neurons, or between neurons and muscles or glands, into which chemical signals are released

synaptic vesicle: Vesicle responsible for the storage, transport, and release of neurotransmitters in the transmission of a nerve impulse

systematics: Classification of organisms into different groups according to evolutionary relatedness

systemic circulation: Circulation of oxygenated blood from the lungs to the heart and around the body

systole: Contraction of the heart

GLOSSARY

T

T cells: See T lymphocytes.

T lymphocytes: White blood cells that mature in the thymus and are involved in immune responses

taproot system: System comprising a large central root and lateral root branches

target cell: Destination cell at which hormones will perform a specified action

template strand: Strand of DNA used by RNA polymerase to build the complementary mRNA

temporal isolation: Mechanism by which different breeding times prevent two or more populations from exchanging genes

temporal lobe: Region of cerebral cortex involved in auditory processing

tendon: Tissue that joins bone to muscle

terminal bud: Undeveloped shoot at the tip of a stem that will either remain dormant or immediately form a new shoot

termination: Point in protein synthesis when the RNA polymerase reaches a stop signal, indicating the completion of translation

tertiary consumer: Carnivore that feeds on secondary consumers; the third level of consumer within the food chain

testes: Primary male sex organ responsible for the production of male gametes and sex hormones

testosterone: Major sex hormone in male mammals that controls reproductive functions

tetrad: Group of four chromatids formed by synapses during the first stages of meiosis

thermoreceptor: Sensory cells that respond to changes in temperature

thigmotropism: Growth of vascular plants in response to touch, such as the growth that allows a vine to extend up a fence

threshold potential: Minimum voltage difference across a membrane required to initiate an action potential

thylakoid: Membrane-bound compartment in chloroplasts that contains the light-absorbing pigments and enzymes required for photosynthesis

thymosin: Hormone released by the thymus that stimulates T-cell production in the immune system

thyroxine (T_4): Major thyroid hormone involved in controlling metabolic rate

tight junction: Tight connection between two cells that forms a barrier to fluid

tissue: Organized group of cells that function together

tolerance: Ability of an organism to withstand the external conditions of an environment

tonsil: Lymphoid tissue found on either side of the throat

trachea: Pipe through which air moves between the throat and the lungs

tracheid: Type of xylem cell in flowering plants that conducts water and dissolves minerals

trait: Characteristic of an organism that can take alternative forms, such as eye color

transcription: Process in which an RNA polymerase assembles a single-stranded RNA complementary to the DNA strand from which it was assembled

transcription factor: Protein in eukaryotes that regulates the binding of RNA polymerase and controls the initiation of transcription

transcriptional control: Process that either allows or prevents the initiation of transcription

transduction: Process by which a cell rapidly converts external signals or stimuli into another signal through biochemical processes

transfer RNA (tRNA): Molecule that transfers amino acids to the growing protein chain during translation

transgenic biotechnology: Field of research in which genes are transferred from one genetically-engineered organism to another

translation: Process in protein synthesis in which the coded sequence information in mRNA is used to assemble a polypeptide chain

transmission: Passage of a nerve signal along neurons within the nervous system

transmission electron microscopes (TEMs): Microscopes that transmit a beam of electrons through a sample to create an image

transpiration: Evaporative loss of water from plant leaves or stems that occurs when stomata open to take in carbon dioxide

transport (channel) protein: Protein that controls the movement of molecules in and out of a cell

triiodothyronine (T_3): Thyroid hormone that is involved in the development of the skeleton and nervous system in embryos and acts to increase basal metabolic rate in adults

trophoblast: Outermost cell layer in the early stage of embryo development that provides the embryo with nutrients and later develops into the placenta

tropism: Growth of vascular plants in response to external environmental stimuli

true-breeding: Lineage of offspring among sexually reproducing animals in which biological traits are uniformly and consistently passed on to subsequent generations

tube feet: Ventral protrusions on starfish arms and some other echinoderms that pass food from the water to the mouth

turgor pressure: Internal pressure of the cell contents on the cell wall following osmosis

U

uniform spacing: Distribution where individuals in a population develop their own space, or territory, as a result of competition for resources in the area

ureters: Tubes through which urine is passed from the kidneys to the bladder for storage

urethra: Tubular structure through which urine is passed from the bladder and expelled from the body

urinary system: Organ system that controls blood volume and composition through the process of filtration

urine: Liquid waste produced through the processes of filtration, reabsorption, and secretion in the kidneys

uterus: Major female reproductive organ that accommodates the developing embryo during pregnancy

V

vacuole: Membrane-bound cell compartment that stores excess food, water, minerals, cellular fluid, and other matter for later use

vagina: Part of the female reproductive system that allows entry of sperm, forms part of the birth canal, and channels menstrual flow out of the body

valence electron: Electrons occupying the outer energy shell and involved in chemical reactions

vas deferens: Ducts in the male reproductive system that transport sperm from the testes to the penis during ejaculation

vascular cambium: Layer of meristematic tissue from which all secondary growth in plants originates

vascular tissue: Plant cells that are arranged to form tubes through which water, minerals, and sugars can flow

vector: Substance used to transfer genes from one organism to another

veins: Blood vessels that return blood to the heart

vertebrate: All members of the phylum Chordata that have a backbone

vesicle: Small membrane-bound compartment within a cell that functions in the transport, storage, or digestion of substances

vessel element: Short, square xylem cells found in the majority of angiosperms that transport water effectively through the plant body

villi: Absorptive structures protruding from the epithelial surface that act to increase surface area

virion: An inert virus particle existing outside of a host cell

virus: Unicellular organism containing only its genome surrounded by a protein coat and capable of replicating only after infecting a host cell

viviparity: Embryonic development occurring inside the body of the mother, from which the embryo gains nourishment

W

water cycle: Process driven by solar energy in which water moves between the atmosphere, land, and oceans

water table: Upper section of groundwater formed by the seepage of ponds, lakes, streams, and rainfall through soil

water-vascular system: System used by echinoderms for locomotion, food, waste transportation, and respiration

white blood cell: Cell in the vertebrate immune system that defends against infection and disease; also known as a leukocyte

X

xylem: One of the two types of vascular tissue found in vascular plants. Xylem transports water and solutes throughout the plant.

Y

yellow marrow: Fatty tissue located in the long bones of adults that produce red blood cells

yolk sac: Membrane that provides nourishment to a growing fetus

Z

zygote: Single, fertilized egg from which all animals are produced

α **helix:** Coil that is the first form in the secondary structure in

ß sheet proteins: Pleated fold that is the second form of regular secondary structure in proteins

Index

INDEX

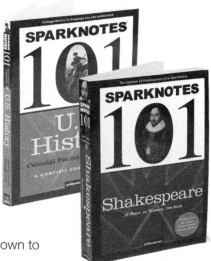